METHODS IN MOLECULAR BIOLOGY™

Series Editor
**John M. Walker
School of Life Sciences
University of Hertfordshire
Hatfield, Hertfordshire, AL10 9AB, UK**

For other titles published in this series, go to
www.springer.com/series/7651

Molecular and Cell Biology Methods for Fungi

Edited by

Amir Sharon

Department of Plant Sciences, Tel Aviv University, Tel Aviv, Israel

Editor
Amir Sharon, Ph.D.
Department of Plant Sciences
Tel Aviv University
Tel Aviv
Israel
amirsh@tauex.tau.ac.il

ISSN 1064-3745   e-ISSN 1940-6029
ISBN 978-1-60761-610-8 e-ISBN 978-1-60761-611-5
DOI 10.1007/978-1-60761-611-5
Springer New York Dordrecht Heidelberg London

Library of Congress Control Number: 2010923232

© Springer Science+Business Media, LLC 2010
All rights reserved. This work may not be translated or copied in whole or in part without the written permission of the publisher (Humana Press, c/o Springer Science+Business Media, LLC, 233 Spring Street, New York, NY 10013, USA), except for brief excerpts in connection with reviews or scholarly analysis. Use in connection with any form of information storage and retrieval, electronic adaptation, computer software, or by similar or dissimilar methodology now known or hereafter developed is forbidden.
The use in this publication of trade names, trademarks, service marks, and similar terms, even if they are not identified as such, is not to be taken as an expression of opinion as to whether or not they are subject to proprietary rights.
While the advice and information in this book are believed to be true and accurate at the date of going to press, neither the authors nor the editors nor the publisher can accept any legal responsibility for any errors or omissions that may be made. The publisher makes no warranty, express or implied, with respect to the material contained herein.

Printed on acid-free paper

Springer Science+Business Media (www.springer.com)

Preface

The kingdom Fungi constitutes an independent group equal in rank to that of plants and animals. It is a diverse clade of heterotrophic eukaryotic organisms that shares some characteristics with animals and includes mushrooms, molds, yeasts as well as many other types of less well known organisms. Approximately 100,000 species have been described, which comprise less than 10% of the estimated number of fungal species in nature. Fungi can be found in every place wherever adequate moisture, temperature, and organic substrates are available; however, they also occupy extreme habitats, from hot volcanoes to arctic zones, arid deserts, and deep oceans.

The importance of fungi as a group is tremendous; most species are saprobes and play prime roles in decomposition and the recycling of organic matter and nutrients, and many of them produce enzymes and metabolites with important applications in pharmacology, biotechnology, and other industries. Alongside the positive aspects, fungi also cause huge damage, primarily as plant pathogens. Fungi are highly amenable to molecular work, and a few fungal species serve as model systems to study basic processes with results that are applicable to many organisms, including humans.

Fungal research has made enormous progress in the past two decades. Two main stages can be identified: (1) the development of fungal transformation systems in the 1980s, which represents the shift of fungal research to the molecular era, and (2) sequencing of the *Saccharomyces cerevisiae* genome in the late 1990s, which represents the entrance of fungal research to the genomic era. Currently over 70 fungal genomes have been published and numerous others are being sequenced. This wealth of genomic information along with high throughput methods has revolutionized fungal research. The combination of the genomic approaches with the more traditional molecular and cell biology methods, particularly the readily available transformation and gene knockout procedures, open new horizons in fungal research, and we may expect that the coming years will bring exciting discoveries and developments.

Lessons from advanced model systems, particularly the yeast *S. cerevisiae*, show that the increased use of high throughput methods will be paralleled by increased research of single genes and proteins, products of such high throughput analyzes. Indeed, molecular, biochemical, and cellular-type work is now practiced in many fungal research laboratories to investigate the function of newly identified genes/proteins/metabolic pathways/signal cascades, etc.

The primary purpose of *Molecular and Cell Biology Methods for Fungi* is to provide readers with an up-to-date set of practical protocols that cover the range of frequently used methods to study molecular and cellular aspects of fungal biology. The set of protocols described in this book includes classical protocols such as transformation systems and traditional protein analysis methods, which have been widely used for many years, alongside the most advanced techniques such as genome amplification, whole genome knockout methods, sophisticated in vivo imaging techniques, etc.

The methods in each chapter describe protocols for specific fungal species. However, the described methods should be useful in a wide range of species; detailed *Notes* are provided in each protocol, which should allow easy adaptation of the protocol to various other species. It is my hope that this book will be helpful to both experienced fungal research laboratories as well as to those that are interested in using fungi as hosts to study their favorite genes.

Finally, I would like to take this opportunity to thank all authors for their commitment, cooperation, and contributions. I am also thankful to John Walker, the series editor, for his professional guidance during all stages and for making this process an enjoyable and educating experience.

Tel Aviv, Israel *Amir Sharon*

Contents

Preface v
Contributors ix

PART I FUNGAL TRANSFORMATION AND GENE KNOCKOUT

1 Protoplast Transformation of Filamentous Fungi 3
   B. Gillian Turgeon, Bradford Condon, Jinyuan Liu, and Ning Zhang

2 Electroporation and *Agrobacterium*-Mediated Spore Transformation 21
   Anna Minz and Amir Sharon

3 High-Throughput Construction of Gene Deletion Cassettes
   for Generation of *Neurospora crassa* Knockout Strains 33
   *Patrick D. Collopy, Hildur V. Colot, Gyungsoon Park, Carol Ringelberg,
   Christopher M. Crew, Katherine A. Borkovich, and Jay C. Dunlap*

4 Development of *Impala*-Based Transposon Systems for Gene
   Tagging in Filamentous Fungi 41
   Marie Dufresne and Marie-Josée Daboussi

5 DelsGate: A Robust and Rapid Method for Gene Deletion 55
   *María D. García-Pedrajas, Marina Nadal, Timothy Denny,
   Lourdes Baeza-Montañez, Zahi Paz, and Scott E. Gold*

6 Gene Silencing for Functional Analysis: Assessing RNAi
   as a Tool for Manipulation of Gene Expression 77
   Carmit Ziv and Oded Yarden

PART II DETECTION AND QUANTIFICATION OF FUNGI

7 Analysis of Fungal Gene Expression by Real Time Quantitative PCR 103
   Shahar Ish-Shalom and Amnon Lichter

8 Identification of Differentially Expressed Fungal Genes
   In Planta by Suppression Subtraction Hybridization 115
   Benjamin A. Horwitz and Sophie Lev

9 Quantification of Fungal Infection of Leaves with Digital Images
   and Scion Image Software 125
   Paul H. Goodwin and Tom Hsiang

10 Expression Profiling of Fungal Genes During Arbuscular Mycorrhiza
   Symbiosis Establishment Using Direct Fluorescent *In Situ* RT-PCR 137
   *Pascale M. A. Seddas-Dozolme, Christine Arnould, Marie Tollot,
   Elena Kuznetsova, and Vivienne Gianinazzi-Pearson*

11 Application of Laser Microdissection to Study Plant–Fungal
   Pathogen Interactions 153
   John Fosu-Nyarko, Michael G. K. Jones, and Zhaohui Wang

12 Multiplex Gene Expression Analysis by TRAC in Fungal Cultures 165
    Jari J. Rautio

13 Amplification of Fungal Genomes Using Multiple
    Displacement Amplification 175
    Simon J. Foster and Brendon J. Monahan

PART III MICROSCOPY AND PROTEIN ANALYSIS

14 Biochemical Methods Used to Study the Gene Expression
    and Protein Complexes in the Filamentous Fungus *Neurospora crassa* 189
    Jinhu Guo, Guocun Huang, Joonseok Cha, and Yi Liu

15 Measuring Protein Kinase and Sugar Kinase Activity in Plant
    Pathogenic *Fusarium* Species 201
    Burton H. Bluhm and Xinhua Zhao

16 A Detailed Protocol for Chromatin Immunoprecipitation
    in the Yeast *Saccharomyces cerevisiae* 211
    Melanie Grably and David Engelberg

17 A Method to Visualize the Actin and Microtubule Cytoskeleton by Indirect
    Immunofluorescence 225
    Flora Banuett

18 Fluorescence In Situ Hybridization for Molecular Cytogenetic
    Analysis in Filamentous Fungi 235
    Dai Tsuchiya and Masatoki Taga

19 Live-Cell Imaging of Microtubule Dynamics in Hyphae
    of *Neurospora crassa* 259
    Maho Uchida, Rosa R. Mouriño-Pérez, and Robert W. Roberson

20 Methods to Detect Apoptotic-Like Cell Death in Filamentous Fungi 269
    Camile P. Semighini and Steven D. Harris

21 Evaluation of Antifungal Susceptibility Using Flow Cytometry 281
    Cidália Pina-Vaz and Acácio Gonçalves Rodrigues

22 Preparation of Fungi for Ultrastructural Investigations
    and Immunogoldlabelling 291
    Gerd Hause and Simone Jahn

23 Split-EGFP Screens for the Detection and Localisation
    of Protein–Protein Interactions in Living Yeast Cells 303
    Emma Barnard and David J. Timson

Index *319*

Contributors

CHRISTINE ARNOULD • *UMR INRA/CNRS/Université Bourgogne Plante-Microbe-Environnement, Dijon Cedex, France*
LOURDES BAEZA-MONTAÑEZ • *Estación Experimental "La Mayora", CSIC, Algarrobo-Costa, Málaga, Spain*
FLORA BANUETT • *Department of Biological Sciences, California State University, Long Beach, CA, USA*
EMMA BARNARD • *School of Biological Sciences, Queen'sec University, Belfast, UK*
BURTON H. BLUHM • *Department of Plant Pathology, Division of Agriculture, University of Arkansas, Fayetteville, AR, USA*
KATHERINE A. BORKOVICH • *Department of Plant Pathology and Microbiology, University of California, Riverside, CA, USA*
JOONSEOK CHA • *Department of Physiology, The University of Texas Southwestern Medical Center, Dallas, TX, USA*
PATRICK D. COLLOPY • *Department of Genetics, Dartmouth Medical School, Hanover, NH, USA*
HILDUR V. COLOT • *Department of Genetics, Dartmouth Medical School, Hanover, NH, USA*
BRADFORD CONDON • *Department of Plant Pathology and Plant-Microbe Biology, Cornell University, Ithaca, NY, USA*
CHRISTOPHER M. CREW • *Department of Plant Pathology and Microbiology, University of California, Riverside, CA, USA*
MARIE-JOSÉE DABOUSSI • *Univ. Paris-Sud 11, CNRS, UMR8621, Institut de Génétique et Microbiologie, Orsay, France*
TIMOTHY DENNY • *Department of Plant Pathology, University of Georgia, Athens, GA, USA*
MARIE DUFRESNE • *Université Paris-Sud 11, CNRS, UMR8621, Institut de Génétique et Microbiologie, Orsay, France*
JAY C. DUNLAP • *Department of Genetics, Dartmouth Medical School, Hanover, NH, USA*
DAVID ENGELBERG • *The Department of Biological Chemistry, The Institute of Life Sciences, The Hebrew University of Jerusalem, Jerusalem, Israel*
SIMON J. FOSTER • *The Sainsbury Laboratory, John Innes Center, Norwich, UK*
JOHN FOSU-NYARKO • *Plant Biotechnology Research Group, School of Biological Sciences and Biotechnology, Murdoch University, Perth, WA, Australia*
MARÍA D. GARCÍA-PEDRAJAS • *Estación Experimental "La Mayora", CSIC, Algarrobo-Costa, Málaga, Spain*
VIVIENNE GIANINAZZI-PEARSON • *UMR INRA/CNRS/Université Bourgogne Plante-Microbe-Environnement, Dijon Cedex, France*

SCOTT E. GOLD • *Department of Plant Pathology, University of Georgia, Athens, GA, USA*
PAUL H. GOODWIN • *Department of Environmental Biology, University of Guelph, Guelph, ON, Canada*
MELANIE GRABLY • *The Department of Biological Chemistry, The Institute of Life Sciences, The Hebrew University of Jerusalem, Jerusalem, Israel*
JINHU GUO • *Department of Physiology, The University of Texas Southwestern Medical Center, Dallas, TX, USA*
STEVEN D. HARRIS • *Department of Plant Pathology and Center for Plant Science Innovation, University of Nebraska, Lincoln, NE, USA*
GERD HAUSE • *Central Microscopy, Biocenter of the Martin-Luther-University Halle-Wittenberg, Halle/Saale, Germany*
BENJAMIN A. HORWITZ • *Department of Biology, Technion – Israel Institute of Technology, Haifa, Israel*
TOM HSIANG • *Department of Environmental Biology, University of Guelph, Guelph, ON, Canada*
GUOCUN HUANG • *Department of Physiology, The University of Texas Southwestern Medical Center, Dallas, TX, USA*
SHAHAR ISH-SHALOM • *Department of Post Harvest Science, ARO, The Volcani Center, Bet Dagan, Israel*
SIMONE JAHN • *Biocenter of the Martin-Luther-University Halle-Wittenberg, Halle/Saale, Germany*
MICHAEL G.K. JONES • *Plant Biotechnology Research Group, School of Biological Sciences and Biotechnology, Murdoch University, Perth, WA, Australia*
ELENA KUZNETSOVA • *UMR INRA/CNRS /Université Bourgogne Plante-Microbe-Environnement, Dijon Cedex, France*
SOPHIE LEV • *Department of Biology, Technion – Israel Institute of Technology, Haifa, Israel*
AMNON LICHTER • *Department of Post Harvest Science, ARO, The Volcani Center, Bet Dagan, Israel*
JINYUAN LIU • *Department of Plant Pathology and Plant-Microbe Biology, Cornell University, Ithaca, NY, USA*
YI LIU • *Department of Physiology, The University of Texas Southwestern Medical Center, Dallas, TX, USA*
ANNA MINZ • *Department of Plant Sciences, Tel-Aviv University, Tel Aviv, Israel*
BRENDON J. MONAHAN • *CSIRO Molecular and Health Technologies, Ian Wark Laboratories, Clayton, VIC, Australia*
ROSA R. MOURIÑO-PÉREZ • *Departmento de Microbiologia. Centro de Investigacion Cientifica y, Ensenada, Mexico*
MARINA NADAL • *Department of Plant Pathology, University of Georgia, Athens, GA, USA*
GYUNGSOON PARK • *Department of Plant Pathology and Microbiology, University of California, Riverside, CA, USA*
ZAHI PAZ • *Department of Plant Pathology, University of Georgia, Athens, GA, USA*

CIDÁLIA PINA-VAZ • *Department of Microbiology, Faculty of Medicine, University of Porto, Porto, Portugal*
JARI RAUTIO • *PLEXPRESS Oy, Helsinki, Finland*
CAROL RINGELBERG • *Department of Genetics, Dartmouth Medical School, Hanover, NH, USA*
ROBERT W. ROBERSON • *School of Life Sciences, Arizona State University, Tempe, AZ, USA*
ACÁCIO GONÇALVES RODRIGUES • *Department of Microbiology, Faculty of Medicine, University of Porto, Porto, Portugal*
PASCALE M.A. SEDDAS-DOZOLME • *UMR INRA/CNRS/Université Bourgogne Plante-Microbe-Environnement, Dijon Cedex, France*
CAMILE P. SEMIGHINI • *Department of Plant Pathology and Center for Plant Science Innovation, University of Nebraska, Lincoln, NE, USA; Department of Molecular Genetics & Microbiology, Duke University Medical Center, Durham, NC, USA*
AMIR SHARON • *Department of Plant Sciences, Tel-Aviv University, Tel Aviv, Israel*
MASATOKI TAGA • *Department of Biology, Faculty of Science, Okayama University, Okayama, Japan*
DAVID J. TIMSON • *School of Biological Sciences, Queen'sec University, Belfast, UK*
MARIE TOLLOT • *UMR INRA/CNRS/Université Bourgogne Plante-Microbe-Environnement, Dijon Cedex, France*
DAI TSUCHIYA • *Department of Biology, Indiana University, Bloomington, IN, USA*
B. GILLIAN TURGEON • *Department of Plant Pathology and Plant-Microbe Biology, Cornell University, Ithaca, NY, USA*
MAHO UCHIDA • *Department of Anatomy, University of California, San Francisco, CA, USA*
ZHAOHUI WANG • *Plant Biotechnology Research Group, School of Biological Sciences and Biotechnology, Murdoch University, Perth, WA, Australia*
ODED YARDEN • *Department of Plant Pathology and Microbiology, The Robert H. Smith Faculty of Agriculture, Food and Environment, The Hebrew University of Jerusalem, Rehovot, Israel*
NING ZHANG • *Department of Plant Pathology and Plant-Microbe Biology, Cornell University, Ithaca, NY, USA*
XINHUA ZHAO • *AureoGen Biosciences, Kalamazoo, MI, USA*
CARMIT ZIV • *Department of Plant Pathology and Microbiology, The Robert H. Smith Faculty of Agriculture, Food and Environment, The Hebrew University of Jerusalem, Rehovot, Israel*

Part I

Fungal Transformation and Gene Knockout

Chapter 1

Protoplast Transformation of Filamentous Fungi

B. Gillian Turgeon, Bradford Condon, Jinyuan Liu, and Ning Zhang

Abstract

The protoplast method for the transformation of filamentous fungi is described in detail, as is the Restriction Enzyme-Mediated Integration (REMI) procedure for introducing tagged mutations into the fungal genome. A split marker method for generating PCR fragments for targeted integration and deletion of genes of interest is also detailed.

Key words: Selectable marker, Drug resistance, Split-marker, Gene-deletion, Cotransformation, Hygromycin B, Geneticin, Bialaphos, Phosphothricin, REMI, Integration

1. Introduction

1.1. Transformation

Transformation, the introduction and incorporation of exogenous DNA into cells, was first achieved experimentally when Griffith (1928) showed that a nonvirulent strain of *Streptococcus pneumoniae* could be "transformed" into a virulent one by exposure to strains of heat-killed, virulent, *S. pneumoniae* (1). In 1944, Avery, MacLeod, and McCarty identified the transforming factor as DNA (2). The first eukaryotic organism to be transformed was *Saccharomyces cerevisiae* in 1978 (3), and the first filamentous fungus transformed was the saprobe *Neurospora crassa* (4). These achievements were followed by the transformation of the saprobe, *Aspergillus nidulans* in 1984 (5), the first plant pathogenic ascomycete, *Cochliobolus heterostrophus* in 1985 (6), and the first basidiomycete, *Ustilago maydis* in 1988 (7). The first demonstration of bacterial transformation proved that *S. pneumoniae* could incorporate the transforming material naturally, while all of the foregoing fungal transformations were artificial, involving exogenous DNA and protoplasts (i.e., cells were stripped of their walls by enzymatic digestion) stabilized in an osmoticum. Subsequent chilling of the protoplasts in the presence of divalent cations such as Ca^{2+} (in $CaCl_2$) renders protoplasts permeable to DNA and thus "competent" to take up DNA. Although there are other methods of transformation, such as cocultivation with *Agrobacterium*, electroporation (*see* Chapter 2), treatment of whole cells with cations, or use of a particle gun, except for the *Agrobacterium* method, none is as widely used as protoplast transformation for introduction of DNA.

Efficiency of transformation, i.e., the fraction of cells that take up DNA is low, therefore strategies are needed to identify those cells that have been transformed. There are two basic approaches: 1) introduction of a dominant selectable marker that transforms the original strain from drug sensitive to drug resistant and 2) introduction of DNA that transforms an auxotroph to a prototroph. The former is much easier to work with than the latter, assuming that strains of the fungus in question can be found that display drug-sensitivity. The latter is more cumbersome since each strain to be transformed must carry a mutation in a gene conferring prototrophy and subsequent manipulations may require the same genetic background (e.g., a wild-type gene in the leucine pathway may be used to transform a leu- strain).

1.2. Cotransformation and Complementation

For fungi such as *C. heterostrophus* (used here as a model), which have a very efficient homologous integration mechanism (8), an entire gene or part of a gene can be deleted with precision and replaced by a selectable marker with close to 100% efficiency. This targeted integration efficiency also allows subsequent complementation of the transgenic strain at the native site, with e.g., the wild-type gene or a version of the gene that carries site-directed mutations, if cotransformation with a second selectable marker is performed.

1.3. REMI Transformation

The Restriction Enzyme-Mediated Integration (REMI) procedure, first described for *S. cerevisiae* (9) and refined for use with *Dictyostelium discoideum* (10, 11) offers the prospect of introducing random tagged mutations into the fungal genome at a relatively high rate, and is the functional equivalent of transposon tagging in prokaryotes. This was first demonstrated to work well with filamentous fungi in 1994 (12) and was quickly adopted by researchers in the filamentous fungal community as a mean of introducing "random" tagged mutations (13).

The procedure is essentially the same as the regular transformation procedure described above and below, except that the transforming DNA is a linearized vector carrying a selectable marker that has no homology to the fungal genome. The DNA in the original digestion mixture plus extra restriction enzyme (usually the same one used for vector linearization) are added to protoplasts. The assumption is that genomic DNA will be cut at sites corresponding to the restriction enzyme used and that the introduced, linearized, transforming DNA, cut with the same enzyme, will insert randomly at these sites and is in a "race" to insert before DNA repair mechanisms repair the genomic sites. In an ideal situation, a single random tagged insertion results, without any other alteration in the genome. In reality, this is not a random process, since the transforming DNA inserts at designated restriction enzyme sites. Assuming these sites are common in fungal genomes, chances of hitting any given gene can be calculated. To increase chances, several restriction enzymes can be used.

1.4. Uses of Transformation

1. Deletion of genes by targeted integration.
2. Introduction of mutations, for example those made by site-directed mutagenesis.
3. Complementation of mutants.
4. Introduction of extra copies of genes.
5. Overexpression of genes.
6. Introduction of conditionally expressed genes.
7. Introduction of heterologous genes.
8. Creation of marked strains.

We recommend the following primary and review articles for a complete history of transformation tools and their uses: (3, 5, 6, 14–19).

2. Materials

2.1. Protoplast Production

1. Glucanex. We have been using a liquid preparation, kindly provided by C. M. Hjort, Novo Nordisk, Bagsvaerd, Denmark (see ref. (20)) (see Note 1).
2. Driselase (Sigma D9515).
3. iProof High Fidelity DNA Polymerase (BioRad, B172-5301).
4. Nylon fabric (SEFAR Filtration Inc., Depew, NY14043, 25 μm pore size lab Pak 03-25/14).
5. Filter unit PES 50 MM (0.2 μm) (NALGENE, 565-0020, VWR 73520-982).
6. Cheesecloth.
7. Centrifuge (Sorvall RC-5C or equivalent).
8. Screw-capped centrifuge tubes (40 mL).
9. 50 mL Erylenmeyer flasks.

10. Optional: ampicillin.
11. 13 × 100 mm pyrex test tubes.
12. Misc. glassware – flasks, beakers.
13. Eppendorf pipettes, tips.
14. Petri dishes.

2.2. Drugs for Selection of Transformants

1. Hygromycin B (Calbiochem, California, USA, 400051).
2. Geneticin (G418 sulphate, GIBCO Invitrogen, 11811-031).
3. PPT (DL-Phosphinothricin, Research Products International Corp. Illinois, USA, P40100) or Bialaphos (Gold Biotechnology, B0178-500).

2.3. Solutions

All solutions must be autoclaved unless otherwise indicated.

1. *Minimal medium (MM) solid or liquid*: 10 mL Solution A, 10 mL Solution B, 0.5 mL Srb's micronutrients, 10 g glucose, 20 g Agar (only for solid media), H_2O to 1000 mL.
2. *Complete medium (CM) or (CMX) solid or liquid*: 10 mL Solution A, 10 mL Solution B, 0.5 mL Srb's micronutrients, 1 g yeast extract, 0.5 g casein hydrolysate (enzymatic), 0.5 g casein hydrolysate (acidic), 10 g glucose (for CM) OR 10 g xylose (for CMX), 20 g Agar (for solid media), H_2O to 1000 mL. For CMNS, leave out Solutions A and B.
3. *Solution A*: 100 g $Ca(NO_3)_2 \cdot 4H_2O$, 1,000 mL H_2O. Store in the dark, shake before use.
4. *Solution B*: 20 g KH_2PO_4, 25 g $MgSO_4 \cdot 7H_2O$, 15 g NaCl, H_2O to 1000 mL. Adjust to pH 5.3 with NaOH, filter sterilize, do not autoclave. Store in the dark, shake before use.
5. *Srb's micronutrients*: 57.2 mg H_3BO_3, 393 mg $CuSO_4 \cdot 5H_2O$, 13.1 mg KI, 60.4 mg $MnSO_4 \cdot H_2O$, 36.8 mg $(NH_4)_6Mo_7O_{24} \cdot 4H_2O$, 5,490 mg $ZnSO_4 \cdot H_2O$, 948.2 mg $FeCl_3 \cdot 6H_2O$, H_2O to 1,000 mL.
6. *Enzyme-osmoticum*: 3.27 g NaCl (0.7 M), 1.6 mL Glucanex, 0.8 g Driselase, H_2O to 80 mL. Stir at room temperature 5 min, filter sterilize with 0.2 µm filter unit.
7. *0.7 M NaCl*: 8.18 g NaCl, H_2O to 200 mL.
8. *STC*: Sorbitol, 21.86 g (1.2 M), 1 mL of 1 M Tris–HCl pH 7.5, (10 mM), 0.735 g $CaCl_2 \cdot 2H_2O$ (50 mM), H_2O to 100 mL.
9. *Polyethylene glycol*: 30 g Polyethylene glycol, MW 3,350 (60% w/v), 0.5 mL of 1 M Tris–HCl pH 7.5 (10 mM), 0.37 g $CaCl_2 \cdot 2H_2O$ (50 mM), H_2O to 50 mL.

10. *Molten regeneration medium*:

    Flask A: *For HygromycinB or Geneticin/G418 Sulfate selection*: 1 g yeast extract, 1 g casein hydrolysate (enzymatic), H_2O to 50 mL.

    For Phosphothricin/Bialaphos selection: 10 mL Solution A, 10 mL Solution B, 0.5 mL Srb's micronutrients, H_2O to 50 mL. Filter sterilize (precipitates form after autoclaving).

    Flask B: For all selection methods: 342 g Sucrose, H_2O to 500 mL.

    Flask C: For all selection methods: 16 g Agar, H_2O to 450 mL.

    Autoclave Flasks A, B and C separately. Combine after autoclaving and hold at 60°C until ready to use.

3. Methods

3.1. Generation and Transformation of Fungal Protoplasts

The following protocol is known to work for a number of filamentous fungi and is based on the use of a gene conferring resistance to an antibiotic e.g., hygromycin B, (*see* ref. (16)). A number of choices for drugs are available and sensitivity of protoplasts to the drug should be experimentally determined on solid medium for each fungus. All steps in this protocol should be performed using sterile technique.

3.1.1. Growth of Cultures for Protoplasting

Start fresh cultures from stock cultures for each experiment. For *Cochliobolus*, scrape a few ice crystals containing conidia and mycelium fragments from −80°C glycerol stock onto CMX medium in 100 mm Petri dishes. Grow at 22–26°C for about 1 week, under warm white lights (General Electric, F40WW-RS.WM) in a 16 h light/8 h dark regime, until nicely conidiated (Fig. 1). Use these plates for Subheading 3.1.2.

3.1.2. Preparation of Protoplasts

1. For *Cochliobolus*, inoculate 100 mL complete medium (CM) in a 500 mL flask with 10^6–10^7 conidia. Alternatively, mycelium from solid or liquid medium may be blended for 5–10 s in a sterile blender cup and then used as inoculum. For other fungi, grow as usual on medium known to produce conidia. Ampicillin (100–150 µg/mL) may be used to prevent bacterial contamination.

2. Grow 12–18 h at room temperature (25–28°C) with shaking at 150–250 rpm (see Note 2).

3. Transfer the culture to 40 mL centrifuge tubes and centrifuge at 5,000 rpm for 5 min at 4°C in a SS34 rotor (Sorvall) or its equivalent. Discard the supernatant and try not to disturb the pellet.

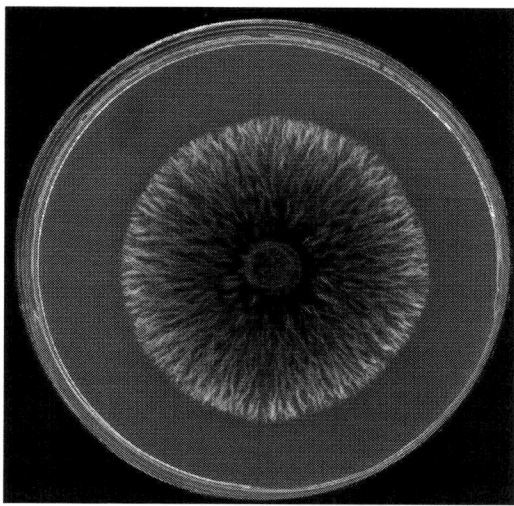

Fig.1. Conidiating colony of *C. heterostrophus* on CMX medium for use in setting up overnight liquid culture of young mycelium for use in transformation.

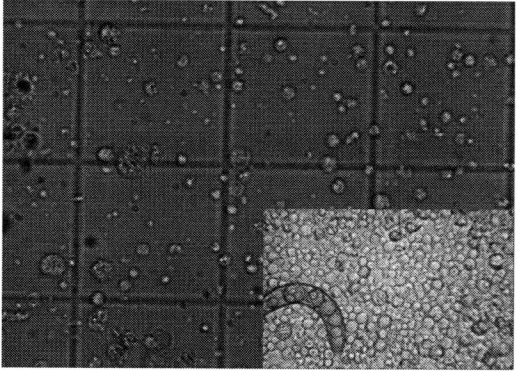

Fig. 2. *C. heterostrophus* protoplasts released in enzyme-osmoticum visualized on a hemocytometer. Sizes are variable. *Inset* shows relationship to single conidium.

4. Scoop 0.5–0.8 g portions (see Notes 3 and 4) of the pellet and resuspend well in 10 mL of enzyme-osmoticum in a 50 mL Erlenmeyer flask (use 80 mL total of enzyme-osmoticum, 8×10 mL) (see Note 5).
5. Place flasks at 30–32°C and shake gently (50 rpm) for 1.5–2 h. 10^8–10^9 protoplasts should be released in this time (count with a hemocytometer) (Fig. 2).
6. Separate the protoplasts from intact mycelium and cell wall debris by filtering the protoplast suspension through four layers of sterile cheesecloth, then through sterile nylon fabric with a 20–25 µm pore size (see Note 6).

7. Pellet the protoplasts by centrifuging the filtrate from Step 6 in 40 mL screw-capped centrifuge tubes at 5,000 rpm for 5–6 min in a SS-34 rotor at 4°C. Discard the supernatant very carefully, trying not to disturb the protoplast pellet (see Note 7). The enzyme-osmoticum can be reused twice if stored at −20°C (see Note 8).

8. Resuspend the pellets gently in a total of 10 mL 0.7 M NaCl, combine into one tube. Recentrifuge as in step 7. Discard the supernatant.

9. Wash the pellet three times with 10 mL STC, pelleting the cells between washes by centrifuging as in step 7. Resuspend protoplasts in 500 μL STC after final wash.

10. Count protoplasts and adjust their density to approximately 10^8/mL with STC.

3.1.3. Transformation

Use at least $0.5–1 \times 10^7$ protoplasts for each transformation. Perform the transformation steps on ice. Always include a control treatment with no DNA.

1. Add 100 μL of protoplast suspension to 13×100 mm test tubes.

2. Add 25 μL of DNA (5–10 μg PCR product or 10–15 μg linearized plasmid DNA) in STC or Tris–EDTA (TE: 10 mM Tris–HCl, 1 mM EDTA, pH 8.0). Mix gently by rolling the tube. Incubate for 2–10 min. For cotransformation, use 5–10 μg PCR product or 10–15 μg linearized plasmid DNA and 2–3 μg of a linearized plasmid carrying a second selectable marker.

3. Add freshly prepared PEG in three aliquots of 200, 200, and 800 μL each, the last at room temperature. Mix well after each addition by rolling the tube. Incubate the tube on ice for 2–10 min after the first two additions and at room temperature after the last addition.

4. Dilute with 1 mL of STC and plate for viability determination and Hygromycin B, Geneticin, or PPT selection as described below.

3.1.4. Plating of Transformed Protoplasts

Determination of protoplast viability (Optional, see Note 9):

1. Remove a 10 μL aliquot from one of the samples. Determine the concentration of the protoplast suspension with a hemocytometer and record.

2. Prepare dilutions from the same sample of $10^2–10^6$/mL in STC.

3. Plate 100 μL aliquots of each dilution in molten regeneration medium by pouring approximately 20 mL of medium into a Petri plate, then adding protoplasts and swirling with the

pipette tip to mix. Allow the medium to solidify, then incubate at 30°C. Estimate protoplast viability as the number of colonies observed/the number of protoplasts plated. Viability of 5–20% is typical.

For Hygromycin B, Geneticin/G418, or Phosphothricin (PPT)/Bialaphos selection:

1. Plate 200–500 μL aliquots in 20 mL molten regeneration medium as described above. After overnight incubation at 30°C, overlay with 10 mL of 1% agar containing 150 μg/mL Hygromycin B, 1,000 μg/mL Geneticin, or 150 μg/mL PPT. The final concentration of Hygromycin B, Geneticin, or PPT in the plate will be 50, 333, and 50 μg/mL, respectively. Use one plate/transformation as no overlay control.

2. Incubate plates at 30°C (see Note 10), transformants generally appear after 3–7 days. Verification that hygromycin B-resistant colonies are indeed transformants requires additional genetic and molecular analyses (see Subheading 3.3.2).

3.1.5. Selectable Markers

A variety of genes conferring resistance to antibiotics is available. Among these are genes for resistance to Hygromycin B, Geneticin/G418, Bialaphos/Phosphothricin, Nourseothricin, Blasticidin, and Phleomycin. Hygromycin B is the most widely used.

3.2. Protocol for Split-Marker PCR (Fig. 3)

Reference: Catlett et al. (21).

3.2.1. Target Fragments and Primers (see Notes 11–14)

Fragment	Primers	Length	Template
First round PCR (four PCR fragments will be amplified in this step)			
I. 5′ Flank region of your gene	FP1 + RP1	(600–1,000 bp)	Genomic DNA containing your gene
II. 3′ Flank region of your gene	FP2 + RP2	(600–1,000 bp)	Genomic DNA containing your gene
III. 5′ Region of *hygB* gene (HY)	M13RHYG + NLC37	(1,147 bp)	pUCATPH (12)
IV. 3′ Region of *hygB* gene (YG)	M13FHYG + NLC38	(1,778 bp)	pUCATPH
Second round PCR (two fused PCR fragments will be amplified in this step)			
V. 5′ Flank fused to HY	FP1 + NLC37	(~2,500 bp)	I + III
VI. 3′ Flank fused to YG	NLC38 + RP2	(~2,500 bp)	II + IV

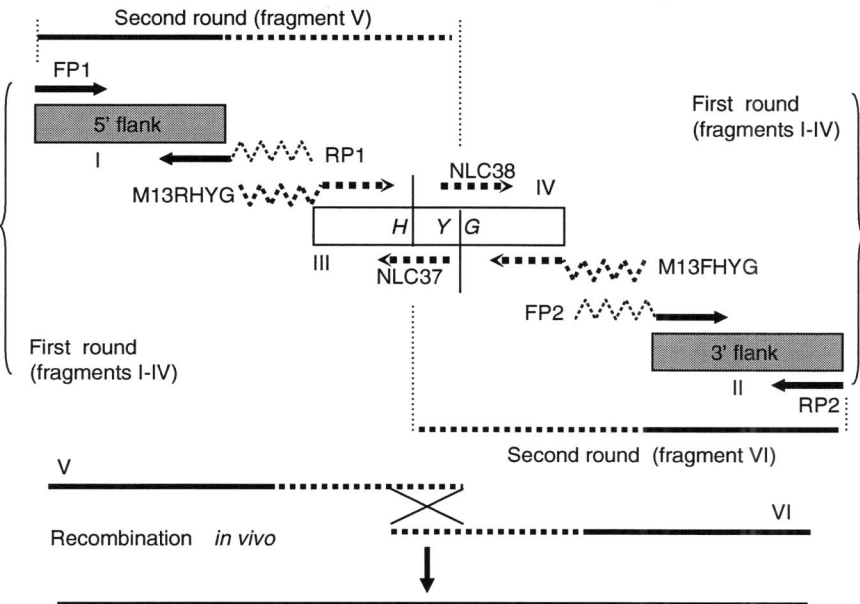

Fig. 3. Split marker procedure (21) as described in text (Subheading 3.2) with primers mapped on fragments.

3.2.2. PCR Conditions

1. First round amplification. 0.4 mM dNTPs, 50–100 nM Primer 1 and Primer 2, 10 µL 5× buffer, 0.5 µL iProof Taq (2 U/µL), 0.5 µg genomic DNA (or 50 ng plasmid DNA), H_2O to 50 µL.
2. Second round fusion. 0.4 mM dNTPs, 50–100 nM Primer 1 and Primer 2, 10 µL 5× buffer, 0.5 µL iProof Taq (2 U/µL), 1–2 µL fragments I, III (or II, IV), H_2O to 50 µL.
3. PCR program: 1 cycle of {98°C, 2 min}, 32 cycles of {98°C 20 s, 59–65°C 30 s, 72°C 0.5–3 min} 72°C for 10 min.

3.2.3. Primer Sequences

Pair I (for the 5′ flank of your gene)

1. FP1: 5′-xxxxxxxxxxxxxxxxxxx-3′ = forward primer to your 5′ flank
2. RP1: 5′-TCCTGTGTGAAATTGTTATCCGCT-xxxxxxxxxxxxxxxxxxxx-3′
(complementary to M13R) reverse primer to your 5′ flank

Pair II (for the 3′ flank of your gene)

1. FP2: 5′-GTCGTGACTGGGAAAACCCTGGCG-xxxxxxxxxxxxxxxxxxx-3′
(complementary to M13F) forward primer to your 3′ flank
2. RP2: 5′-xxxxxxxxxxxxxxxxxxx-3′ = reverse primer to your 3′ flank

Pair III (for 5' region of *hygB*, "HY")

1. M13RHYG   AGCGGATAACAATTTCACACAGGA (=pUCATPH bp 2865–2888RC)
2. NLC37 GGATGCCTCCGCTCGAAGTA (=pUCATPH bp 1685–1704)

Pair IV (for 3' region of *hygB*, "YG")

1. M13FHYG   CGCCAGGGTTTTCCCAGTCACGAC (=pUCATPH bp 352–375)
2. NLC38 CGTTGCAAGACCTGCCTGAA (=pUCATPH bp 2132–2150RC)

3.3. REMI

For *Cochliobolus*, we use the vector pUCATPH (Fig. 4), which has no detectable homology to the *C. heterostrophus* genome. This vector was constructed by ligating the 2.4 kb *Sal*I fragment from pDH25, containing *hygB* fused to the *A. nidulans trpC* promoter and terminator (*12*) into the *Sal*I site of pUC18.

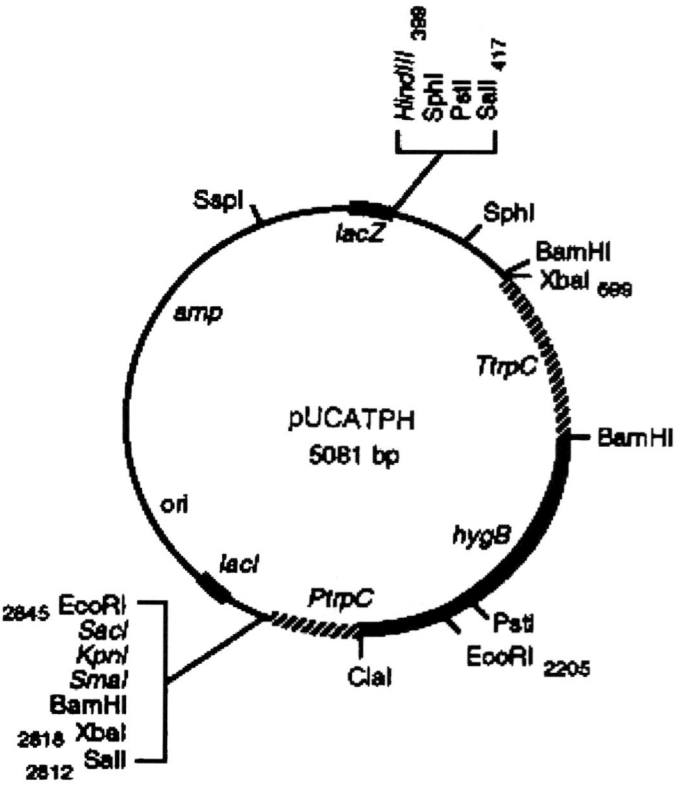

Fig. 4. pUCATPH and REMI vector pUCATPH, constructed as described in (*12*). Six base-pair restriction enzyme sites are shown. Italicized sites are unique. *amp*, Ampicillin resistance; *hygB*, hygromycin B resistance; *PtrpC*, *A. nidulans trpC* promoter; *TtrpC*, *A. nidulans trpC* terminator; *ori*, *Escherichia coli* origin of replication.

3.3.1. Transformation

1. Plasmid DNA is digested with a restriction enzyme that cuts the vector once e.g., *Hind* III and used directly in the original digestion solution (100 µL) or after phenol extraction, ethanol precipitation, and resuspension in 100 µL of sterile TE.
2. Protoplasts are prepared from an overnight culture of *C. heterostrophus* as above (Subheading 3.1.2).
3. Approximately 5×10^6 protoplasts in 100 µL of STC are gently mixed with 30 ug of linearized plasmid in 100 µL of solution with or without restriction enzyme.
4. The mixture is kept on ice for 5–10 min before addition of polyethylene glycol and completion of the usual transformation and plating protocols (Subheadings 3.1.3 and 3.1.4).
5. Control protoplast mixtures include uncut plasmid and phenol extracted, ethanol precipitated linearized plasmid resuspended in 100 µL of sterile TE.
6. After 10 h to overnight incubation at 30°C, plates (each containing 20 mL of regeneration medium) are overlaid with 10 mL of 1% agar containing e.g., hygromycin B at 150 µg/mL and incubated.

3.3.2. Verifying Transformants

1. Purification and confirmation of drug resistance. Under selection, the expectation is that candidate transformants will be obvious against a background of 'fuzz' (Table 1, Figs. 5 and 6). The next step is to transfer plugs of putative transformants to fresh selective medium (CMNS plus Hygromycin B) to test if they are truly resistant to the drug in question. Plugs of these colonies are then transferred to both CMX and to fresh selective medium [CMNSHygB (80–120 µg/mL), CMNSG418 (400 µg/mL), or MM + PPT (50 µg/mL)] to check for drug resistance and to compare the growth of candidates on CMX to that of wild type. Single, drug-resistant conidia are isolated from putative transformants for the analysis of integration events.
2. PCR verification of targeted insertion. To determine the type of integration event, PCR can be performed with selected pairs of primers on DNA from candidate purified transformants

Table 1
Transformation plates

Sample	Expected result	Interpretation
– DNA	No colonies on selection medium	Selection is tight, antibiotic is active
– selection	Many colonies, or lawn	Protoplasts are viable
+ selection	Some colonies	Transformation procedure working

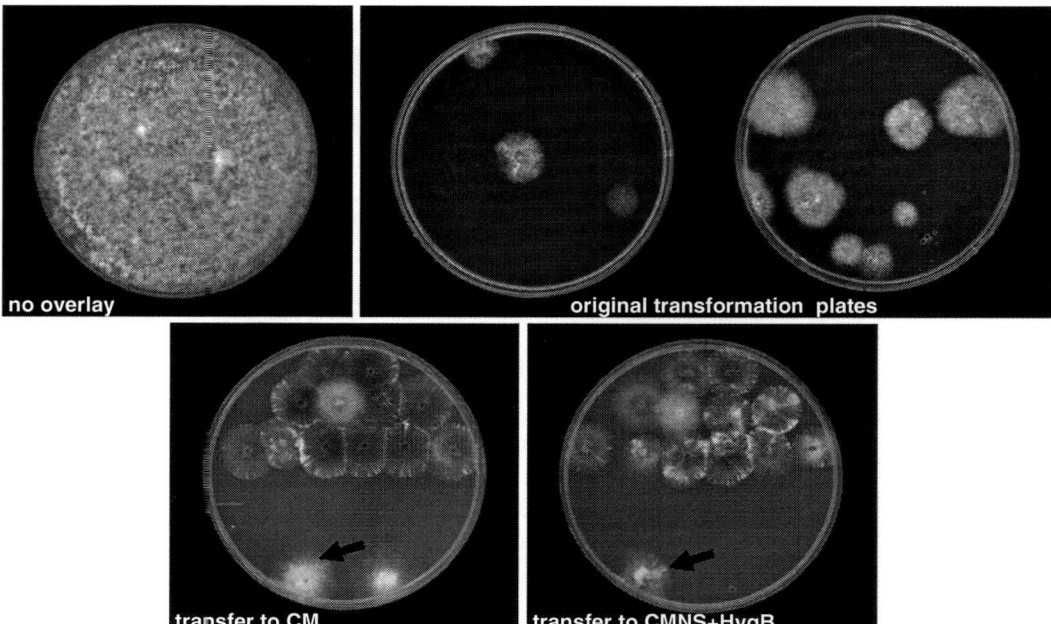

Fig. 5. *C. heterostrophus* on transformation plates. *Top left* protoplasts plated on regeneration medium with no subsequent selective medium overlay. Lush growth indicates that protoplasts were viable. *Top right* protoplasts plated on regeneration medium with subsequent Hygromycin B medium overlay. Candidate transformants appear as fast growing large colonies that grow up through the overlay. If selection is tight, the sparsely growing background of untransformed protoplasts does not break through the overlay. *Bottom* transfer of plugs of candidate transformants (*top rows*) from original transformation plates (e.g., from *Top, right*) to CM and CMNS + hygromycin B. Note that the positive (hygromycin B resistant) and negative (hygromycin B sensitive) control samples, *bottom row* on each plate, both grow on nonselective CM medium, but only the control (hygromycin B resistant) sample grows on the selective medium (*arrows*).

(Fig. 7a, b). It is essential to also include negative control DNA, such as that from a wild-type, drug sensitive strain, and positive control DNA such as that from a known drug resistant strain (Tables 2 and 3).

3. PCR verification of complementation. To determine the type of complementation event (whether or not complementing DNA has integrated at the native locus), PCR can be performed with selected pairs of primers on DNA from candidate purified transformants (Fig. 7c). Include negative control DNA, such as that from a strain resistant to the first selectable marker antibiotic, and a positive control DNA such as that from a strain known to be resistant to the second drug (Fig. 7c, Tables 2 and 3).

4. Southern hybridization. It is our opinion that, in 2009, it is not necessary to determine/prove integration events by DNA–DNA hybridization, if PCR results are clear-cut. There

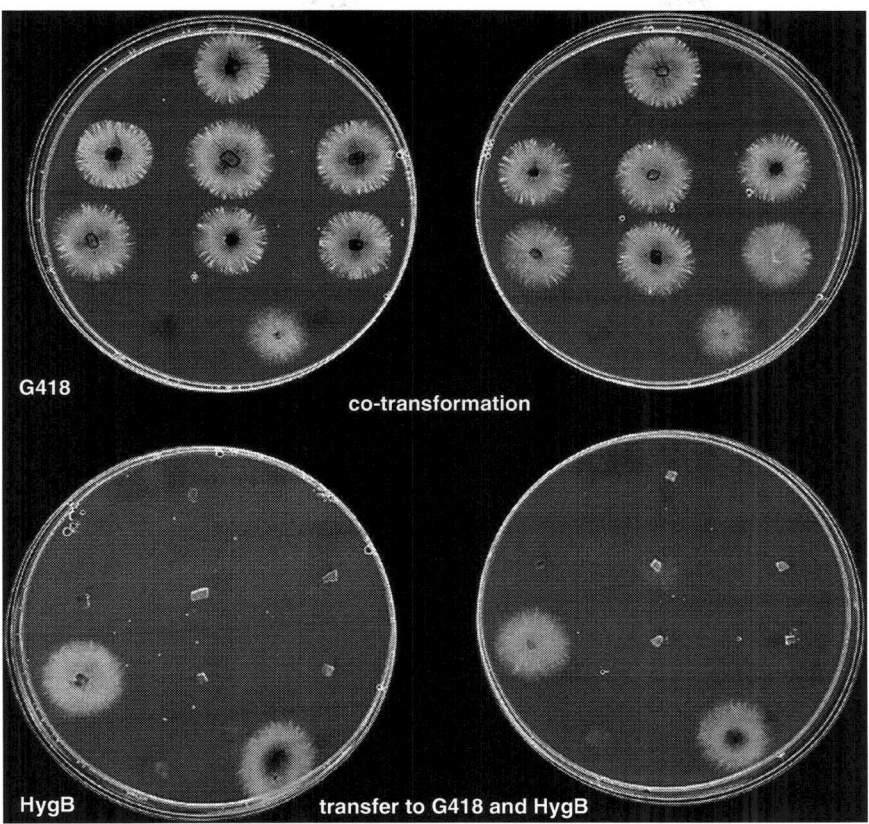

Fig. 6. *C. heterostrophus* cotransformation transfer plates. Strain transformed was hygromycin resistant. Second selectable marker used was *gen^R* conferring resistance to Geneticin (G418). Linearized vector carrying *gen^R* plus a fragment expected to replace the hygB marker were used for transformation. *Top rows*, plugs of candidate transformants. *Bottom* two plugs are the positive (Geneticin/G418 resistant, *right*) and negative (Geneticin/G418 sensitive, *left*) control samples. Note the latter fail to grow on Geneticin. *Bottom row*, same colonies as on *Top plates*. Most colonies fail to grow on Hygromycin B, indicating that the marker was replaced by the introduced DNA.

are only two possible types of integration event – by homology or at an ectopic site. Similarly, there are only two possible configurations of the introduced DNA – as a single copy or as multiple copies. Finally, there are only two possible outcomes in terms of integration location – at a single or at multiple sites. If the integration events are not clear when analyzed by PCR, then Southern hybridizations can be performed. This is best done by using an enzyme that does not cut the introduced DNA, as this type of digestion when probed with the introduced DNA can resolve all three aforementioned integration issues. Additional enzymes that cut the introduced DNA can be used to further sort out details (*see* for example ref. (6, 16)).

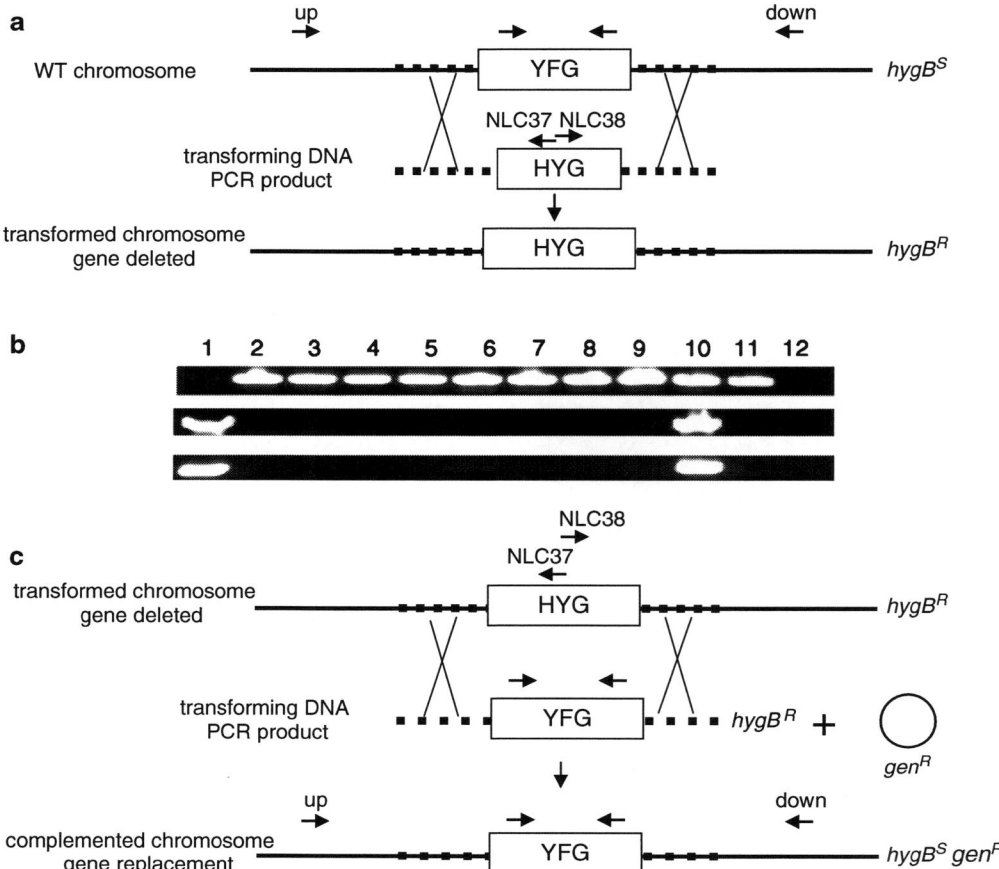

Fig. 7. Diagram of integration events and primer locations. (a) Chromosomal configurations if a gene deletion by double crossover occurs (YFG = your favorite gene). *Top line*, wild-type chromosome (hygB^S). *Middle line*, transforming DNA generated by split marker PCR, carrying the *hygB* gene. *Bottom line*, transformed chromosome, YFG is replaced by the *hygB* gene (HYG), strain is resistant to hygromycin B (hygB^R). NLC37, NCL38 = primers to the *hygB* gene (21), up and down = primers to the chromosome on the *left* and *right* flanks of YFG, outside the introduced PCR fragment. Unmarked arrows (*top line*) are gene (YFG)-specific primers. *See* Table 2. (b) Agarose gel image of PCR results using primers shown in (a) and Table 2. and DNA from transformants and control strains. Lanes: 1 = WT, 2–11 = candidate transformants, and 12 = negative control. Top panel shows products using *hygB* primers NLC37 and NLC38, panel in the middle, up and YFG gene-specific primers, bottom panel, YFG gene-specific primers. Isolate in lane #10 is ectopic, isolates in lane 2–9, 11 show targeted integration and YFG gene loss. (c) Chromosomal configuration if complementation of a hygB^R strain occurred at the native YFG site. *Top line*, wild-type chromosome (hygB^R). *Middle line*, YFG DNA generated by PCR, plus a separate vector (linearized) fragment carrying *gen*^{R}. Bottom line, transformed complemented chromosome, *hygB* is replaced by YFG and strain is resistant to geneticin (gen^S). NLC 37, NCL38 = primers to the *hygB* gene (21), up and down = primers to the chromosome on the left and right flanks of YFG, outside the PCR fragment. Unmarked arrows (*middle* and *bottom lines*) are YFG-specific primers. *See* Table 3.

Table 2
PCR verification of gene deletion

Strain	Primer combination[a]					HygB[R]
	Up/NLC37	Down/NLC38	NLC37/38	Gene specific	Up/Down	
WT	–	–	–	+	+	–
knockout	+	+	+	–	+ (Different size from WT)	+
ectopic	–	–	+	+	+ (WT size)	+

[a]Up and down primers = matching left and right flanks, respectively, outside of introduced PCR product p, NLC37 and NLC38 primers match the Hygromycin B gene see Catlett et al. (21), gene-specific primers match the gene of interest (Fig. 7)

Table 3
PCR verification of complementation

Strain	Primer combination[a]						First marker e.g., HygB[R]	Second marker e.g., Gen[R]
	Up/ NLC37	Down/ NLC38	NLC37/ 38	Gen[R] specific	Gene specific	Up/Down		
Starting strain (HygB[R])	+	+	+	–	–	Different from WT	+	–
Complemented (HygB[S], Gen[R])	–	–	–	+	+	+ (Same size as WT)	–	+
Ectopic (HygB[R], Gen[R])	–	–	+	+	+	Different from WT	+	+

[a]Assuming e.g., cotransformation and complementation of a hygB-resistant strain with a second selectable marker Gen[R]. Up and down primers = matching left and right flanks, respectively, outside of introduced PCR product, NLC37 and NLC38 primers match the HygB[R] gene see Catlett et al. (21), Gen[R]-specific match the gene for resistance to Geneticin, gene-specific primers match the introduced gene of interest

4. Notes

1. Glucanex (Novozyme) is widely used, however it does not always work well. Various other enzyme preparations might be tested in this case.
2. For some fungi, it may be difficult to digest cell walls if grown too long. If so, use only young (~12 h) germlings.
3. It is important not to have too much mycelium per unit volume of enzyme-osmoticum. We usually use a stainless steel spatula with a "spoon" tip and use about half a spoonful per 10 mL enzyme osmoticum in a 50 mL flask.

4. When scooping pellets (Subheading 3.1.2), keep in mind that the best protoplasts will come from the youngest tissue. Try to collect only the white part of the pellet, avoiding the black part at the very bottom of the tube. This is often the case if the overnight growth was inoculated with mycelium rather than conidia.
5. To suspend scooped pellets in enzyme osmoticum, gently shake the 50 mL flask.
6. Cheesecloth and nylon fabric should be stretched over 100 mL beakers, separately, covered in foil and autoclaved. Alternatively, cheesecloth and nylon may be preassembled as a unit in a funnel supported in a beaker and autoclaved.
7. Pellets tend to be loose. Decant carefully and stop if protoplasts are being resuspended. Osmoticum will be diluted out in subsequent washes.
8. Quality of protoplasts deteriorates upon storage. We do not recommend using frozen protoplasts for important experiments, since the entire transformation procedure is labor intensive. Use 5–10× more protoplasts per transformation if using frozen protoplasts.
9. It is a good idea to determine protoplast viability when first trying to transform a previously untransformed fungus. With success, this step is no longer necessary.
10. For *Cochliobolus*, plates can be kept at 30°C until colonies appear or moved to conditions described in Subheading 3.1.1.
11. Primers must be correct (pay attention to complementary sequences, *see* Fig. 3).
12. Concentration of primers (always use minimum).
13. PCR tips for generating fragments V and VI: Use longer melting and annealing temperatures as specified by TAQ to help reduce nonspecific binding and amplification. Annealing temperatures should also be set higher than normal to reduce nonspecific binding.
14. Purifying fragments I–IV is essential for removing trace amounts of primers and obtaining the correct fragments V + VI. Purifying I–IV 2× can be important. Alternatively, increasing the amount of template and reducing the primer concentration can reduce the formation of extra bands for V + VI.

Acknowledgments

We thank previous lab members (Shun-Wen Lu, Natalie Catlett) who contributed to developing transformation, REMI, and split marker protocols.

References

1. Lorenz MG, Wackernagel W (1994) Bacterial gene transfer by natural genetic transformation in the environment. Microbiol Rev 58:563–602
2. Avery OT, MacLeod CM, McCarty M (1979) Studies on the chemical nature of the substance inducing transformation of pneumococcal types. Inductions of transformation by a desoxyribonucleic acid fraction isolated from pneumococcus type III. J Exp Med 149:297–326
3. Hinnen A, Hicks JB, Fink GR (1978) Transformation of yeast. Proc Natl Acad Sci U S A 75:1929–1933
4. Case ME, Schweizer M, Kushner SR, Giles HN (1979) Efficient transformation of *Neurospora crassa* by utilizing hybrid plasmid DNA. Proc Natl Acad Sci U S A 76:5259–5263
5. Yelton MM, Hamer JE, Timberlake WE (1984) Transformation of *Aspergillus nidulans* by using a *trpC* plasmid. Proc Natl Acad Sci U S A 81:1470–1474
6. Turgeon BG, Garber RC, Yoder OC (1985) Transformation of the fungal maize pathogen *Cochliobolus heterostrophus* using the *Aspergillus nidulans amdS* gene. Mol Gen Genet 201:450–453
7. Wang J, Holden DW, Leong SA (1988) Gene transfer system for the phytopathogenic fungus *Ustilago maydis*. Proc Natl Acad Sci U S A 85:865–869
8. Turgeon BG, Oide S, Bushley K (2008) Creating and screening *Cochliobolus heterostrophus* non-ribosomal peptide synthetase mutants. Mycol Res 112:200–206
9. Schiestl RH, Petes TD (1991) Integration of DNA fragments by illegitimate recombination in *Saccharomyces cerevisiae*. Proc Natl Acad Sci U S A 88:7585–7589
10. Kuspa A, Loomis WF (1992) Tagging developmental genes in *Dictyostelium* by restriction enzyme-mediated integration of plasmid DNA. Proc Natl Acad Sci U S A 89:8803–8807
11. Dynes JL, Clark AM, Shaulsky G, Kuspa A, Loomis WF, Firtel RA (1994) *LagC* is required for cell-cell interactions that are essential for cell-type differentiation in *Dictyostelium*. Genes Dev 8:948–958
12. Lu SW, Lyngholm L, Yang G, Bronson C, Yoder OC, Turgeon BG (1994) Tagged mutations at the *Tox1* locus of *Cochliobolus heterostrophus* using restriction enzyme-mediated integration. Proc Natl Acad Sci U S A 91:12649–12653
13. Sweigard J (1996) A REMI primer for filamentous fungi. IS-MPMI Rep Spring, 1996:3–5
14. Fincham JRS (1989) Transformation in fungi. Microbiol Rev 53:148–170
15. Griffin D (1981) Fungal physiology. Wiley, New York
16. Turgeon BG, Garber RC, Yoder OC (1987) Development of a fungal transformation system based on selection of sequences with promoter activity. Mol Cell Biol 7:3297–3305
17. Turgeon G, Yoder O (1985) Genetically engineered fungi for weed control. In: Cheremisinoff P, Ouellette R (eds) Biotechnology handbook. Technomic, Lancaster PA, pp 220–229
18. Case M (1982) Transformation of *Neurospora crassa* utilizing recombinant plasmid DNA. In: Hollander A, DeMoss RD, Kaplan S, Konisky J, Salvage D, Wolfe RS (eds) Genetic engineering of microorganisms for chemicals. Plenum, NY, pp 87–100
19. Bennett JW, Lasure LL (1991) More gene manipulations in Fungi. Academic, San Diego
20. Ganem S, Lu SW, Lee BN et al (2004) G-protein beta subunit of *Cochliobolus heterostrophus* involved in virulence, asexual and sexual reproductive ability, and morphogenesis. Eukaryot Cell 3:1653–1663
21. Catlett N, Lee B-N, Yoder O, Turgeon B (2003) Split-marker recombination for efficient targeted deletion of fungal genes. Fungal Genet Newsl 50:9–11

Chapter 2

Electroporation and *Agrobacterium*-Mediated Spore Transformation

Anna Minz and Amir Sharon

Abstract

Genetic transformation is a key technology in modern fungal research. Most commonly, protoplasts are transformed using the polyethylene glycol-mediated transformation protocols. Because protoplasts are generated by treatment of mycelia with a crude enzyme preparation, the results tend to be inconsistent. Furthermore, some species cannot be transformed by this method. Electroporation (EP) and *Agrobacterium tumefaciens*-mediated transformation (AMT) are two alternative methods. These methods allow the transformation of spores or mycelia, they are simple to perform and provide consistent results. In this chapter, we describe EP and AMT protocols for the fungus *Colletotrichum gloeosporioides* f. sp. *aeschynomene* (*C. gloeosporioides*). These protocols can be used as baseline for the calibration of similar transformation protocols in other species.

Key words: Transformation, *Agrobacterium*, AMT, Electroporation, *Colletotrichum*

1. Introduction

Genetic transformation is an essential part of modern fungal research. Polyethylene glycol (PEG)-mediated protoplasts transformation has been the method of choice for many years and transformation protocols were developed in numerous species according to the original *Neurospora crassa* protocol (1). For a detailed protocol of protoplast transformation, see Chapter 1. However, although widely used, this system has several drawbacks. Primarily, protoplasts production and recovery might be problematic, especially in slow growing species. It is also well known that results vary considerably between different enzyme batches resulting in inconsistent transformation efficiencies. Alternative methods for direct transformation of spores or hyphae can provide efficient solution to these inherent problems. EP and AMT are the most commonly used alternatives. Both methods can be used to transform spores or hyphae, once working they are highly reproducible, and they have been used to transform a range of fungal species.

During EP, a high-voltage electric pulse creates a population of small, aqueous pores in the cell membrane through which DNA can enter the cell by diffusion or electrophoretically (2). The technique can be used to transform protoplasts as well as mycelia or spores (3–6). For spore transformation, the spores are usually pregerminated or incubated with a mild concentration of cell wall degrading enzymes (7–9).

Transformation of *Saccahromyces cerevisiae* by *Agrobacterium tumefaciens* was first reported in 1995 (10). Subsequently, it was demonstrated that AMT can be used to transform protoplasts as well as spores or mycelia of filamentous species (11). AMT is based on the natural ability of *A. tumefaciens* to transfect plant cells with a specific part of DNA (T-DNA). The discovery that *A. tumefaciens* can also transfect fungal cells attracted much interest in this method, and protocols were developed for numerous species, including member of Ascomycota, Basidiomycota, and Zygomycota, as well as Oomycetes (12). AMT solved transformation problems in a number of species such as *Agaricus bisporus* and *Sclerotinia sclerotiorum* and has become the method of choice when developing a new transformation protocol. AMT has also been reported to help solving problems of low homologous integration rates in certain species (13).

In this chapter, we describe an EP and AMT spore-transformation protocols. The protocols are specific to the fungus *C. gloeosporioides*, however, these protocols can be used as baseline for the calibration of similar procedures in various fungal species.

2. Materials

2.1. Agrobacterium-Mediated Transformation

2.1.1. Fungal Spores

2.1.2. Agrobacterium

1. Emerson's YpSs (EMS) agar medium. For 1,000 mL: yeast extract 4 g, soluble starch 2.5 g, $K_2HPO_4 \cdot 3H_2O$ 1 g, $MgSO_4 \cdot 7H_2O$ 0.5 g, agar 16 g (see Note 1).

1. Antibiotics. Ampicillin (Amp), kanamycin (Kan), carbenicillin (Carb, Duchefa), and clafuran (Cla) are prepared in water as 100 mg/mL stock solutions and stored at −20°C. The final concentration for all antibiotics is 100 μg/mL (see Note 2).
2. Luria-Bertani (Miller) bacterial growth medium (LB) 25 g/L.

3. MES buffer. Prepare 1 M stock solution, pH 5.5: 625 g/L MES, titrate to pH 5.5 with 40% NaOH, store at 4°C under dark conditions.

4. Microelements solution. For 1,000 mL: 100 mg $ZnSO_4 \cdot 7H_2O$, 100 mg $MnSO_4 \cdot H_2O$, 100 mg $CuSO_4 \cdot 5H_2O$, 100 mg $Na_2MoO_4 \cdot 7H_2O$, 100 mg H_3BO_3. Store solution in a light protected bottle at 4°C.

5. 1% $CaCl_2$. Dissolve 0.3 g $CaCl_2$ in 30 mL water, autoclave and store at Room Temperature (RT).

6. 0.01% $FeSO_4$. Dissolve 0.02 g $FeSO_4$ in 200 mL water. Filter sterilize and store at RT in a light protected bottle.

7. 50% Glycerol. Dissolve 25 g in 40 mL water. Sterilize before use.

8. MN buffer. Dissolve 9 g $MgSO_4 \cdot 7H_2O$ and 4.5 g NaCl in 300 mL water and autoclave. Store at RT.

9. 20% NH_4NO_3. Dissolve 8 g of NH_4NO_3 in 40 mL water. Autoclave and store at RT.

10. 20% glucose. Dissolve 20 glucose in 100 mL water and autoclave. Store at RT.

11. Minimal medium (MM). For 800 mL (add in the following order): 0.64 mL K_2HPO_4 buffer 1.25 M, pH 4.8, 0.8 mL $CaCl_2$ 1%, 4 mL microelements solution, 8 mL $FeSO_4$ 0.01%, 32 mL MES buffer (1 M, pH 5.5), 8 mL glycerol 50%, 16 mL MN buffer, 2 mL NH_4NO_3 20% (w/v), 8 mL glucose 20% (w/v), water to 800 mL. Divide into 50 and 750 mL volumes. To the 50 mL volume, add 0.5 mL of 20% glucose and autoclave. To the 750 mL, add 11 g agar and autoclave.

12. Acetosyringone (Sigma). For 200 mM stock solution: dissolve 40 mg acetosyringone into 1 mL DMSO. Store in a light protected tube at −20°C. The final concentration is 200 µM.

13. Induction medium (IM): MM + 200 µM acetosyringone (see Note 3).

14. YENB. For 1,000 mL: 7.5 g yeast extract, 8 g nutrient broth, complete to 1 L with water.

2.1.3. Transformation of Agrobacterium

1. Plasmid DNA. A binary vector with T-DNA insertion sites should be used. Any commercial vector such as pBin19, pCAMBIA or equivalent can be used (see Note 4).

2. *Escherichia coli*. Propagate plasmid DNA in an *E. coli* commercial strain such as DH5α (Bethesda Research Laboratories) or equivalent.

3. Electroporator. All types of electroporators should be suitable. Use 0.2 cm cuvettes.

2.1.4. Fungal Transformation

1. Hemacytometer (Neubauer Improved Counting Chamber, 0.1 mm depth).
2. IM plates. To every 100 mL of MM agar (750 mL bottle, see Subheading 2.1.2, item 11) add 100 µL acetosyringone from stock solution. Keep plates in complete darkness.
3. Cellophane membranes. Use cellulose-based cellophane membranes (see Note 5). Cut 90 mm discs, place between two layers of wet filter paper and autoclave.
4. Regeneration medium (REG). For 1,000 mL: mannitol 145.7 g, yeast extract 4 g, soluble starch 1 g, agar 16 g, water to 1 L.

2.1.5. Selection of Transformants

1. Hygromycin B (Hyg). Prepare stock in water, 100 mg/mL (see Note 6).
2. REG + Hyg + Cla plates. To 100 mL of 50°C molten REG medium add 40 µL Hyg and 100 µL Cla. Mix thoroughly before plating. Keep plates in complete darkness.
3. EMS + Hyg. To 100 mL of 50°C molten EMS medium, add 100 µL Hyg. Mix thoroughly before plating. Keep plates in complete darkness.
4. Overlay agar + Cla. Prepare and autoclave 1% (w/v) water agar. To 100 mL of 50°C molten agar, add + 400 µL Cla.

2.2. Electroporation

All solutions must be autoclaved and stored at RT unless otherwise indicated.

2.2.1. Fungal Spores

1. Emerson's YpSs (EMS) agar medium. see Subheading 2.1.1, item 1.
2. Hemocytometer. see Subheading 2.1.4, item 1.
3. Pea extract (PE). Cook 800 g of frozen peas in a pressure cooker for 15 min, in 1.6 L water. Filter through Miracloth (Calbiochem), transfer liquid to bottles and autoclave for 30 min. Can be stored at RT for up to 1 month (see Note 7).

2.2.2. Fungal Transformation

1. Electroporation Buffer. For 1 L: HEPES 238 g, mannitol 9.108 g. Adjust pH to 7.5 with 1 M NaOH. Buffer must be cooled to 4°C before use.
2. Plasmid DNA. Prepare plasmid DNA in distilled water or electroporation buffer at a concentration of 1 µg/µL (see Note 8).
3. Electroporator. Small instruments such as the MicroPulser Electroporator (BioRad), which are used to transform bacteria are not suitable for most fungi. An instrument in which the voltage, resistance, and capacitance can be controlled, [e.g., Gene Pulser Xcell (BioRad)] should be used.

4. Liquid REG. See Subheading 2.1.4, item 4 without the agar.

2.2.3. Selection of Transformants

1. Hyg. see Subheading 2.1.5, item 1.
2. REG. see Subheading 2.1.4, item 4.
3. Overlay agar + Hyg. Prepare and autoclave 1% water agar. To 100 mL of 50°C molten agar, add 100 µL Hyg. Mix thoroughly before plating.
4. REG + Hyg plates. To 100 mL of 50°C molten REG medium, add 100 µL Hyg. Mix thoroughly before plating.

3. Methods

3.1. Agrobacterium-Mediated Transformation

3.1.1. Preparation of Fungal Spores

1. Grow the fungus on EMS plates for 5 days at 28°C (9) (see Note 9).
2. Wash plates with sterile distilled water. Transfer spore suspension to centrifuge tubes and spin at $6,000 \times g$ for 10 min at 4°C.
3. Resuspend spores in 2 mL MM (control) or IM (MM + acetosyringone), count spores with the aid of a hemacytometer (dilute first if too dense) and bring to a final density of 10^8 spores/mL.
4. Dispense 100 µL of the spore suspension into Eppendorf tubes and place on ice.
5. Number tubes according to treatments.

3.1.2. Preparation of Competent Agrobacterium Cells

1. Plate *Agrobacterium* strain AGL1 from glycerol on LB agar plates + Kan + Carb. Incubate at 28°C for 2 days. For longer periods, keep plates at RT.
2. In the morning, inoculate 50 mL culture tubes containing 5 mL YENB + Kan + Carb with single colonies from plates. Grow cells at 28°C with agitation at 220 rpm.
3. In the evening, use the resulting starter to inoculate two 1 L flasks containing 200 mL of YENB + Kan + Carb. Grow cells at 28°C with agitation at 220 rpm overnight or until O.D$_{A595}$ 0.6–0.9.
4. Chill flasks on ice and transfer content to sterile centrifuge tubes.
5. Centrifuge at $4,000 \times g$ for 10 min at 4°C, discard supernatant, combine the resulting pellets with the remaining supernatant and recentrifuge.
6. Resuspend cells in 50 mL of ice cold sterile distilled water, centrifuge at $4,000 \times g$ for 10 min at 4°C, and discard the supernatant. Repeat this washing step one more time.

7. Resuspend pellet in 10 mL of 10% ice cold glycerol, centrifuge at 4,000×g for 10 min at 4°C and discard the supernatant.
8. Resuspend cells in 2–3 mL of 10% ice cold glycerol, dispense 100 µL aliquots of the competent cells into Eppendorf tubes and store at −80°C.

3.1.3. Transformation of Agrobacterium with Plasmid DNA by Electroporation

1. Prepare sufficient amount of growth medium such as YENB (without antibiotics) for recovery of the electroporated cells. Adjust the electroporator settings to 2.5 kV, 25 µF, and 400 Ω, or follow the instructions of the electroporation apparatus if it has specific settings for *Agrobacterium*.
2. Chile cuvettes on ice, add 1 µg of plasmid DNA into 100 µL of competent cells and transfer the mixture to a cuvette (keep on ice). Dry the cuvette with a paper towel, place in the electroporation chamber and pulse once. When using an *Agrobacterium* specific program, simply follow the instructions.
3. Remove the cuvette from the chamber, immediately add 700 µL of sterile YENB medium and mix gently. Transfer the bacteria to an Eppendorf tube and incubate at 28°C for 3 h with agitation.
4. Plate 50–100 µL of cells on solid LB medium with the appropriate antibiotics to select for positive transformants. Incubate plates at 28°C for 2 days and then transfer colonies to separate plates.

3.1.4. PCR Analysis

Presence of the transforming vector in resulting colonies can be verified by "colony PCR." Prepare typical PCR reaction mix and aliquot 25 µL into each PCR tube (use each tube to analyse an individual colony). Place tubes on ice and transfer a small amount of colony to each tube using a toothpick or a 100 µL tip. After transferring of bacteria, leave tip inside the tube for 5–10 min (keep tubes on ice during the entire process). The amount of cells needed is small and sufficient mixing will result in complete cell lysis. Set the first cycle of PCR program conditions (single first cycle operating at 95°C) to 5–10 min to achieve initial cell breakage in addition to DNA denaturation.

3.1.5. Fungal Transfection

1. Pick a single colony from fresh plates and inoculate into 5 mL of LB + Kan + Carb. Culture overnight at 28°C with agitation at 220 rpm.
2. Centrifuge the culture, remove medium and resuspend in 2 mL LB *without antibiotics*. Divide the culture into two 50 mL tubes (2.5 mL each), dilute the bacteria to $OD_{A595} = 0.1$ with LB (control) or LB + 200 µM acetosyringone.

Incubate overnight under dark conditions with agitation at 220 rpm. Centrifuge the cells, remove the medium and dilute with MM (control) or IM to $OD_{A595} = 0.25–0.3$ (approximately 10^8 cells/mL).

3. Place sterile cellophane discs on MM plates (control) and on IM plates. Prepare four plates per each transformation treatment.

4. Mix 100 µL of bacteria suspension with 100 µL of spore suspension in an Eppendorf tube. Plate on the cellophane discs, 50 µL of the suspension per plate. Spread the suspension with glass beads or with a sterile dispensing stick. Incubate under dark conditions at 28°C for 48 h (see Note 10).

5. After the first incubation (48 h), transfer the cellophane discs onto REG plates without selection. Incubate plates for 18 h under light at 28°C (see Note 11).

6. Transfer the cellophane discs onto REG + Hyg (40 µg/mL) + Cla (100 µg/mL) (see Note 12). Incubate under dark conditions and watch for the appearance of colonies over 3–8 days (see Note 13).

7. When colonies appear on the plates, transfer a small piece (approximately 2 × 2 mm) onto fresh EMS + Hyg (100 µg/mL) + Cla (100 µg/mL) (see Note 14).

8. After 2 days, apply 1% agar + Cla (400 µg/mL) onto the plates. Incubate in light for 24 h, and then cut tips of hyphae that grow on the top agar layer and transfer to REG + Hyg (100 µg/mL) plates.

9. Check for insert using PCR or Southern blot according to conventional methods.

3.2. Electroporation

All steps in this protocol are carried under sterile conditions. It is also important to use large orifice tips in order to minimize damage to the conidia. Ampicillin (100 µg/mL) may be added to REG or EMS solid media to prevent bacterial contamination.

3.2.1. Preparation of Fungal Spores

1. Grow fresh fungal cultures on EMS plates for 5 days at 28°C under continuous fluorescent light. At this stage, the cultures sporulate profusely and spores germinate at high rates (see Note 15). After 5 days, harvest conidia by washing the plates with sterile distilled water.

2. Collect the liquid with spores from the plate, determine spore concentration with the aid of a hemacytometer (dilute first if too dense) and bring to a final density of 10^6 spores/mL in 50 mL of PE (see Note 16).

3. Dispense 50 mL spore suspension into 250 mL Erlenmeyer flasks and incubate for 2.5 h at 28°C with agitation at 190 rpm (see Note 17).

4. After 2.5 h, determine germination rate (see Note 18). Proceed only if germination rates exceed 50%.

5. Transfer the flasks to ice for 2 min to stop germination, and then transfer the cultures into 40 mL centrifuge tubes and centrifuge at $5,000 \times g$ for 4 min at 4°C (see Note 19).

6. *Carefully* discard supernatant without disturbing the pellet. When using several tubes, combine all pellets (with small amount of supernatant) into a single tube, centrifuge again and discard the remaining supernatant.

7. Resuspend the pellet gently but thoroughly in 30 mL of ice cold sterile distilled water by pipetting (*do not vortex*) and centrifuge at $5,000 \times g$ for 4 min at 4°C. Carefully discard the supernatant.

8. For each sample, prepare a 1.7 mL sterile Eppendorf tube and a 0.2 cm cuvette. Keep tubes and cuvettes on ice. Resuspend conidia by pipetting with sterile, ice cold Electroporation buffer (1 mL per sample).

9. Distribute 1 mL of the cell suspension into each tube (see Note 20). Centrifuge at $5,000 \times g$ for 4 min at 4°C in a micro centrifuge, carefully remove the supernatant and then add 100 µL of sterile, ice cold Electroporation buffer to each tube.

10. Add 1 µg DNA to each tube and mix gently (see Note 21).

11. Incubate on ice for at least 10 min (can be extended to several hours if necessary). Always include negative (no DNA) and positive (a known plasmid) control treatments.

3.2.2. Transformation of Germinated Spores with Plasmid DNA

1. Prepare plates containing 20 mL of solid REG medium and a sterile, ice cold, REG with 5% PE.

2. Adjust the electroporator to the desired parameters; the optimal conditions for *C. gloeosporioides* 1.4 kV, 25 µF, 800 Ω (see Note 22).

3. Pipette the cells carefully inside the Eppendorf tube and transfer the mixture of cells and DNA to a cold, 0.2 cm electroporation cuvette.

4. Tap the spore suspension to the bottom of the cuvette, place the cuvette in the electroporation chamber and pulse once. Remove the cuvette from the chamber and immediately add 1 mL of sterile, ice cold REG with 5% PE medium.

5. Quickly but gently resuspend the cells, transfer back to the Eppendorf tube, and place on ice.

6. After 10 min, apply the content of each tube onto four plates of solid REG medium (250 µL per plate) and spread the cells using a sterile dispensing stick. Leave the plates in the sterile

hood until they are completely dry (about 30 min), and then incubate with lid facing down at 28°C (see Note 23).

7. After an overnight incubation, overlay with 10 mL of 1% agar with 100 µg/mL Hyg. Overlay half of the negative control plates with agar without the selection marker. Allow to dry and then incubate plates with lid facing down at 28°C. Colonies appear on top of the overlay after 4–6 days.

3.2.3. Analysis of Transformants

1. Transfer plugs of putative transformants to fresh selective medium (e.g., REG with 100 µg/mL Hyg) to verify if they are resistant to the drug. Always include wild type strain as control.
2. When using a reporter gene such as GFP, transformants can be identified by fluorescent microscopy. Otherwise, extract DNA from colonies and verify the presence of the transforming vector by PCR or Southern blot using conventional methods.

4. Notes

1. In other species, use appropriate sporulation medium.
2. Clafuran is a common drug and can be usually purchased from local pharmacies under the trade name Cefotaxime.
3. Prepare fresh before use.
4. For calibration of a new transformation protocol, it is recommended to use a plasmid with a GFP-expression cassette. This will allow easy and fast screening of practically unlimited number of colonies, to differentiate between transformants and background.
5. It is also possible to use nitrocellulose membranes; however, they are very expensive and have no advantage over cellophane. When using cellophane membranes, it is important to use only the type made up of natural cellulose which are used for backing and can be found in food stores. The synthetic cellophane (that is usually obtained in office supply stores) does not transfer the nutrients and should not be used.
6. This protocol describes only the use of the *HPH* (Hygromycin phosphotransferase) gene that confers resistance to hygromycin B as selectable marker. Other markers such as phleomycin or nourseothricin can be used in a similar way.
7. Pea extract (PE) can be prepared using either a pressure cooker or an autoclave for 15 min. Filtrate through Miracloth

and autoclave again for sterilization. Do not use immediately and allow to stand over-night for precipitation. It is advisable to use the medium for up to 1 month as well as avoiding its cooling due to crystallization and precipitation.

8. Linear or circular plasmid DNA can be used, although higher (up to 70%) transformation rates are obtained with linear DNA. Co-transformation with two or three plasmids is also possible.

9. When using other species, optimal conditions (medium, light, days) for the production of high-rate germinating spores must be determined.

10. Time of incubation may vary and should be determined experimentally. For most species 48 h should be fine, however fast growing species might be removed after 24 h to avoid overgrowth.

11. Normally, filters are transferred directly from the transformation plates (IM) to selection plates. In cases of weak growth on the selection plates and lack of colonies, it is recommended to add this extra step, which allows recovery of the fungus and expression of the resistance gene before exposure to the selective drug.

12. Hyg concentration is critical; too low concentration will result in high background and inability to distinguish transformants from the background, whereas at too high concentrations there will be no colonies. The optimal concentration should allow the development of low background after 24 h, but such that will not further develop after 48 h.

13. Colonies may appear already after 2 days, but should be transferred only when a clear colony has been produced, which usually takes at least 3 days. Additional colonies may appear during several days. After more than 8 days, colonies may develop that are not transgenic, and it is recommended to avoid such late developed colonies. The size of colonies may vary considerably, e.g., due to the number of colonies per plate. Ideally, there should be no more than ten colonies per 90 mm plate.

14. Colonies will appear as round, pink to dark color and should be clearly distinguished from background. When GFP is used, the colonies can be easily scanned using a fluorescent stereoscope and true transformants can be identified without opening the plate lids.

15. Achieving high rates of uniform germination is critical for this method. Therefore, the optimal stage for the production of high numbers of readily germinating spores must be determined.

16. Uniform germination is very important. If for any reason the spores are not germinated directly after harvest, they must be kept on ice until transferred to PE in order to prevent initiation of germination during preparations.

17. The amount of Erlenmeyers needed is half the amount of samples (cuvettes) used (e.g., for eight samples, four Erlenmeyers are needed). However, at least five Erlenmeyers should be used in order to pellet the cells properly.

18. In order to determine germination rates, 20 μL samples are taken out of the Erlenmeyers and viewed with light microscope. High germination rates are expected (up to 80% of conidia form aggregates and develop short germ tubes). If no or little germination is observed, incubation may be extended for an additional 20 min. Below 50% germination, the transformation will not work and it is recommended not to continue.

19. It is important to perform the following steps in sterile and ice cold conditions, thus utilization of sterile precooled solutions, cuvettes, and maintenance of spores on ice at all times are required.

20. The amount (mL) of electroporation buffer needed is the amount of samples (e.g., for ten samples, pelleted cells should be resuspended in 10 mL of buffer).

21. The highest number of transformants (up to 80 transformants/cuvette) is obtained with 1 μg DNA (9). When transforming with more than a single plasmid (co-transformation), use 1 μg of each plasmid. It is advisable to use as high concentration of the plasmid DNA as possible, and in any case the volume of plasmid DNA that is added to each cuvette should not exceed 4 μL in order to keep salt concentration low.

22. Transformation efficiency is affected by the rate of DNA uptake and cell viability. However, these parameters are in opposite correlation: DNA uptake increases with enhanced energy, whereas cell viability decreases. The optimal conditions for this protocol were determined after testing a wide range of conditions, in which each variable (voltage and resistance) was modified (see (9) for the complete set of conditions tested). It has been found that when parallel resistance was kept at 800 Ω, and voltage was varied between 1.25 and 1.75 kV, a primary peak in efficiency was observed at 1.4 kV.

23. Incubation period must be calibrated according to the selection marker. For Hyg, the optimal time is 12 h. Deviation from this time will result in either lack of growth (too short) or intense background (too long) and inability to isolate colonies.

References

1. Case ME, Schweizer M, Kushner SR, Giles NH (1979) Efficient transformation of *Neurospora crassa* by utilizing hybrid plasmid DNA. Proc Natl Acad Sci U S A 76:5259–5263
2. Weaver JC (1995) Electroporation theory. Concepts and mechanisms. Methods Mol Biol 47:1–26
3. Chakraborty BN, Kapoor M (1990) Transformation of filamentous fungi by electroporation. Nucleic Acids Res 18:6737
4. Kothe GO, Free SJ (1995) Protocol for the electroporation of *Neurospora* spheroplasts. Fungal Genet Newsl 43:31–32
5. Kuo CY, Huang CT (2008) A reliable transformation method and heterologous expression of beta-glucuronidase in Lentinula edodes. J Microbiol Methods 72:111–115
6. Richey MG, Marek ET, Schardl CL, Smith DA (1989) Transformation of filamentous fungi with plasmid DNA by electroporation. Phytopathology 79:844–847
7. Chakraborty BN, Patterson NA, Kapoor M (1991) An electroporation-based system for high-efficiency transformation of germinated conidia of filamentous fungi. Can J Microbiol 37:858–863
8. Sanchez O, Aguirre J (1996) Efficient transformation of *Aspergillus nidulans* by electroporation of germinated conidia. Fungal Genet Newsl 43:48–51
9. Robinson M, Sharon A (1999) Transformation of the bioherbicide *Colletotrichum gloeosporioides* f. sp. *aeschynomene* by electroporation of germinated conidia. Curr Genet 36:98–104
10. Bundock P, den Dulk-Ras A, Beijersbergen A, Hooykaas PJ (1995) Trans-kingdom T-DNA transfer from *Agrobacterium tumefaciens* to Saccharomyces cerevisiae. EMBO J 14:3206–3214
11. de Groot MJ, Bundock P, Hooykaas PJ, Beijersbergen AG (1998) *Agrobacterium tumefaciens*-mediated transformation of filamentous fungi. Nat Biotechnol 16:839–842
12. Michielse CB, Arentshorst M, Ram AF, van den Hondel CA (2005) *Agrobacterium*-mediated transformation leads to improved gene replacement efficiency in *Aspergillus awamori*. Fungal Genet Biol 42:9–19
13. Zeilinger S (2004) Gene disruption in *Trichoderma atroviride* via *Agrobacterium*-mediated transformation. Curr Genet 45:54–60

Chapter 3

High-Throughput Construction of Gene Deletion Cassettes for Generation of *Neurospora crassa* Knockout Strains

Patrick D. Collopy, Hildur V. Colot, Gyungsoon Park, Carol Ringelberg, Christopher M. Crew, Katherine A. Borkovich, and Jay C. Dunlap

Abstract

The availability of complete genome sequences for a number of biologically important fungi has become an important resource for fungal research communities. However, the functions of many open reading frames (ORFs) identified through annotation of whole genome sequences have yet to be determined. The disruption of ORFs is a practical method for loss-of-function gene analyses in fungi that are amenable to transformation. Unfortunately, the construction of knockout cassettes using traditional digestion and ligation techniques can be difficult to implement in a high-throughput fashion. Knockout cassettes for all annotated ORFs in *Neurospora crassa* were constructed using yeast recombinational cloning. Here, we describe a high-throughput knockout cassette construction method that can be used with any fungal transformation system.

Key words: Gene targeting, Filamentous fungi, *ku70*, *ku80*, *mus-51*, *mus-52*

1. Introduction

Many ORFs identified through whole-genome sequencing of numerous fungi cannot be functionally characterized based on sequence alone. Gene disruption serves as a fundamental approach for understanding the role of many ORFs that do not have an assigned function. We have undertaken the task of disrupting all ~10,000 ORFs in the annotated genome of *Neurospora crassa* as part of a functional genomics consortium (1). This project aims to provide fungal researchers with a set of individual gene knockouts that covers the whole *N. crassa* genome similar to the set of knockouts that encompasses the entire genome of the yeast *Saccharomyces cerevisiae* (2–4). While the scale of a whole-genome knockout project may not be feasible for all fungal researchers, systematic disruption of groups of functionally relevant ORFs can be quite informative. For example, the disruption of 103 putative transcription factors in *N. crassa* (5) demonstrated the usefulness of a high-throughput knockout procedure for fungi with completed genome sequences on a scale that is realistic for many suitably equipped labs.

One of the first hurdles to overcome with a large-scale knockout project like this is the construction of the knockout cassettes. The use of traditional restriction digestion and ligation, for example, is too cumbersome to be a favorable option for high-throughput knockout cassette assembly. For this reason, we employed a method that utilizes *S. cerevisiae* to assemble cassette components that possess short complementary overlapping sequence ends (6, 7). Components are synthesized in separate PCR reactions and cotransformed into *S. cerevisiae* for assembly by the host cell's own recombination machinery. An overview schematic for the construction of knockout cassettes using the yeast recombinational system is presented in Fig. 1.

Crucial to the success of this project was the work from Ninomiya et al. (8) demonstrating that mutation of single genes (*mus-51* and *mus-52*) involved in nonhomologous end-joining (NHEJ) leads to a high rate of homologous recombination in *Neurospora*. Prior to this work, high-throughput gene disruption in *N. crassa* was extremely difficult due to the low frequency of homologous recombination relative to ectopic insertions. We created two disruption strains, Δmus-51 and Δmus-52, using the selectable marker *bar*, which confers resistance to phosphinothricin (9, 10). Knockout cassettes are transformed into one of the *mus* mutants to produce primary transformants, which are then crossed to wild type to generate knockout mutants in a genetic background lacking the *mus* mutation. Subsequent to our work, mutants deficient in the NHEJ-pathway have been shown to display high homologous recombination frequencies in a number of filamentous fungi (11). Therefore, assessment of targeting efficiency and generation of recipient strains that are deficient in the NHEJ-pathway are important areas of consideration prior to undertaking a large-scale gene deletion effort.

The basic protocol for the high-throughput construction of knockout cassettes in *N. crassa* genes is described here with some aspects of the procedure elaborated upon in greater detail. Additional details of the entire *N. crassa* process (including *N. crassa* electroporation and verification of targeted gene disruptions) can be found in the methods and Supplementary Methods of Colot et al. (5). Many of the protocols that were developed for *N. crassa* are easily transferred to other filamentous fungi. However, depending on the fungal organism of interest, *Agrobacterium*-mediated, biolistic or protoplast transformation methods may be more suitable than electroporation using this high-throughput knockout cassette construction protocol.

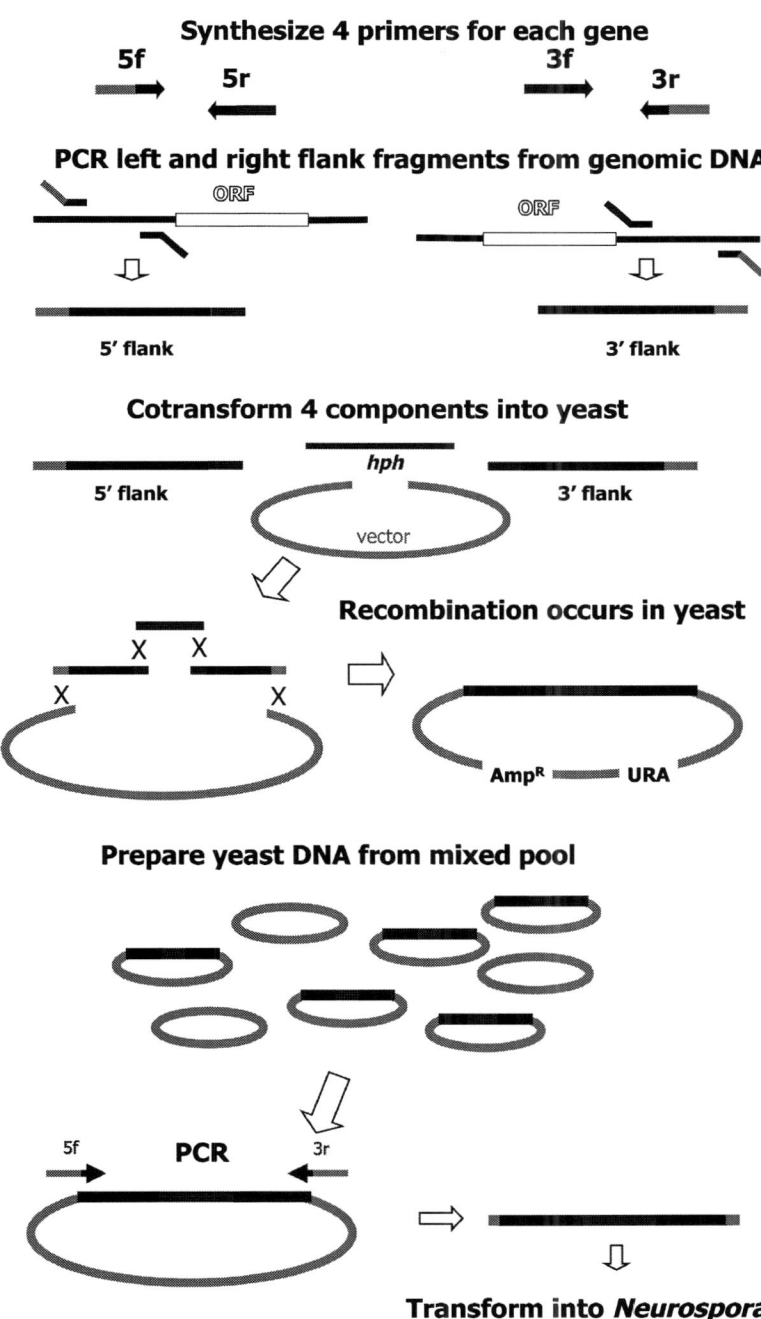

Fig. 1. Schematic of KO cassette construction using yeast recombinational cloning techniques. Primers were designed to amplify flanks both upstream and downstream of the ORF. Each primer also contained 29 nucleotides of "common" sequence to overlap with the yeast cloning vector (*light grey*) or the hygromycin B phosphotransferase (*hph*) resistance cassette (*dark grey*).

2. Materials

2.1. PCR of Knockout Cassette Components

1. *Gene-specific PCR products to be cloned.* Typically, these consist of three fragments: the 5′ and 3′ flanks of the gene(s) of interest and a resistance cassette. There should be approximately 30 bp of overlapping complementary sequence between fragments to be assembled. Four gene-specific primers are needed for each knockout cassette flank: 5′ flank forward (5f), 5′ flank reverse (5r), 3′ flank forward (3f), and 3′ flank reverse (3r). The following 5′ common sequences were added to the respective primers for each gene to allow cassette assembly in the yeast recombinational system:

   - 5f, GTAACGCCAGGGTTTTCCCAGTCACGACG...
   - 5r, ATCCACTTAACGTTACTGAAATCTCCAAC...
   - 3f, CTCCTTCAATATCATCTTCTGTCTCCGAC...
   - 3r, GCGGATAACAATTTCACACAGGAAACAGC...

2. Plasmid pCSN44 (12, 13). This plasmid contains the dominant drug-resistance marker gene hygromycin B phosphotransferase (*hph*) from *E. coli*, flanked by the promoter and terminator of the *trpC* gene from *Aspergillus nidulans*.

3. Primers for the amplification of the *hph* resistance cassette:

   hphF (GTCGGAGACAGAAGATGATATTGAAGGAGC)
   hphR (GTTGGAGATTTCAGTAACGTTAAGTGGAT)

4. A high-fidelity, high-yield Taq polymerase to be used for the generation of gene flanks and resistance cassettes that minimizes errors in the final knockout cassettes. We use LA Taq by TaKaRa (Fisher Scientific, Pittsburgh, PA).

2.2. Yeast Transformation and Cassette Assembly

1. Cloning vector pRS426 (14). This is a 2 μm vector that contains a *URA3* marker for selection in *S. cerevisiae*.

2. *S. cerevisiae* strain FY834 (*MATα his3 Δ200 ura3-52 leu2Δ1 lys2Δ202 trp1Δ63*) (15).

3. *YPD medium.* Dissolve 10 g yeast extract, 20 g peptone and 20 g dextrose in dH_2O to a final volume of 1 L. Add 15 g agar and autoclave.

4. *96PEG solution.* Dissolve 45.6 g PEG (MW 3350) in dH_2O to a final volume of 88 mL. Add 10 mL 1 M lithium acetate and 2 mL 50× TE buffer. Autoclave or filter-sterilize.

5. *Sheared salmon sperm DNA, 10 mg/mL (Fisher Scientific, Pittsburgh, PA).* Dilute to 2 mg/mL in ddH_2O, boil for 5 min and store at −20°C. This does not need to be boiled each time it is thawed for use.

6. *SC-Ura medium*. To prepare 1 L, mix 26.7 g Drop-out Base with Glucose (US Biologicals, Swampscott, MA) and 2 g Drop-out Mix Synthetic Minus Uracil w/o Yeast Nitrogen Base (US Biologicals, Swampscott, MA) in H_2O. Autoclave.

7. Airpore tape sheets (Qiagen, Valencia, CA).

2.3. Yeast DNA Purification

1. Gentra Puregene Tissue Kit (Qiagen, Valencia, CA).
2. Zymolyase T-100 (Seikagaku, Tokyo, Japan).
3. CleanSeq beads (Agencourt, Beverly, MA).
4. Agencourt SPRIPlate® 96R – Ring Magnet Plate (Agencourt, Beverly, MA).

3. Methods

Although many of the cassette construction steps were performed using a Biomek NX robot, these methods may also be easily scaled-down for small groups of genes.

3.1. PCR of Knockout Cassette Components

1. Gene-specific knockout cassettes should be designed to have 1–1.3 kb of gene-specific 5′ and 3′ flanks and a selectable marker (the hygromycin B phosphotransferase gene, *hph*, in our case). To ensure appropriate expression of the selectable marker, we use the *A. nidulans* promoter *trpC* (13).

2. For the primer design of 5′ and 3′ flanks, software written for us by John Jones (John Jones Consulting, Moreno Valley, CA) was used to retrieve regions adjacent to each ORF in the annotated genome of *N. crassa*. This information was then passed to PRIMER3 (http://primer3.sourceforge.net/), which would automatically select a list of candidate primers based on defined conditions (5).

3. Primers for the *Neurospora* knockout project are ordered from Illumina (San Diego, CA) in 96-well plates at 50 µM. These primers are mixed in three pairs (5f+5r, 3f+3r and 5f+3r) using a Beckman Biomek NX robot to a final concentration of 10 µM each. (The 5f+3r primer combination is required for the amplification of the final knockout cassette to be used for transformation.)

4. Perform separate 25 µL PCR reactions for the 5′ and the 3′ flanks using primer pair mixtures 5f+5r or 3f+3r. PCR reactions consist of: 2.5 µL 10× buffer (LA Taq), 4 µL dNTP's (LA Taq), 1 µL 10 µM primer mixture, 0.25 µL genomic *N. crassa* DNA (prepared using the Puregene DNA kit; ~150 ng/µL), 17 µL water, and 0.25 µL LA Taq (5 units/µL).

5. Generate the *hph* cassette fragment in a 50 µL PCR reaction containing: 5 µL 10× buffer (LA Taq), 8 µL dNTP's (LA Taq), 1 µL 10 µM primer hphF, 1 µL 10 µM primer hphR, 0.5 µL pCSN44 DNA (from 1:100 dilution of plasmid miniprep), 34 µL water, and 0.5 µL LA Taq (5 units/µL).

6. PCR cycling parameters for both flanks and the *hph* resistance marker are: 94°C for 1:00 min followed by 35 cycles of: {94°C for 30 s, 60°C for 30 s, 72°C for 2:00 min}, and a final extension of 72°C for 10:00 min.

7. Digest several µg of pRS426 vector with *Eco*RI and *Xho*I and dilute to a final concentration of 100 ng/µL.

3.2. Yeast Transformation and Cassette Assembly

The PCR products and digested vector do not need to be cleaned up prior to yeast transformation. Yeast transformation was adapted from a 96-well protocol described in http://depts.washington.edu/sfields/protocols/cloning_protocol.html

1. Inoculate 50 mL of YPD with 0.3 mL of a FY834 saturated culture. Grow overnight at 30°C with shaking (up to 300 rpm).

2. Pellet cells in a 50 mL conical tube ($2,530 \times g$ for 3 min). Discard supernatant, resuspend cells in 2 mL 100 mM lithium acetate and transfer to two microcentrifuge tubes.

3. Spin at top speed in a microcentrifuge for 30 s and discard the supernatant. Resuspend cell pellets in 100 mM lithium acetate to a total final volume of 1.8 mL (approximately 700 µL to each microcentrifuge tube). At this point, it is safe to keep cells on bench at room temperature until use.

4. Prepare fresh CT110 mixture (16). For 100 transformations mix 20.7 mL 96PEG, 580 µL boiled salmon sperm (2 mg/mL), 210 µL *hph* cassette (directly from PCR reaction of Section 3.1, step 5), 105 µL cut vector pRS426 (100 ng/µL) and 2.62 mL DMSO. Mix thoroughly for 30 s, then add 1.8 mL yeast cells and mix thoroughly for an additional 60 s.

5. Pipette 200 µL of CT110 mixture into each well of a 96-well deep-well plate.

6. Add 4 µL of 5′ flank and 4 µL of 3′ flank PCR reaction for each gene. Seal plate, vortex 4 min, and incubate at 42°C for 30 min.

7. Spin plate at $2,000 \times g$ for 7 min and aspirate off supernatant.

8. Add 200 µL SC-Ura to each well and resuspend by pipetting up and down.

9. Transfer 80 µL of resuspended transformed cells to 1 mL of SC-Ura liquid in each well of a 96-well deep-well plate. Seal the plate with Airpore tape (Qiagen) and grow 3 days at 30°C with shaking.

3.3. Yeast DNA Purification

The following steps are described as they were performed using the Biomek NX robot, with the exception of centrifugation and vortexing steps. The Gentra Puregene™ Tissue kit (Qiagen) was used on the robot with a few modifications that are described below.

1. After 3 days, centrifuge the plate at $2,000 \times g$ for 3 min.
2. Remove supernatant and resuspend pellet in 300 µL of Gentra cell suspension solution (Qiagen) containing 120 µg/mL Zymolyase T-100 (Seikagaku).
3. Incubate the plate at 37°C for 40 min, then centrifuge at $830 \times g$ for 3 min.
4. Remove the supernatant and add 300 µL Gentra cell lysis solution (Qiagen) to the pellet and mix.
5. Add 100 µL Gentra protein precipitation solution (Qiagen) and mix.
6. Vortex the plate for 20 s and centrifuge at $3,070 \times g$ for 12 min.
7. Transfer 90 µL of supernatant into a round-bottomed 96-well plate, and add 10 µL CleanSeq beads and 143 µL fresh 85% ethanol. Mix components and incubate statically for 3 min.
8. Place the plate on an SPRI magnetic plate (Agencourt) for 10 min and aspirate the cleared solution.
9. With the plate remaining on the magnet, add 200 µL of 85% ethanol to each well. After 30 s, remove the ethanol and repeat this rinse with 200 µL of 85% ethanol.
10. Allow the plate to air-dry for 10 min and remove from the magnet.
11. Add 40 µL 1× TE to each well and mix.
12. Return the plate to the magnet for 5 min and elute DNA to a new 96-well plate.

3.4. PCR of Final Full-Length Knockout Cassette

1. 50 µL PCR reactions are performed to generate each deletion cassettes. PCR reactions consist of: 5 µL 10× buffer (LA Taq), 8 µL dNTP's (LA Taq), 2 µL 10 µM primer 5f+3r primer mixture, 4 µL yeast DNA, 30.5 µL water, and 0.5 µL LA Taq.
2. PCR cycling parameters for the final full-length knockout cassette is: 94°C for 1:00 min followed by 35 cycles of: {94°C for 30 s, 60°C for 30 s, 72°C for 5:00 min} and a final extension of 72°C for 10:00 min.
3. Estimate DNA yields by agarose gel electrophoresis.
4. At this point, the cassettes we use for creating the *Neurospora* Genome Project knockout strains are cleaned up with the Qiagen QIAquick 96 PCR Purification Kit #28181 according

to the manufacturer's protocol prior to *N. crassa* transformation. 5 µL of purified PCR product usually yields a sufficient number of *N. crassa* transformants by electroporation, but this amount will vary depending on the recipient organism and the transformation technique.

Acknowledgments

This work was supported by grant P01 GM068087 from the National Institute of General Medical Sciences. We would like to thank John Jones for software design and LIMS implementation.

References

1. Dunlap JC, Borkovich KA, Henn MR, Turner GE, Sachs MS, Glass N et al (2007) Enabling a community to dissect an organism: overview of the *Neurospora* functional genomics project. Adv Genet 57:49–96
2. Winzeler EA, Shoemaker DD, Astromoff A, Liang H, Anderson K, Andre B et al (1999) Functional characterization of the S. cerevisiae genome by gene deletion and parallel analysis. Science 285:901–906
3. Martin AC, Drubin DG (2003) Impact of genome-wide functional analyses on cell biology research. Curr Opin Cell Biol 15:6–13
4. Ooi SL, Pan X, Peyser BD, Ye P, Meluh PB, Yuan DS et al (2006) Global synthetic-lethality analysis and yeast functional profiling. Trends Genet 22:56–63
5. Colot HV, Park G, Turner G, Ringelberg C, Crew CM, Litvinkova L et al (2006) A high-throughput gene knockout procedure for *Neurospora* reveals functions for multiple transcription factors. Proc Nat Acad Sci USA 103:10352–10357
6. Oldenburg KR, Vo KT, Michaelis S, Paddon C (1997) Recombination-mediated PCR-directed plasmid construction in vivo in yeast. Nucleic Acids Res 25:451–452
7. Raymond CK, Pownder TA, Sexson SL (1999) General method for plasmid construction using homologous recombination. Biotechniques 26:134–141
8. Ninomiya Y, Suzuki K, Ishii C, Inoue H (2004) Highly efficient gene replacements in *Neurospora* strains deficient for nonhomologous end-joining. Proc Natl Acad Sci USA 101:12248–12253
9. Pall ML (1993) The use of Ignite (basta; glufosinate; phosphinothricin) to select transformants of bar-containing plasmids in *Neurospora crassa*. Fungal Genet Newsl 40:58
10. Pall ML, Brunelli JP (1993) A series of six compact fungal transformation vectors containing polylinkers with multiple unique restriction sites. Fungal Genet Newsl 40:59–62
11. Meyer V (2008) Genetic engineering of filamentous fungi – Progress, obstacles and future trends. Biotechnol Adv 26:177–185
12. Gritz L, Davies J (1983) Plasmid-encoded hygromycin B resistance: the sequence of the hygromycin B phosphotransferase gene and its expression in *Escherichia coli* and *Saccharomyces cerevisiae*. Gene 25:179–188
13. Staben C, Jensen B, Singer M, Pollock J, Schechtman M, Kinsey J et al (1989) Use of a bacterial hygromycin B resistance gene as a dominant selectable marker in *Neurospora crassa* transformation. Fungal Genet Newsl 36:79–81
14. Sikorski RS, Hieter P (1989) A system of shuttle vectors and yeast host strains designed for efficient manipulation of DNA in *Saccharomyces cerevisiae*. Genetics 122:19–27
15. Winston F, Dollard C, Ricupero-Hovasse SL (1995) Construction of a set of convenient *Saccharomyces cerevisiae* strains that are isogenic to S288C. Yeast 11:53–55
16. Rajagopala SV, Titz B, Uetz P (2007) Array-based yeast two-hybrid screening for protein-protein interactions. Method Microbiol 36:139–163

Chapter 4

Development of *Impala*-Based Transposon Systems for Gene Tagging in Filamentous Fungi

Marie Dufresne and Marie-Josée Daboussi

Abstract

Genome sequences of many filamentous fungi are now available and additional genomes are currently being sequenced. One of the next strategic goals is to generate collections of tagged genes in order to establish a link between the several thousands of predicted genes and their function. Transposable elements have been invaluable for the identification and isolation of genes of interest as insertion of a transposon both disrupts and tags a gene. In an effort to exploit active transposons identified in the genome of *Fusarium oxysporum* as insertional mutagens, a binary system including the tagging element, MITE, and the transposase of a *Tc1* element has been established as an efficient tool for gene-tagging in *Fusarium graminearum*. In this chapter, we provide an overview of the techniques used and highlight some of the critical steps for the application of this tool to other fungal species.

Key words: *Fusarium*, *Tc1/mariner* elements, MITEs, Transposon-tagging

1. Introduction

With the availability of complete genome sequences of many filamentous fungi, the next challenging phase of research is to link sequence information to gene function. One of the best ways to ascertain function is to disrupt genes in a random and genome wide fashion. Transposon insertional mutagenesis has been used for large-scale analysis of the genome in a wide variety of organisms including microorganisms, flies, worms, mice, and plants (1, 2). A range of tagging strategies by using endogenous transposons when available or transposons from heterologous species have been developed. The identification of active transposons from the phytopathogenic fungus *Fusarium oxysporum* by trapping them in the *niaD* target gene was the first step in the development of tagging strategies (3). Among them, *impala*, a *Tc1* member, was demonstrated to transpose

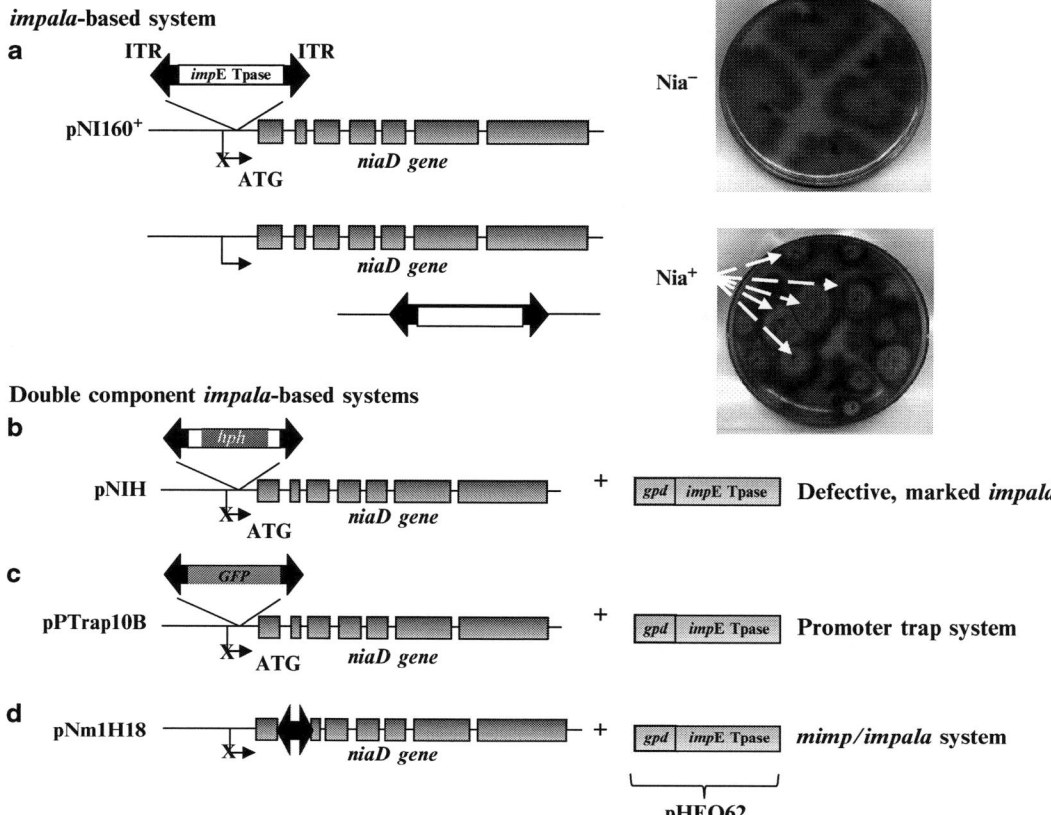

Fig. 1. Structure of the constructs and principle of the phenotypic assay for excision used in the different *impala*-based transposon-tagging systems. Plasmids pNI160+, pNIH, pPTrap10B and pNm1H18 have the same skeleton consisting in a plasmid carrying the wild-type *niaD* gene of *Aspergillus nidulans* (exons are represented as open *grey boxes*, non-coding regions as a *black line*, the arrow below the line corresponds to the transcription inititation site). Large *black* opposite arrows indicate the terminal inverted repeats (TIRs) of the different transposons. (**a**) Plasmid pNI160+ carries the natural transposon *impala160*, inserted in the 5′ untranslated region of *niaD*. When present, it prevents transcription of the gene leading to a *Nia*− phenotype (see upper Petri plate). Upon excision of *impala160*, *niaD* transcription is restored leading to a *Nia*+ phenotype (see lower Petri plate). The principle of this phenotypic assay for excision is the same for all the transposons described in this figure. (**b**) Plasmid pNIH carries at the same position of the *niaD* gene a modified *impala* transposon in which an internal 1-kbp containing most of the *impala* coding sequence has been replaced by the bacterial gene *hph* (hygromycin B resistance) driven by the *A. nidulans trpC* promoter (9). (**c**) The vector pPTrap10B containing the *imp160::gfp* element was constructed by introducing an 862-bp *Sma*I fragment containing the coding region of the *egfp* gene into the *Sma*I site of a transposase-deleted *impala160* version containing 61 bp and 31 bp of the 5′ and 3′ border sequences, respectively. The construct was then used to replace the *impala* element previously cloned into the promoter region of the *A. nidulans niaD* gene in the pNI160+ plasmid (11). (**d**) Plasmid pNm1H18 was constructed by introduction of a *mimp1* element into the first intron of the *niaD* gene (12). When defective, the transposons have to be mobilized in trans by the *impala* transposase brought by the pHEO62 construct carrying both the open reading frame encoding the *impalaE* transposase cloned between the *gpdA* promoter and the *trpC* terminator of *A. nidulans* (10) and an expression cassette conferring resistance to hygromycin B (for more detail, see ref. (12)).

through a phenotypic assay for excision (Fig. 1a) based on the restoration of the activity of the nitrate reductase after *impala* excision (4). Similar to other *Tc1/mariner* elements (5), *impala* has already been demonstrated to transpose in several phylogenetically distant species (3). In some of them *impala* was successfully used to tag genes of interest (6, 7). However, *impala* reinsertion frequency, in the range of 50–75%, hampers its use as a high throughput gene-tagging system (8, 9). To overcome these difficulties, different two-component systems including the *impala* transposase and versions of the tagging element have been developed. First, *impala* has been engineered by insertion of the *hph* gene conferring resistance to the drug hygromycin B (9). This defective element is transactivated by the *impala* transposase under the control of its own promoter (9) or fused to the constitutive *Aspergillus nidulans gpdA* promoter (10) on a separate vector. This selectable marker facilitates the recovery of strains with a reinserted element on condition that only one copy of this construct has been integrated (Fig. 4.1b). Second, we establish a promoter-trap system that utilizes a promoterless *GFP* gene within the transposon so that this reporter gene should be expressed only when the excised transposon reinserts near or into an active gene (11) (Fig. 4.1c). The difficulties to detect GFP expression (weak expression, variable among replicates) among a collection of 1,000 strains tested limit the use of such a strategy and new trapping vectors have to be developed. Third, we demonstrated the ability of the *impala* transposase to transactivate *mimp1*, which shows features of MITEs (12) and developed a novel double component system based on these two partners (Fig. 1d). We assessed the *mimp1* mutagenic potential by generating *mimp*-transposed strains in *F. graminearum* isolate Fg820. We mapped the insertion sites of the tagging element in a collection of transposed strains by comparison with a published genome sequence and screened for mutant phenotypes (13). The high rate of excision of some strains, the distribution of the reinserted element throughout the genome, and the high proportion of insertions within or close to genes make this system very promising. The tools and technologies based on the *mimp1/impala* double component system we developed are presented here in order to provide a valuable resource to accelerate research on gene function in filamentous fungi.

2. Materials

2.1. Isolation of Nitrate-Reductase Deficient Strains

1. Petri plates.
2. Minimal medium agar (MM): 2% (w/v) glucose, 7.5 mM KH_2PO_4, 2 mM $MgSO_4$, 6.5 mM KCl, 6 µM $FeSO_4$ (1 mL from a 1% stock solution prepared extemporaneously). Add

the appropriate nitrogen source (5 mM urea, 23 mM sodium nitrate, 10 mM sodium nitrite or 0.7 mM hypoxanthine), adjust to pH 5.8, and solidify with 1.5% purified agar (Bacteriological agar No 1, Oxoïd).

2.2. Recovery of Cotransformants Giving Rise to Excision Events

1. Potato Dextrose Agar (PDA): We use "home-made" PDA medium: peal 200 g potatoes and let them boil in 1 L tap water for 1 h. Filtrate over gaze and cotton, add 20 g glucose, 1 g yeast extract and 1 g casein hydrolysates. Solidify with 1.5% agar and autoclave at 120°C for 15 min. A commercial product may also be used (e.g., from Difco).
2. Petri plates.
3. Cellophane disks (Focus Packaging).
4. $MgSO_4$: 1.2 M $MgSO_4$, adjust to pH 5.8 with H_3PO_4, filter sterilize on a 0.45 µm unit. Store at 4°C.
5. MS, pH 6.3: 0.6 M sorbitol, 10 mM MOPS, adjust pH to 6.3 with 0.5 N sodium hydroxide, autoclave for 15 min. Store at 4°C.
6. TMS: 1 M sorbitol, 10 mM MOPS, adjust to pH 6.3 with 0.5 N sodium hydroxide and autoclave for 15 min. Store at 4°C.
7. TMSC: 1 M sorbitol, 10 mM MOPS, 40 mM $CaCl_2$, adjust pH 6.3 with 0.5 N sodium hydroxide and autoclave for 15 min. Store at 4°C.
8. Tris–EDTA-$CaCl_2$: 10 mM Tris, pH 7.5, 1 mM EDTA, 40 mM $CaCl_2$. Store at 4°C.
9. PEG solution: 60% (w/v) PEG 4000 in 0.6 M MS. Store at RT for no more than 1 month.
10. Glucanex® (Novo Nordisk, Denmark).
11. PDAS: made as for the PDA medium except that 20% sucrose is added.
12. Hygromycin B (Fermentas).
13. MMS Top agar: MM with 20% (w/v) sucrose, 23 mM sodium nitrate, and 0.4% (w/v) purified agar.
14. Disposable 50 mL and 5 mL sterile tubes.
15. Nylon N disks (89 mm diameter, GE Healthcare).
16. Denaturation solution: 0.5 M NaOH, 5× SSC.
17. Neutralization solution: 0.5 M Tris–HCl, pH 7.5, 10× SSC. For optimal treatments, prepare the denaturation solution extemporaneously. Can be kept for weeks at room temperature (RT).
18. CL-1000 UV crosslinker® (UVP).

2.3. Optimization of the System for Large Scale Mutagenesis

1. PDA plates.
2. Cellophane disks (Focus Packaging).
3. Lysis solution: 1% sarkosyl, 100 mM Tris–HCl, pH 9.0, 10 mM EDTA.
4. Fastprep® Cell Disrupter (QBiogene).
5. Phenol solution equilibrated at pH 7.0.
6. RNase: Dissolve in water at 100 mg/mL.
7. All standard components that are needed to set up PCR reactions.

2.4. Generation of a Collection of Revertants

1. Petri plates (89 mm).
2. 96-well microtiter plates.
3. MM agar supplemented with 23 mM sodium nitrate.

3. Methods

As described above, transposition events are selected using a phenotypic assay for excision based on the restoration of the expression of the *A. nidulans niaD* gene upon excision of the mobilizable *mimp1* element (Fig. 1a, d). The successful application of our transposon-tagging system relies first on the recovery of mutants impaired in the nitrate-reductase structural gene (*nia*⁻ mutants), without the use of a mutagen, through positively screening for resistance to chlorate (14). This selection system appears to be possible for many fungal species (15). The second step consists of introduction of the two constructs into a *nia*⁻ mutant strain: one construct (plasmid pHEO62) contains the source of *impala* transposase together with the *hph* selectable marker conferring resistance to hygromycin B, and the second carries the tagging element inserted into the first intron of the *A. nidulans niaD* gene (plasmid pNm1H18, Fig. 1d). Cotransformants, that is transformed strains carrying both constructs, giving rise to excision events will then be selected. The most critical step will consist of the selection of the optimal cotransformants for the generation of a collection of revertants. These strains will have to fulfill a number of criteria: (1) show high rates of excision and reinsertion of the mobilizable element *mimp1*, (2) carry a single copy of this element to ensure an easy and reliable molecular analysis, and (3) give rise to reinsertion events distributed randomly throughout the recipient genome. The following part describes all these different steps.

3.1. Isolation of nia⁻ Strains

3.1.1. Isolation of Chlorate-Resistant Strains

1. Plate conidia on minimal medium (MM) with 5 mM urea and 200 mM chlorate at 10^3 conidia per plate. Prepare at least three platees (see Note 1) and incubate at 26°C.
2. Chlorate resistant colonies appear after 1 or 2 weeks as fast growing colonies (see Note 1). Pick each of colony onto fresh MM urea (5 mM) chlorate (200 mM) plates both to individualize them and to verify their resistance to chlorate.

3.1.2. Identification of nia⁻ Mutants Among Chlorate-Resistant Strains

1. Single-spore each individual chlorate-resistant mutant.
2. Transfer plugs of each of the purified chlorate-resistant mutants to MM amended with different nitrogen sources: 23 mM nitrate, 10 mM nitrite, or 0.7 mM hypoxanthine (see Note 1) and grow at 26°C for 5–7 days to determine which gene of the nitrate-reductase activity is affected. *nia⁻* mutants will exhibit sparse mycelium on MM containing nitrate as sole nitrogen source but will develop normal aerial mycelium on the two other media.
3. Stable *nia⁻* mutants can be screened by plating conidia (10^6 per plate, at least five plates) on minimal medium (MM) containing nitrate (23 mM) as sole nitrogen source and incubating at 26°C for 8–10 days. Reversion of the *nia* mutation can be monitored through the appearance of aerial colonies. A reversion frequency of selected *nia⁻* mutants less than 10^{-6} viable spores is acceptable for the following steps.

3.2. Recovery of Cotransformants Giving Rise to Excision Events

3.2.1. Protoplast Transformation of a Stable nia⁻ Mutant with the Two Plasmids

1. Preparation of fungal protoplasts:
   - Prepare six PDA Petri plates with a cellophane disk, and plate 10^6 spores of the *nia⁻* recipient strain on each of them. Incubate 17 h at 26°C.
   - Scrape the mycelium from the surface of the plates, and resuspend it in 10 mL of 1.2 M $MgSO_4$. Add Glucanex® (20 mg/mL) and incubate 2 h at 26°C with agitation at 150 rpm.
   - Purify the protoplasts by adding gently and slowly 10 mL MS on top of the digestion solution (see Note 2) and centrifugating 10 min at $750 \times g$.
   - Collect the protoplasts from the interface using a Pasteur pipette (see Note 3).
   - Add 5 mL TMS and centrifuge 5 min at $750 \times g$.
   - Resuspend the pellet of protoplasts in 200 µL TMSC. Protoplasts are ready to be counted using a 1:100 dilution in TMSC and a Thoma's cell (see Note 4).
2. Transformation of fungal protoplasts:

- Pour Petri plates of PDAS+hygromycin B at a final concentration of 50 μg/L, dry them well (see Note 5), and place at 37°C until use.
- Adjust protoplasts density to 10^7 protoplasts/100 μL (in TMSC) and incubate on ice for at least 20 min.
- Prepare a solution of the plasmids to be transformed: 5 μg of each plasmid in a total volume of 60 μL of Tris–EDTA–$CaCl_2$, pH 7.5, for each transformation reaction.
- Preheat a water bath under the flow hood at 42°C, and preheat 5 mL tubes for MMS Top agar. Melt MMS Top Agar in a microwave, and dispatch 3 mL per tube (plan four tubes per transformation). Equilibrate at 42°C until use.
- Mix gently DNA (60 μL) and protoplasts (100 μL) and incubate 20 min on ice.
- Remove from the ice and add very gently, drop by drop, 160 μL of PEG solution (using a blue tip, P1000). Mix by "gentle vortexing" (half the maximal shaking), and incubate 15 min at RT.
- Add 1 mL of TMSC, mix gently by inverting the tube then by "gentle vortexing", and centrifuge for 5 min (not more than 5,500 g, in a bench top centrifuge).
- Discard the supernatant without drying the protoplasts pellet, and resuspend in 200 μL of TMSC, using a blue tip.
- Add 50 μL of the transformation reaction per tube of MMS Top Agar. Vortex quickly to mix well, and pour the content of each tube onto a PDAS+hygromycin B plate (see Note 6). Incubate at 26°C. Potential transformants usually appear after 3–5 days as fast growing hygromycin-resistant colonies.
- Transfer a plug from each hygromycin-resistant colony to a fresh PDA+hygromycin B plate. Use the recipient strain as a negative control. Purify each hygromycin-resistant transformant by single-spore isolation.

3.2.2. Identification of the Cotransformants

1. Apply Nylon N disks (GE Healthcare) onto PDA plates (one per plate).
2. Transfer a small plug (1 mm²) of each hygromycin-resistant transformant as well as of the *nia⁻* recipient strain (usually, up to 15 strains can be placed on a single 89 mm diameter plate). Incubate at 26°C for 24–48 h until colonies reach 2 cm diameter.
3. Remove the Nylon N membranes, and treat them by immersion into the denaturation solution for 30 min at room temperature under shaking (150 rpm). Remove the solution and replace

it with the same volume of the neutralization solution. Incubate under the same conditions. Repeat these two treatments once again.

4. Dry the Nylon N disks on Whattman 3 MM paper. Remove any piece of agar.

5. Fix the membranes under UV using a crosslinker (3 min at 1,200 µJ/mm/s).

6. Perform molecular hybridization under standard conditions using a probe specific for the *niaD::mimp1* construct (for example, a *niaD*-specific probe). The proportion of cotransformants, that is strains giving a positive hybridization signal, is usually in the range of 50–75% in *Fusarium* strains. (See Note 7).

3.2.3. Phenotypic Assay for Excision

1. Grow each cotransformant onto a PDA plate for 8–10 days at 26°C.

2. Transfer four plugs (4–6 mm²) from each strain to a plate of MM containing nitrate as sole nitrogen source and incubate at 26°C.

3. Check the plates weekly for the appearance of colonies exhibiting aerial growth ([*Nia*⁺]), corresponding to potential excision events, above a mat of sparse mycelium ([*Nia*⁻]), over a period of 5–6 weeks. (See Note 8).

4. Transfer a plug of each aerial colony to a fresh MM nitrate plate to verify its [*Nia*⁺] phenotype.

3.3. Optimization of the System for Large Scale Mutagenesis

This is the most critical step. The identification of the most adequate cotransformant(s) is essential to guarantee the quality of the resulting collection of revertants that is mutant strains resulting from true transposition events of the mobilizable element. Such strains must fulfill several criteria. First, in order to ensure the feasibility of obtaining a large collection of revertants, the original cotransformant(s) must exhibit a high rate of excision. This parameter can be estimated through the phenotypic assay described above (see Subheading 3.2.3), in which, if possible, we try to select cotransformants giving rise to at least 15–20 aerial [*Nia*⁺] colonies per plate. Second, among these cotransformants, it is essential to determine which ones carry a single copy of the tagging element. This is one of the most important points as it ensures that most mutations will be tagged (only one transposon moves into the genome) and greatly facilitates molecular analyses. Last, it is important to estimate both quantitatively and qualitatively the reinsertion step. Quantification of reinsertion events will allow estimation of the proportion of collection that is efficient. Molecular analysis of sequences flanking the element in a sample of revertants (usually 24) will give some indication about the randomness of reinsertion events and prevent any strong bias in the collection (see Note 9). The following paragraph describes the procedure for the second and third steps.

3.3.1. Identification of Cotransformants Carrying a Single Copy of the Tagging Element

1. Select two or three revertants from each cotransformant giving rise to a good number of aerial colonies (15–20 per plate) in the phenotypic assay described above (see Subheading 3.2.3), and purify them by single-spore isolation.
2. Spread conidia of each strain (approximately 10^5 or 10^6) on a sterile cellophane disk applied on a PDA plate, and incubate 16–24 h at 26°C.
3. Recover the mycelium by scraping the cellophane disk, place the mycelium in a 2 mL Eppendorf tube, and freeze in liquid nitrogen. Store at –80°C until use.
4. Add 0.6 mL lysis solution and a few glass beads to each tube. Grind in a Fastprep twice for 30 s at 3,500 g, let settle down a few minutes, and then spin shortly.
5. Extract successively with 500 µL phenol, 500 µL phenol:chloroform (1/1 v/v), and finally with 500 µL chloroform.
6. Add 0.05 volume 3 M NaAc (usually 25 µL), 0.6 volume isopropanol (usually 320 µL), invert the tubes five times, and centrifuge 15 min at 13,000 g at RT.
7. Remove supernatant and wash the pellet with 70% ethanol (approx. 200 µL), vortex and centrifuge again 5 min at RT and at 13,000 g.
8. Vacuum dry 5 min (optional) and resuspend in 100 µL sterile water + RNase A at a final concentration of 1 mg/mL. Incubate 1–2 h at 37°C. Genomic DNA is ready for use, either for PCR (0.5 or 1 µL per PCR reaction) or for digestion by restriction endonucleases for Southern blot analysis (usually 60–80 µL per digest depending on the efficiency of the extraction). Tubes can be stored either at 4°C or at –20°C until use.
9. Perform a PCR test using 50 ng of genomic DNA and primers *niaD*144 (5′-GTTCATGCCGTGGTCGCTGC-3′) and *niaD*754r (5′-AGTTGGGAATGTCCTCGTCG-3′), which are specific for the *A. nidulans niaD* gene and surrounding the insertion site of the tagging element within the construct. PCR reactions should be performed under the following conditions: 4 min at 94°C, then 30 cycles of 1 min at 94°C, 1 min at 59°C, and 2 min at 72°C. A schematic representation of typical results is presented in Fig. 2a. A 717-bp PCR product corresponding to a *niaD* locus still carrying the tagging element is expected for cotransformants (Fig. 2a, COTR1 and COTR2). In revertants, two situations might be encountered. A single

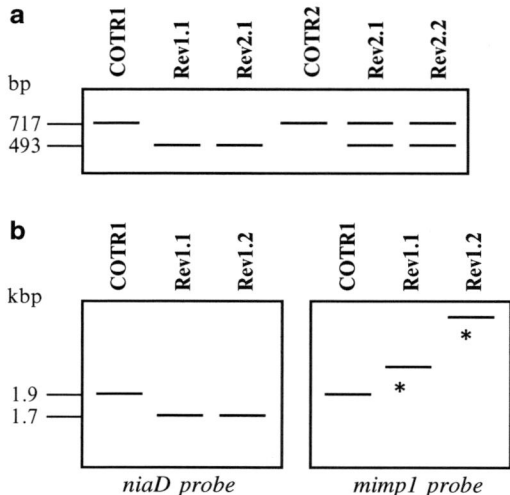

Fig. 2. Schematic representation of typical results of the molecular analysis of different sets of strains each corresponding to a cotransformant and the derived revertants. (**a**) PCR using the *niaD*144/*niaD*754r primer pair: numbers correspond to sizes in base pairs expected for a *niaD* donor site either empty (493 bp) or still carrying *mimp1* (717 bp). (**b**) Southern blot analysis of cotransformant 1 (COTR1) and two corresponding revertants (Rev1.1 and Rev1.2). Genomic DNA was digested with *Xba*I and membranes were successively hybridized with a *niaD* probe (*left*), and a *mimp1* probe (*right*). The stars indicate the reinsertion of excised copies.

493 bp PCR product corresponding to the size of the wild-type *niaD* fragment plus the typical excision footprints left by the element (Fig. 2a, Rev1.1 and Rev1.2) indicates excision of tagging element from the *niaD* gene in these revertants and that the cotransformant (COTR1) they derive from is supposed to contain a single copy of the tagging element. Besides this 493 bp fragment, in other revertants (Fig. 2a, Rev2.1 and Rev2.2), an additional PCR product that corresponds to a *niaD* locus still carrying the tagging element (717 bp) can be observed. These revertants also result from an excision event but at least one additional copy of the *mimp1*-interrupted *niaD* construct was present in the original cotransformant COTR2. Strains corresponding to the latter case should be excluded from further analysis.

10. To further characterize the strains at the molecular level, Southern blot experiments are highly recommended. Genomic DNA of each cotransformant (carrying a single copy of the tagging element) and of 2–3 corresponding revertants is digested using the *Xba*I restriction enzyme, separated electrophoretically on 0.7% agarose gels and transferred onto nylon membranes. Molecular hybridizations are then performed using successively two probes on each membrane: a probe matching the *niaD* gene and overlapping the initial

insertion site of the tagging element and a probe specific for the tagging element itself. As illustrated in Fig. 2b (*left panel*), using the *niaD* probe, cotransformants are characterized by the presence of a single 1.9 kb *Xba*I hybridization band while revertants show a single 1.7 kb *Xba*I signal. This first molecular hybridization thus confirms the results obtained by PCR (see previous step). Using the *mimp1* element as a probe, reinsertion is observed as hybridizing bands of different sizes (Fig. 2b, *right panel*). The proportion of revertants for which such a hybridizing signal is detected corresponds to the reinsertion frequency.

3.3.2. Verification of the Randomness of Reinsertion Events

1. On a sample of purified revertants (24), perform thermal asymmetric interlaced (TAIL)-PCR reactions to recover the sequences flanking the tagging element (see Note 10). In the primary PCR (TAIL1-PCR), the arbitrary degenerate (AD) primer AD2, 5′-AG(A/T)GNAG(A/T)ANCA(A/T)AGA-3′, and the specific primer m1Div53F, 5′-GCCTCTTTGAGCCACG-CTAC-3′, are used with a reduced-stringency and a high-stringency annealing temperatures of 44°C and 66°C, respectively. The TAIL1-PCR program is the following: 92°C 2 min, 95°C 1 min, five cycles consisting of 94°C 30 s, 66°C 1 min and 72°C then 2 min 30 s, 94°C 30 s, 30°C 3 min and a ramping to 72°C over 2 min, 72°C 2 min 30 s and finally 15 supercycles consisting of 94°C 30 s, 66°C 1 min, 72°C 2 min 30 s, 94°C 30 s, 66°C 1 min, 72°C 2 min 30 s, 94°C 30 s, 44°C 1 min, 72°C 2 min 30 s. The secondary PCR (TAIL2-PCR) is set up with 2 µL of a 1/100 dilution of the primary PCR, using the same AD2 primer and Div149 (5′-GCAGGCTAAACTCCAAAT-AGGC-3′) as the second specific primer. The TAIL2-PCR program is as follows: 12 supercycles consisting of 94°C 30 s, 66°C 1 min, 72°.
2. C 2 min 30 s, 94°C 30 s, 66°C 1 min, 72°C 2 min 30 s, 94°C 30 s, 44°C 1 min, 72°C 2 min 30 s followed by an elongation step at 72°C for 5 min.
3. Sequence the resulting PCR products and search for homology with the genome sequence of the species you are working on. If reinsertion occurs randomly, matches at different genomic loci should be obtained.

3.4. Generation of a Collection of Revertants

After selection of strains exhibiting a high rate of excision and verification that reinsertion is not biased, saturation mutagenesis of *Fusarium* genes can be performed. To cover the genome thousands of transposed *mimp*-strains have to be recovered. Two strategies can be used depending on the rate of excision. Revertants can be picked on plates (see Subheading 3.2.3.) or they can be obtained on microtiter plates ensuring that they result from independent excision events.

1. For the microtiter plate method, inoculate single wells of a 96-well plate containing nitrate minimal agar medium (200 µL per well) with a single germinating spore per well.
2. Collect revertants as they appeared, through the appearance of dense mycelium (Fig. 1a).
3. Purify by single spore isolation and store either on microtiter plate with PDA at −80°C or in eppendorf tubes (containing 500 µL PDA) at 4°C.

4. Notes

1. In *Fusarium* strains, we observed an influence of the nitrogen source in the minimal medium containing chlorate on the occurrence of *nia−* mutants among other nitrate non-utilizing mutants. For other fungal species, urea might have to be replaced by other nitrogen sources such as glutamine, asparagine, uric acid or proline. Because some fungi may exhibit a low sensitivity to chlorate, concentration might be increased up to 500 mM chlorate. Finally, the detection of chlorate-resistant as well as of *nia−* mutants is not always obvious. In these cases, mutants are isolated and identified using morphological differences as compared with the wild-type strain, such as the formation of sectors or colonies with more aerial mycelium or faster growth on the selection media.
2. Alternatively, tubes containing 10 mL MS each can be prepared. The digestion solution can then be slowly poured at the bottom of the tube as it is denser than the MS itself.
3. We usually bend the Pasteur pipette in a flame. Alternatively a 1 mL tip can be used.
4. When resuspended in an osmotic solution such as TMSC, protoplasts can be kept on ice for at least 2 h.
5. Proper drying of the plates under a flow hood prevents condensation and will ensure a correct adhesion of the Top agar layer on the solid PDAS bottom plate.
6. Hygromycin B concentration should be adjusted, depending on the sensitivity of the recipient strain. For *F. graminearum* strains, we usually use hygromycin B at a final concentration of 50 µg/mL. For more sensitive strains, adding the antibiotic directly into the bottom plates might be too vigorous. Another selection method consists in pouring the transformed protoplasts into MMS Top agar on PDAS plates and incubate

the plates without any selection pressure for 16–24 h. This step will allow protoplast regeneration. Only after this incubation time, a second MMS Top agar containing the antibiotic at the adequate concentration is then poured onto the first one, allowing the selection of transformants.

7. In order to increase the proportion of cotransformants carrying a single-copy of the tagging element, the amount of plasmid DNA used in the transformation experiment (particularly the one carrying the tagging element) might be lowered to 2 µg.

8. Two criteria, necessary but not sufficient, are used to estimate the quality of excision events. First, aerial colonies should appear in the first 3 weeks; later events might be enriched in spontaneous mutations corresponding to the reversion of the endogenous *nia* mutation. Second, appearance of colonies should be progressive along the incubation period, which indicates that events are independent.

9. In the case of integration of both constructs at the same genomic locus, we observed two categories of bias: (1) aberrant transposition events resulting from excision of the tagging element together with genomic sequences (16) or (2) transposition of the tagging element at proximal sites, mainly in the *A. nidulans niaD* promoter sequences.

10. Alternatively, flanking sequences can be recovered using inverse PCR (12).

References

1. Mátés L, Izsvak Z, Ivics Z (2007) Technology transfer from worms and flies to vertebrates: transposition-based genome manipulations and their future perspectives. Genome Biol 8:S1.1–S1.19
2. Maes T, De Keukeleire P, Gerats T (1999) Plant tagnology. Trends Plant Sci 4:90–96
3. Daboussi MJ, Capy P (2003) Transposable elements in filamentous fungi. Annu Rev Microbiol 57:275–299
4. Langin T, Capy P, Daboussi MJ (1995) The transposable element *impala*, a fungal member of the *Tc1/mariner* superfamily. Mol Gen Genet 246:19–28
5. Plasterk RH, Izsvak Z, Ivics Z (1999) Resident aliens: the *Tc1/mariner* superfamily of transposable elements. Trends Genet 15:326–332
6. Villalba F, Lebrun MH, Hua-Van A, Daboussi MJ, Grosjean-Cournoyer MC (2001) Transposon *impala*, a novel tool for gene tagging in the rice blast fungus *Magnaporthe grisea*. Mol Plant Microbe Interact 14:308–315
7. Firon A, Villalba F, Beffa R, D'Enfert C (2003) Identification of essential genes in the human fungal pathogen *Aspergillus fumigatus* by transposon mutagenesis. Eukaryot Cell 2:247–255
8. Migheli Q, Steinberg C, Davière JM, Olivain C, Gerlinger C et al (2000) Recovery of mutants impaired in pathogenicity after transposition of *Impala* in *Fusarium oxysporum* f. sp. *melonis*. Phytopathology 90:1279–1284
9. Hua-Van A, Pamphile JA, Langin T, Daboussi MJ (2001) Transposition of autonomous and engineered *impala* transposons in *Fusarium oxysporum* and a related species. Mol Gen Genet 264:724–731
10. Li Destri Nicosia MG, Brocard-Masson C, Demais S, Van Hua A, Daboussi MJ, Scazzocchio C (2001) Heterologous transposition in *Aspergillus nidulans*. Mol Microbiol 39:1330–1344
11. Lopez-Berges MS, DiPietro A, Daboussi MJ, Wahab Ha, Vasnier C, et al (2009) Identification

of virulence genes in *Fusarium oxysporum* f. sp. *lycopersici* by large-scale transposon-tagging. Mol Plant Pathol 10(1):95–107. doi: 10.1111/J.1364-3703.2008.00512.X

12. Dufresne M, Hua-Van A, Abd el Wahab H, Ben M'Barek S, Vasnier C et al (2007) Transposition of a fungal MITE through the action of a *Tc1*-like transposase. Genetics 175:441–452

13. Dufresne M, van der Lee T, Ben M'Barek S, Xu X, Zhang X et al (2008) Transposon-tagging identifies novel pathogenicity genes in *Fusarium graminearum*. Fungal Genet Biol. doi:10.1016/j.fgb.2008.09.004

14. Cove DJ (1976) Chlorate toxicity in *Aspergillus nidulans*: the selection and characterization of chlorate resistant mutants. Heredity 36:191–203

15. Daboussi MJ, Djeballi A, Gerlinger C, Blaiseau PL, Bouvier I et al (1987) Transformation of seven species of filamentous fungi using the nitrate-reductase gene of *Aspergillus nidulans*. Curr Genet 15:453–456

16. Hua-Van A, Langin T, Daboussi MJ (2002) Aberrant transposition of a *Tc1-mariner* element, *impala*, in the fungus *Fusarium oxysporum*. Mol Genet Genomics 267:79–87

Chapter 5

DelsGate: A Robust and Rapid Method for Gene Deletion

María D. García-Pedrajas, Marina Nadal, Timothy Denny, Lourdes Baeza-Montañez, Zahi Paz, and Scott E. Gold

Abstract

Gene deletion is one of the most powerful tools to study gene function. In the genomics era there is great demand for fast, simple high-throughput methods for gene deletion to study the roles of the large numbers of genes that are being identified. Here we present an approach that speeds up the process of generation of deletion mutants by greatly simplifying the production of gene deletion constructs. With this purpose we have developed a method, which we named DelsGate (Deletion via Gateway), that combines PCR and Gateway cloning technology together with the use of the I-*Sce*I homing endonuclease to generate precise deletion constructs in a very simple, universal and robust manner in just 2 days. DelsGate consists of standard PCR of only the 5′ and 3′ 1 kb gene flanks directly followed by in vitro Gateway cloning and final generation of the circular deletion construct by in vivo recombination in *Escherichia coli*. For use in DelsGate we have modified a Gateway cloning vector to include selectable markers for the transformation of Ascomycetes and the Basidiomycete fungus *Ustilago maydis*. The PCR and transformation steps of DelsGate should be well suited for high-throughput approaches to gene deletion construction in fungal species. We describe here the entire process, from the generation of the deletion construct with DelsGate to the analysis of the fungal transformants to test for gene replacement, with the Basidiomycete fungus *Ustilago maydis*. Application of DelsGate to other fungal species is also underway. Additionally, we describe how this basic approach can be adapted to other genetic manipulations with minor changes. We specifically describe its application to create unmarked deletions in *Ralstonia solanacearum*, a Gram-negative phytopathogenic bacterium.

Key words: Gene deletion, DelsGate, Fungal transformation, *Ustilago maydis*, Ascomycetes, Verticillium, *Ralstonia solanacearum*

1. Introduction

A central tool to study gene function is deletion of genes of interest. Numerous fungal genomes have been or are being sequenced and annotated producing a large number of putative ORFs encoding proteins whose functions need to be studied.

Additionally, transcriptomic technology is generating large sets of candidate genes with potential roles worthy of functional exploration. To be able to develop high-throughput approaches to gene deletion, fast and simple methods to generate gene deletion constructs are required. The method we describe here, named DelsGate (1), can be used universally for fungal species and in general for any organism with an efficient homologous recombination system. In our laboratories we routinely use DelsGate to produce deletion mutants in the basidiomycete *U. maydis* and we are currently extending its use to produce mutants in the Ascomycete *Verticillium dahliae*. Additionally, this method is also being applied in another Ascomycete plant pathogen, *Colletotrichum graminicola* (S. Sukno, personal communication).

DelsGate combines PCR with Gateway cloning technology. Gene flanking regions are produced by PCR, allowing for precise deletion of the ORF. The Gateway system developed by Invitrogen (2) is used for very efficient cloning of the amplified gene flanks. In this step, the *attB1* and *attB2* recombination sites, introduced in the PCR products during the amplification, promote in vitro recombination with a pDONR vector containing the *attP1* and *attP2* recombination sites. In the pDONR vector the *attP1* and *attP2* sites flank the *ccdB* gene and the in vitro recombination by the BP clonase replaces this gene with the PCR products generating an entry clone. The *ccdB* gene is lethal for most *E. coli* strains including the commonly used strain employed in this study, DH5α, thus only transformants harbouring the PCR products are highly favored to produce a colony. To speed up the process for production of deletion constructs for fungi we modified the Invitrogen pDONR201 vector by introduction of suitable selectable markers. In this way the entry clone produced after Gateway cloning is the deletion construct itself. We also greatly simplified the process by taking advantage of a report by Suzuki et al. (3) showing that two independent PCR fragments, each carrying an *attB1* or *attB2* site on one end, can simultaneously recombine with a single pDONR vector. This generates a linear construct that is then circularized in vivo via *E. coli* transformation, provided that there are homologous sequences at the free ends of the PCR fragments to promote recombination. In the DelsGate method the sequence added to the 5' and 3' flanks to promote homologous recombination in vivo is the 18 bp recognition site for the homing endonuclease I-*Sce*I, absent in most fungal genomes. Thus, in addition to promoting homologous recombination, this site is then universally used to generate the linear DNA for fungal transformation without concern for inadvertent digestion of the gene flanks.

Although the frequency of homologous integration can vary among fungal species, and even within species among genes, we regularly use 1 kb of the 5' and 3' gene flanks in the deletion construct to promote homologous recombination, which should be sufficient in most cases. The requirement for only the amplification of 1 kb of sequences flanking the ORF of the gene of interest prior to the BP clonase reaction, combined with the high efficiency of the Gateway cloning system makes DelsGate not only a very fast method, able to produce deletion constructs in just 2 days, but also robust, universal and technically very simple. For the convenience of users we described here the entire process from the production of the deletion construct to transformation of the fungus and confirmation of gene deletion.

This basic approach can also be adapted to other genetic manipulations. For example, in bacteria it is becoming increasingly desirable to create strains in which multiple, dispersed genes have been mutated simultaneously. Creating deletions that incorporate an antibiotic resistance gene (a marked deletion), as with the standard DelsGate strategy, will not work for more than two or three genes, due to limited number of useful antibiotics. Therefore, bacterial geneticists have devised methods to delete genes while introducing little or no foreign DNA (unmarked deletions). In our hands, the most problematic step in generating an unmarked deletion has been fusing the upstream and downstream flaking fragments to make the deletion allele, because neither splice overlap extension PCR (4, 5) nor ligation independent cloning (6) routinely worked for all gene targets. As an alternative, we found that the robust BP clonase reaction and in vivo fragment fusion (steps 2 and 3) of the DelsGate method provided a solution to these limitations in other methods and we describe here the general modifications introduced for this specific application.

2. Materials

2.1. Culture Media

1. Potato dextrose agar supplemented to 2% agar (2PDA): 39 g PDA powder (Difco, Franklin Lakes, NJ), 5 g supplemental agar, 1 L dH$_2$O.

2. Potato dextrose broth (PDB): 24 g PDB powder (Difco), 1 L dH$_2$O. After autoclaving, store at room temperature (RT). (See Note 1).

3. Low Na LB kanamycin plates: 1% bactotryptone, 0.5% yeast extract, 0.1% NaCl, 50 µg/ml kanamycin. To prepare 1 L: 10 g tryptone, 5 g yeast extract, 1 g NaCl, dH$_2$O to 1 L. After autoclaving add 0.5 ml of a 100 mg/ml solution of kanamycin sulfate (Sigma-Aldrich, Saint Louis, MO).

4. YEPS medium: 1% yeast extract, 2% bactopeptone and 2% sucrose. For 500 ml: dissolve 5 g of yeast extract, 10 g of bactopeptone and 10 g sucrose in a final volume of 500 ml dH$_2$O, dispense in aliquots of 100 ml in 500 ml flasks. After autoclaving, store at 4°C.

2.2. PCR Amplification of Gene Flanks and Clean Up of PCR Products

1. Primers: gene-specific primers 1 and 2 to amplify 5′ flanks and 3 and 4 to amplify 3′ flanks (Table 1).
2. *Taq*-polymerases and reaction buffers. Because the bands to be amplified are only 1 kb long any *taq*-polymerase of general use work well for this step. We regularly use home-made *taq*-polymerase. For commercial *taq*-polymerases the reaction buffer provided with them is used. For home-made *taq*-polymerase the 10× buffer we prepare contains: 0.5 M KCl, 100 mM Tris–HCl, pH 8.3, 0.1% gelatine, 1% Triton X-100.
3. 30% PEG 8000/30 mM $MgCl_2$. To prepare 100 ml dissolve 30 g of polyethylene glycol (PEG) 8000 (Sigma-Aldrich) in approximately 70 ml dH_2O, bring volume to 90 ml, after

Table 1
Primers used for DelsGate deletion construct generation, verification of deletion constructs and testing of deletion mutants

Primer	Use	Sequence
Primer 1-(I-SceIF)	Amplification of 5′ flank, primer forward	5′- TAG GGA TAA CAG GGT AAT-(gene-specific sequence, N_{20-22})-3′
Primer 2-(attB1)	Amplification of 5′ flank, primer reverse	5′-GGGG ACA AGT TTG TAC AAA AAA GCA GGC TAA-(gene-specific sequence N_{20-22})-3′
Primer 3-(attB2)	Amplification of 3′ flank, primer forward	5′-GGGG ACC ACT TTG TAC AAG AAA GCT GGG TA-(gene-specific sequence, N_{20-25})-3′
Primer 4-(I-SceIR)	Amplification of 3′ flank, primer reverse	5′- ATT ACC CTG TTA TCC CTA-(gene-specific sequence, N_{20-25})-3′
SceI-F	Verification of deletion construct	5′- TAG GGA TAA CAG GGT AAT-3′
SceI-R	Verification of deletion construct	5′-ATT ACC CTG TTA TCC CTA- 3′
DonrF-C	Verification of deletion construct when using pDONR-Cbx	5′-TCGCGTTAACGCTAGCATGGATCTC -3′
DonrF-H	Verification of deletion construct when using pDONR-A-Hyg and pDONR-Hyg	5′-ATCAGTTAACGCTAGCATGGATCTC-3′
DonrR	Verification of deletion construct for all vectors	5′-GTAACATCAGATTTTGAGACAC-3′
CbxF-DG	Verification of gene deletion when using pDONR-Cbx	5′-GACAGCCTATTGTGGCAGCC- 3′
Hyg-DG	Verification of gene deletion when using pDONR-Hyg	5′-AGAGCTTGGTTGACGGCAATTTCG-3′

autoclaving add 10 ml of 0.3 M $MgCl_2$. Store at 4°C or dispense in aliquots and freeze at −20°C.

4. QIAquick PCR purification kit (QIAgen, Valencia, CA).

2.3. BP Clonase Reaction

1. Gateway BP clonase II enzyme mix (Invitrogen, Carlsbad, CA).
2. Modified Gateway Donor vectors: For protoplast-mediated transformation of *U. maydis*: pDONR-Cbx and pDONR-Hyg; for protoplast-mediated transformation of Ascomycete fungi: pDONR-A-Hyg (Fig. 1); and for production of unmarked deletion mutants in bacteria: pDONR-SacTet.
3. Donor vectors are maintained in *E. coli* strain DB3.1 (Invitrogen) since they contain the *ccdB* gene which is toxic to most other *E. coli* strains used in molecular biology.

2.4. Transformation of E. coli

1. *E. coli* strain DH5α (Bethesda Research laboratories).
2. To increase transformation frequencies, commercial One Shot® MAX Efficiency™ DH5α-T1^R or One Shot® OmniMAX™ 2-T1^R *E. coli* competent cells (Invitrogen) can be used.

2.5. Verification of Deletion Construct

1. Cracking buffer: 100 mM Tris–HCl, pH 7.5, 100 mM EDTA, 5% glycerol, 100 mM NaCl, 0.004% bromophenol blue, 0.1% SDS. For 100 ml mix 5 ml 2 M Tris–HCl, pH 7.5, 20 ml 0.5 M EDTA, 5 ml glycerol, 2.5 ml 4 M NaCl, 4.0 mg bromophenol blue, 1 ml 10% SDS, and dH_2O to 100 ml. Keep at RT.
2. Phenol:chlorophorm:isoamylalcohol (25:24:1).
3. QIAprep spin miniprep kit (QIAgen).
4. Primers: SceI-F and SceI-R (Table 1) combined with gene specific primers from Subheading 2.2, or alternatively with vector primers DonrF-C or DonrF-H (for Donor vectors harbouring carboxin and hygromycin and selectable markers, respectively) and DonrR (Table 1).

2.6. Preparation of Deletion Construct for Fungal Transformation

1. I-*Sce*I (New England Biolab, Ipswich, MA).

2.7. Protoplast-Mediated Fungal Transformation

1. SCS buffer: 20 mM Sodium citrate, pH 5.8, 1 M sorbitol. For 200 ml: dissolve 1.18 g Na_3citrate and 36.44 g sorbitol (Sigma-Aldrich) in approximately 180 ml dH_2O, bring volume to 200 ml and autoclave. Store at 4°C.
2. STC buffer: 10 mM Tris–HCl pH 7.5, 100 mM $CaCl_2$, 1 M sorbitol. For 200 ml: dissolve 36.44 g sorbitol in approximately 160 ml dH_2O, bring volume to 178 ml and autoclave, then add 2 ml 1 M Tris–HCl, pH 7.5 and 20 ml 1 M $CaCl_2$. Store at 4°C.
3. Buffer II: 25 mM Tris–HCl pH 7.5, 25 mM $CaCl_2$, 1 M sorbitol. For 100 ml: Dissolve 18.22 g sorbitol in approximately 80 ml

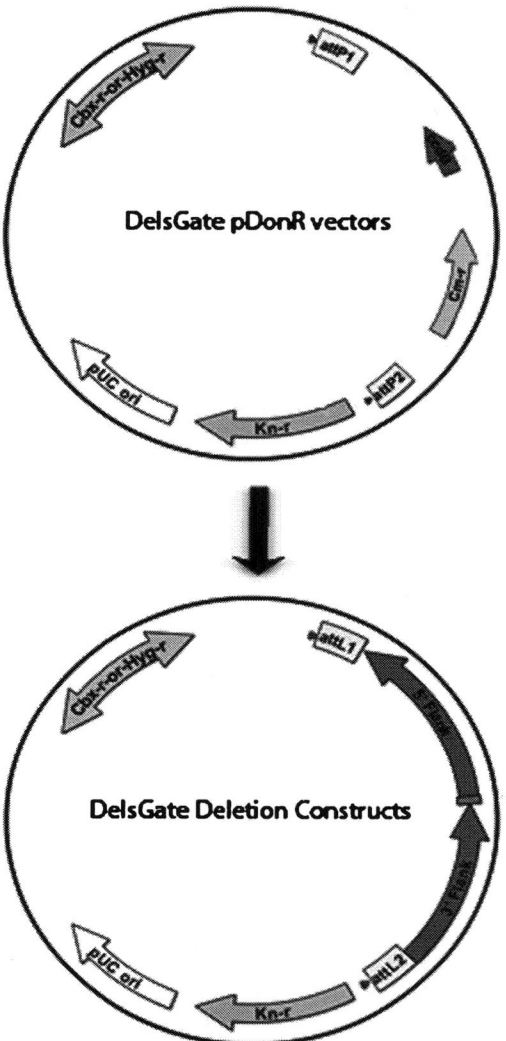

Fig. 1. Schematic representation of the general structure of pDONR vectors adapted to DelsGate and deletion constructs. *U. maydis* pDONR-Cbx and pDONR-Hyg and Ascomycete pDONR-A-Hyg are used in protoplast-mediated transformations. To produce pDONR-Cbx useful for transformation of *U. maydis* a *cbxR* marker was inserted in the blunt ended *Nsp*I site of the Gateway vector pDONR201 (Invitrogen). The *Nsp*I sites have thus been destroyed in pDONR-Cbx. To produce pDONR-Hyg, also for transformation of *U. maydis*, a *Bgl*II site was introduced into pDONR201 at its *Nsp*I site with an *Nsp*I linker carrying the *Bgl*II restriction site to generate pDONR201-BglII. Then, a *hsp70* promoter-*hygR-hsp70* terminator marker cassette was obtained by digesting plasmid pIC19HL with *Bam*HI and *Bgl*II and cloned into pDONR201-BglII at the newly introduced *Bgl*II site. Therefore, in pDONR-Hyg the *hygR* marker is flanked by an *Nsp*I site at both sites of insertion, but the *Bam*HI site was lost at the corresponding cloning site. It should be mentioned, that the *hygR* marker itself contains an additional *Nsp*I site. To produce pDONR-A-Hyg useful for transformation of Ascomycetes a TrpC promoter-*hygR* cassette (8) was inserted in the *Hpa*I site of pDONR201.

dH₂O, bring volume to 95 ml and autoclave. Then add 2.5 ml 1 M Tris–HCl, pH 7.5 and 2.5 ml 1 M CaCl$_2$. Store at 4°C.

4. Lallzyme MMX solution: 500 mg/ml in Buffer II. For 10 ml: dissolve 5 g of Lallzyme MMX (standard activities: 1,840 poly-galacturonase units/g, 24 pectin lyase units/g and 545 pectin esterase units/g) (Lallemand) in a final volume of 10 ml Buffer II by gently pipetting up and down, use fresh or dispense in aliquots and store at −80°C. Thaw at RT upon use.

5. Alternatively Vinoflow® FCI (Novozyme) can be used instead of Lallzyme MMX to digest cell walls. Vinoflow solution: 384 mg/ml in buffer II. For 10 ml: dissolve 3.84 g of Vinoflow in a final volume of 10 ml Buffer II, filter sterilize, use fresh or dispense in aliquots and store at −80°C. Thaw at RT upon use.

6. Forty percent PEG in STC: Autoclave 4 g PEG 4000 (Sigma-Aldrich) and 1.82 g sorbitol with 3 ml dH$_2$O (see Note 2), then add 0.1 ml 1 M Tris–HCl, pH 7.5, 1 ml 1 M CaCl$_2$ and sterile dH$_2$O to 10 ml. Store at 4°C and keep on ice when in use.

7. YEPS with sorbitol (YEPS-S): For 1 L: dissolve 10 g yeast extract, 20 g bactopeptone, 20 g sucrose and 182.2 g sorbitol in approximately 800 ml of dH$_2$O, bring volume to 1 L, add 20 g of agar and autoclave. After autoclaving add 3 µg/ml carboxin (cbx) or 150 µg/ml hygromycin (hyg) depending on the vector you are using.

2.8. Analysis of Fungal Transformants to Confirm Gene Deletion

1. Gene-specific primer 5 and CbxF-DG or Hyg-DG (Table 1 and Fig. 2b) for cbx and hyg vectors, respectively. Additionally, it is useful to use ORF specific primers to confirm deletion.

2. Alternatively use gene-specific primers 2-O and 3-O (Fig. 2b).

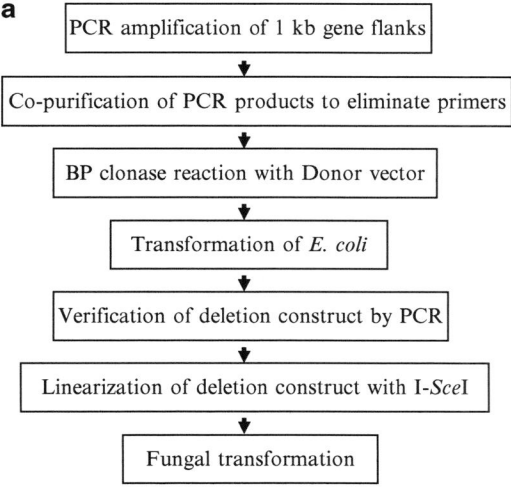

Fig. 2. DelsGate gene deletion methodology. (a) Flowchart of steps in the DelsGate method.

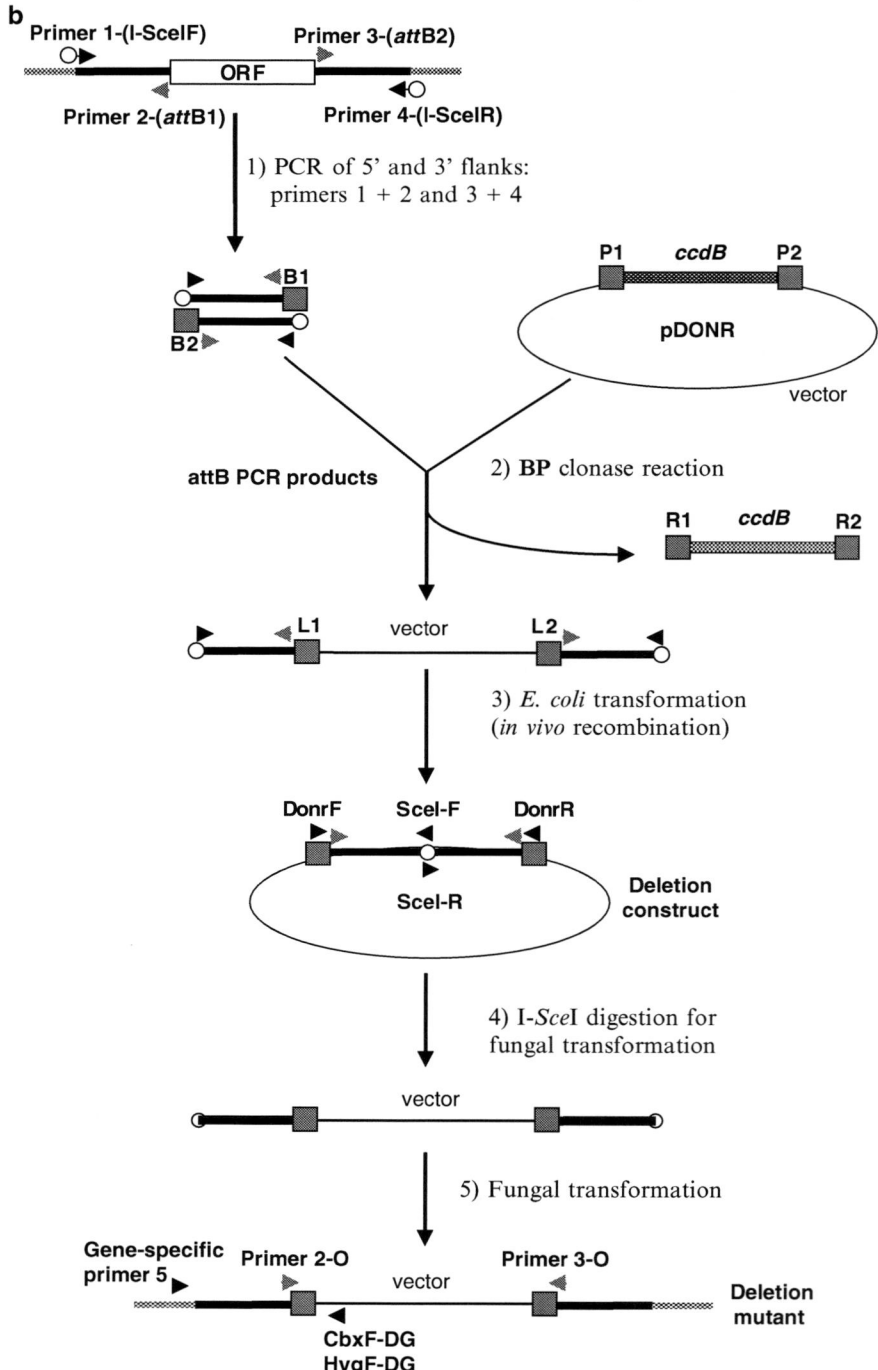

Fig. 2. (continued) (**b**) Schematic representation of DelsGate deletion construction method and generation of deletion mutants using DelsGate constructs.

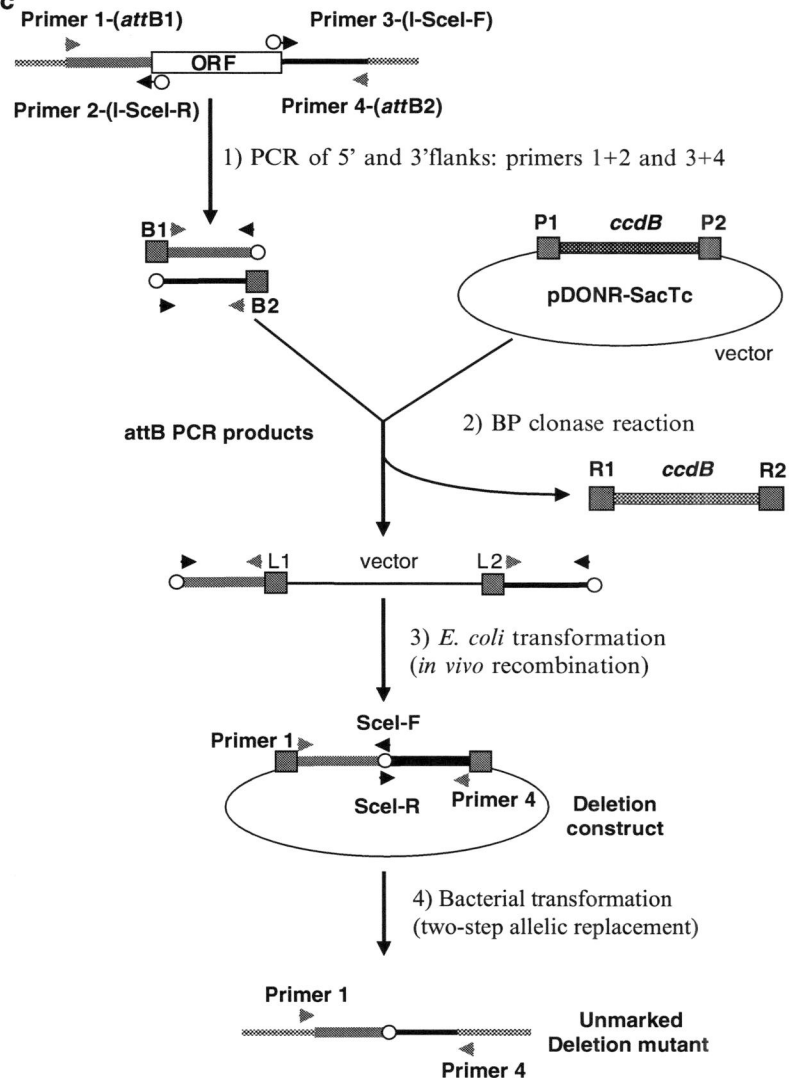

Fig. 2. (continued) (c) Schematic representation of DelsGate approach adapted for the generation of unmarked mutants in bacteria. B1, B2, P1, P2, R1, R2, L1, and L2 represent lambda phage recombination sites, for further details refer to Gateway technology from Invitrogen.

3. Methods

Figure 2a outlines the DelsGate construction steps. Figure 2b shows a schematic representation of the entire procedure to generate deletion mutants: production of deletion constructs by

DelsGate, manipulation of the deletion construct for fungal transformation, and finally analysis of transformants to test for gene replacement. DelsGate deletion construction involves the following three primary steps: (1) simultaneous independent PCRs of the 5′ and 3′ ORF flanks, (2) Gateway BP cloning, and (3) *E. coli* transformation. A number of features have been included in the design of our method to accelerate generation of deletion constructs. For the amplification of the 5′ flank the recognition sequence for the homing endonuclease I-*Sce*I is included at the 5′ end of the forward primer (primer 1), while an *attB1* sequence is included at the 5′ end of the reverse primer (primer 2). For the 3′ flank, an *attB2* sequence is included at the 5′ end of the forward primer (primer 3) and the I-*Sce*I recognition sequence in the reverse orientation is included at the 5′ end of the reverse primer (primer 4). PCR products are then inserted into a deletion plasmid vector via the Invitrogen Gateway BP clonase system. During the BP clonase reaction the co-purified 5′ and 3′ gene flank PCR products recombine with the *attP1* and *attP2* sequences of the donor vector, respectively. This reaction generates a linear molecule harbouring 18 bp of homologous sequences in opposite orientation at the free ends (the I-*Sce*I recognition site). After *E. coli* transformation this homologous sequence recombines in vivo as reported by Suzuki et al. (3) generating a circular construct. DelsGate modified donor vectors pDONR-Cbx and pDONR-Hyg for use with *U. maydis* and pDONR-A-Hyg for use in Ascomycete fungi have been produced by addition of appropriate selectable markers. Therefore the entry clone resulting from the BP clonase reaction and in vivo recombination is the final deletion construct itself, which has the ORF precisely replaced by the vector containing the selectable marker for fungal transformation and both flanks separated by the I-*Sce*I recognition site. Thus, in addition to promoting homologous recombination to generate the circular molecule in vivo, this 18 bp sequence is then used to generate the linear DNA for fungal transformation. The I-*Sce*-I site is extremely rare, it does not exist in the *U. maydis* genome and is likely completely absent from most fungal genomes. The modified pDONR vectors we initially constructed are compatible with protoplast-mediated transformation; we describe here the formation and transformation of protoplasts of the basidiomycete fungus *U. maydis* with DelsGate deletion constructs, and finally the analysis of fungal transformants for gene deletion. We also provide a brief description of the recent adaptation of our methodology for the production of constructs to generate unmarked mutants in bacteria.

3.1. Primer Design for the Amplification of Gene Flanks

1. Primers are designed to separately amplify 1 kb of the 5′ and 3′ sequences flanking the ORF of the gene of interest.
2. Primers 1 and 2 are designed to amplify the 5′ flank and their 5′ ends contain the I-*Sce*I recognition sequence in the forward

orientation and the *attB1* sequence, respectively (Table 1). The gene specific sequence of primer 1 (forward primer) is designed for the primer to anneal approximately 1 kb before the start codon while primer 2 (reverse primer) is designed to anneal as close as practical before the start codon (Fig. 2b).

3. Primers 3 and 4 are designed to amplify the 3′ flank and their 5′ ends contain the *attB2* sequence and the I-*Sce*I recognition sequence (in the reverse orientation), respectively (Table 1). The gene specific sequence of primer 3 (primer forward) is designed for the primer to anneal as close as practical after the termination codon while the gene specific sequence of primer 4 (primer reverse) is designed for the primer to anneal approximately 1 kb distal to primer 3 (Fig. 2b).

4. Gene specific sequences between 20 and 25 nucleotides work well for these primers.

3.2. PCR Amplification of 1 kb Gene Flanks and Clean Up of PCR Products

1. The 1 kb 5′ and 3′ gene flanks are amplified separately. PCRs are performed in a total volume of 50 μl. For each of the two reactions combine in a PCR tube: 32.0 μl PCR-grade ddH$_2$O, 5.0 μl 10× *taq*-polymerase reaction buffer, 5.0 μl MgCl$_2$ 25 mM (see Note 3), 1.0 μl 50× dNTP mix (10 mM each), 2.5 μl primer forward 20 μM, 2.5 μl primer reverse 20 μM, 1.0 μl *taq*-polymerase, 1.0 μl (10–100 ng) fungal genomic DNA.

2. Amplification is performed under the following conditions: an initial denaturation of 1 min at 94°C, 30 cycles of 30 s at 94°C, 30 s at 60°C, and 1 min at 72°C, and completed with a final extension of 5 min at 72°C (see Note 4).

3. After amplification, mix 5 μl of each PCR reaction with 1 μl of loading buffer and load samples on a 0.8% agarose gel to confirm amplification of the desired bands.

4. To eliminate primers from the sample prior to BP clonase reaction, the 5′ and 3′ PCR products are co-purified either by PEG/MgCl$_2$ precipitation or by using the QIAquick PCR purification system (Qiagen) (see Note 5 and 6).

5. For PEG precipitation 90 μl of the combined PCR products (see Note 7) are mixed with 270 μl of TE buffer, pH 7.5–8 and 180 μl of 30% PEG 8000/30 mM MgCl$_2$, vortexed thoroughly and centrifuged 15 min at 18,000 g. Finally the pellet is resuspended in 15 μl of TE buffer or ddH$_2$O.

6. Purification of the combined 5′ and 3′ flanks with QIAquick columns is performed according to the manufacturer's instructions (see Note 7). In the final step DNA is eluted in 30 μl of buffer EB (10 mM Tris–Cl, pH 8.5) or ddH$_2$O.

7. After purification of the combined 5′ and 3′ flanks by either method DNA concentration is determined by A_{260} nm. One micro litter of the purified DNA can be directly used for this measurement when using a NanoDrop spectrophotometer. When using a standard spectrophotometer measure concentration of co-purified PCR products by mixing 2 μl DNA with 58 μl TE buffer or ddH$_2$O (Dilution Factor (DF) = 30) and measuring A_{260} nm. To calculate concentration: $A_{260} \times DF \times 50/1{,}000 =$ concentration (μg/μl).

3.3. BP Clonase Reaction with Donor Vector

1. For BP clonase reaction, add the following components in a 1.5 ml microcentrifuge tube at RT: 1–7 μl (150–250 ng) of combined purified 5′ and 3′ gene flanks (see Note 8), 1 μl (150 ng) of pDONR vector (see Note 9), TE buffer, pH 8, to 8 μl.
2. Thaw BP Clonase II enzyme mix on ice for about 2 min and vortex briefly twice (2 s each time).
3. Add 2 μl of BP Clonase II enzyme mix to the reaction and vortex briefly twice. Spin down briefly.
4. Incubate overnight at 25°C. (See Note 10).
5. To terminate reaction add 1 μl of proteinase K, vortex briefly and incubate sample 10 min at 37°C.

3.4. Transformation of E. coli

1. Five to 10 μl of BP clonase reaction are used to transform *E. coli* competent cells. Generally, although using linear DNA, transforming home-made DH5α competent cells with standard heat shock methods is sufficient to produce a number of colonies to analyze. However, the use of commercial One Shot® MAX Efficiency™ DH5α-T1[R] or One Shot® OmniMAX™ 2-T1[R] *E. coli* competent cells (Invitrogen) can be an option to increase transformation frequencies. (See Note 11).
2. After transformation plate aliquots of 200 μl on low Na kanamycin plates, incubate plates overnight at 37°C. (See Note 12).

3.5. Verification of Deletion Construct

1. Prepare a replica LB kanamycin plate with the transformants to be analyzed for the presence of the correct construct (approximately 20).
2. Plasmid DNA is then directly purified from bacteria grown on this plate using a "cracking" method and loaded on an agarose gel to test for the expected plasmid size.
3. For cracking, take approximately half of the bacterial biomass from each replicated colony with a toothpick or a yellow pipette tip and mix it with 25 μl of cracking buffer by vortexing in a microcentrifuge tube.

4. Add 25 μl of phenol:chlorophorm:isoamylalcohol (25:24:1), mix by thorough vortexing and centrifuge 1 min at maximum speed.

5. Take 7–10 μl of the upper phase being careful not to disturb the inter-phase and load them on a 1% agarose gel.

6. Load the pDONR vector that was used to produce the construct as a control. Since the fragment between the *attP1* and *attP2* sites in the vector, which is replaced by the gene flanks, is roughly 2.4 kb and the size of the flanks combined is 2 kb, the correct construct should generally be very similar in size to the pDONR vector. (See Note 13).

7. Plasmid DNA is then purified to further confirm construct structure by PCR and to prepare DNA for fungal transformation.

8. Select one of the colonies with the desirable construct from the LB kanamycin replica plate, inoculate cells in 5 ml of LB amended with 50 μg/ml kanamycin and grow overnight at 37°C and 200 rpm.

9. Purify plasmid DNA using the QIAprep method from Qiagen column following the manufacturer's manual (see Note 14). Measure DNA concentration as above described.

10. A PCR analysis standardized to test for the presence of the 5′ and 3′ gene flanks by using two primer combinations, SceI-F and gene-specific primer 2, and SceI-R and gene-specific primer 3 (Fig. 2b) is then performed to further confirm construct structure. Primers SceI-F and SceI-R are designed from the I-*Sce*I recognition site in the forward and reverse orientation, respectively (Table 1). Alternatively, this analysis can be performed using primers from the vector in substitution for the gene-specific primers. In this case the primer combinations used are DonrF-C or DonrF-H (for Donor vectors harbouring carboxin and hygromycin and selectable markers, respectively) and DonrR (Table 1) in combination with the SceI-F and SceI-R primers, respectively.

11. Each reaction is performed in a total volume of 25 μl which contains: 11.5 μl PCR-grade ddH$_2$O, 2.5 μl 10× *taq*-polymerase reaction buffer, 2.5 μl MgCl$_2$ 25 mM (see Note 3), 0.5 μl 50× dNTP mix (10 mM each) 1.25 μl primer forward 20 μM, 1.25 μl primer reverse 20 μM, 0.5 μl *taq*-polymerase, approximately 10 ng of plasmid DNA, PCR-grade ddH$_2$O to 25 μl.

12. For each of the two primer combinations prepare a master mix containing all the components except the DNA, for all the reactions you need to perform plus one. For each reaction mix 24 μl of the master mix with 1 μl of plasmid DNA sample.

13. Amplification is then performed under the following conditions: an initial denaturation of 1 min at 94°C, 30 cycles of 30 s at 94°C, 30 s at 60°C, and 1 min at 72°C, and completed with a final extension of 5 min at 72°C.
14. After PCR load 10 μl of each PCR product on a 0.8% agarose gel to test for the presence of a band of the expected size. (See Note 15).

3.6. Preparation of Deletion Construct for Fungal Transformation

1. After confirmation of proper deletion construct structure, plasmids are digested with I-SceI for fungal transformation. This generates a linear vector molecule containing the selectable marker flanked by the 5′ and 3′ flanks of the ORF to be deleted.
2. For I-SceI digestion, combine in a microcentrifuge tube 3–5 μg of plasmid DNA, 5 μl of 10× I-SceI buffer, 1 μl (5 units) of I-SceI and to 50 μl of ddH$_2$O. Incubate 2 h at 37°C.
3. Generally, this incubation is sufficient to fully digest the vector DNA, however, it is a good practice to test if the digestion is completed by taking 5 μl of the digestion reaction, mixing it with 1 μl of loading buffer and running this sample on a 0.8% agarose gel. A single band of about between 6.8 and 5.5 kb should be visible depending on the pDONR vector that was used to make the deletion construct.
4. After digestion is complete precipitate DNA by adding 5 μl of sodium acetate 3 M, pH 5.2 and 100 μl 95% ethanol. Vortex briefly and incubate on ice for 15 min. Centrifuge 15 min at 18,000 g. Discard supernatant and wash pellet with 70% ethanol. Dry pellet and resuspend in 12 μl of ddH$_2$O.
5. Concentration of transforming DNA can then be measured by A$_{260}$ nm as described above. (See Note 16).

3.7. Protoplast-Mediated Fungal Transformation

1. For protoplast-mediated transformation of *U. maydis*, we use a variation of the method described by Tsukuda et al. (7). (See Note 17).
2. Inoculate 5 ml of PDB with one isolated colony of *U. maydis* grown on 2PDA, incubate over night (o/n) at 28°C and 250 rpm.
3. Use approximately 100 μl of the o/n culture to inoculate 100 ml YEPS medium in 500 ml flasks.
4. Incubate o/n at 28°C and 250 rpm until an OD$_{600}$ between 0.6 and 0.8.
5. Spin cells down in two 50 ml conical tubes, fairly gently approximately 1,100 × g, discard supernatant (s/n).

6. Add 10 ml of SCS to each tube, and resuspend with gentle vortexing. Combine the two tubes from each strain into one. Bring the volume to 30 ml with SCS, and spin as in step 5.

7. Resuspend cells in 1 ml SCS, and add 200 µl Lallzyme MMX solution (500 mg/ml in buffer II). Alternatively, add 200 µl Vinoflow solution (384 mg/ml in Buffer II).

8. Incubate with gentle mixing at room temperature (RT) checking protoplast formation on microscope every 15 min. It takes around 45 min for a majority of cells to turn into protoplasts when using Lallzyme solution. With Vinoflow incubation times are shorter, ranging from 10 to 30 min. (See Note 18).

9. When a majority of cells have formed protoplasts spin them down for 10 min at $1,000 \times g$, discard s/n, resuspend pellet in 1 ml SCS and transfer this suspension to a 1.5 ml microfuge tube.

10. Spin protoplasts down in a microcentrifuge for 5 min at 1,000 g at RT (see Note 19), carefully remove s/n and resuspend protoplasts in 1 ml SCS.

11. Spin down as above, resuspend in 1 ml STC and spin down in the same conditions.

12. Resuspend protoplasts, on ice, in 1 ml of ice cold STC. Protoplasts are now ready for transformation, they can either be used fresh or store at −80°C for future transformations. For freezing protoplasts add filtered sterilized DMSO to 7% (70 µl for 1 ml of protoplast suspension) and dispense in aliquots.

13. For transformation, in a 1.5 ml microcentrifuge tube combine approximately 1 µg of transforming linear DNA in a maximum volume of 5 µl (from step 4, Subheading 3.6), 1 µl heparin (15 mg/ml in STC; stored at 4°C) and 50 µl of protoplasts suspension, mix gently and incubate on ice for 20 min.

14. Add 500 µl of PEG 4000 solution, mix by inverting the tube several times or by very gentle pipetting, and incubate on ice for 30 min.

15. Add 500 µl STC, mix by inverting the tubes several times and pellet protoplasts by centrifuging at 600 g for 5 min.

16. Aspirate off the s/n carefully and resuspend protoplasts in 200 µl of STC.

17. Plate cells on YEPS-S plates containing the appropriate selection (carboxin or hygromycin) and incubate plates at 30°C, checking them every 24 h. Usually tranformants start appearing after 4–5 days.

18. When transformants start to appear transfer them to 2PDA with the appropriate selection.

3.8. Analysis of Fungal Transformants to Confirm Gene Deletion

1. Analysis of fungal transformants for gene deletion have been standardized by performing a PCR with a gene-specific primer located approximately 1.1 kb upstream of the start codon (i.e., outside the deletion construct) (gene-specific primer 5), combined with a primer that anneals to the selectable marker (Fig. 2b), CbxF-DG when using the *U. maydis* pDONR-Cbx vector and Hyg-DG when using vectors containing *hygR* as selectable marker (Table 1).

2. The reaction for each transformant to be analyzed is performed in a total volume of 25 µl which contains: 16.0 µl PCR-grade ddH$_2$O, 2.5 µl 10× *taq*-polymerase reaction buffer, 2.5 µl 25 mM MgCl$_2$ (see Note 3), 0.5 µl 50× dNTP mix (10 mM each), 1.25 µl 20 µM gene-specific primer, 1.25 µl selectable marker primer 20 µM, 0.5 µl *taq*-polymerase, 1.0 µl genomic DNA (10–100 ng).

3. Prepare a master mix containing all the components except the DNA, for all the reactions you need to perform plus one. For each reaction, mix 24 µl of the master mix with 1 µl of DNA.

4. Amplification is then performed under the following conditions: an initial denaturation of 1 min at 94°C, 30 cycles of 30 s at 94°C, 30 s at 60°C, and 90 s at 72°C, and completed with a final extension of 5 min at 72°C.

5. After PCR, load 10 µl of each PCR product on a 0.8% agarose gel; only transformants in which the gene has been replaced by the deletion construct through homologous recombination are expected to produce a band with this primer combination.

6. When using this approach to test for gene deletion it is good practice to confirm absence of the gene of interest in those transformants that generated a band in the above PCR by designing primers that anneal within the ORF. Perform this PCR in the manner describe in steps 2–4. Use a wild type strain as positive control in the amplification.

7. Replacement of the gene ORF by the deletion construct through homologous recombination can also be tested using primers with the same sequence as gene-specific primers 2 and 3 but in the opposite orientation, we name these primers 2-O and 3-O. (See Note 20).

8. Prepare reactions like in step 2 but substituting gene-specific and marker primers by primers 2-O and 3-O. Proceed as in step 3.

9. Amplification is then performed under the following conditions: an initial denaturation of 1 min at 94°C, 40 cycles of 30 s at 94°C, 30 s at 60°C, and 5 min at 72°C (see Note 21), and completed with a final extension of 7 min at 72°C.

10. After PCR load 18 μl of each PCR product on a 0.8% agarose gel. In transformants in which gene replacement has taken place you should observed absence of the wild type band and presence only of the deletion construct band, which size ranges from about 3.4 kb to 4.4 depending of which pDONR vector was used.

11. You can further confirmed deletion of the gene of interest by Southern blot hybridization.

12. Table 2 contains a list of genes for which we have generated DelsGate deletion constructs.

3.9. Other Applications

We have also adapted this methodology to the production of unmarked mutants in bacteria. The steps leading to the generation of constructs to produce these unmarked mutants, schematically represented in Fig. 2c, are very similar to those described for generation of deletion mutant constructs for fungi, however a number of modifications were introduced to the basic DelsGate approach to adapt it to this specific application. First, a new donor vector, pDONR-SacTet, was created by introducing a cassette with *sacB* and a tetracycline resistance marker into the *Hpa*I site of pDONR201. A basic step of most strategies to create unmarked deletions requires a method to select for loss of a particular function (rather than the normal selection for gain of antibiotic resistance). A common way to solve this problem is to incorporate the *Bacillus subtilis sacB* gene into the desired cloning vector. *sacB* encodes for levan sucrase, an enzyme that synthesizes a polyfructan from sucrose. When Gram-negative bacteria expressing *sacB* are cultured on sucrose-containing agar media the polyfructan accumulates intracellularly and inhibits multiplication. Consequently, visible colonies arise from cells that have either deleted or mutated the *sacB* gene. The second modification only required reversing which site-specific primers are modified by the addition of *attB*1/2 and the I-*Sce*I recognition sequences (Fig. 2c). For example, primers 1 and 4 carry the *attB* sites when making unmarked deletions, whereas primers 2 and 3 have this addition in the standard DelsGate method. This results in the deletion construct having the upstream and downstream flanks in their natural outside-in orientation with respect to the genome, instead of being oriented inside-out as in the standard DelsGate method. The third modification was simply to eliminate the I-*Sce*I digestion (step 4), which normally is required to linearize the plasmid. Instead, the circular plasmid is introduced into *R. solanacearum* by electroporation.

Table 2
Genes for which deletion constructs have been generated using DelsGate

Protein entry (similarity)	Organism	Gene deleted in target species
UM00116 (related to nicotinamide mononucleotide permease)	*U. maydis*	Yes
UM00118 (probable UDP-glucose 6-dehydrogenase)	*U. maydis*	Yes
UM00496 (cruciform DNA recognition protein Hmp1)	*U. maydis*	Yes
UM01230 (related to NTE1 – serine esterase)	*U. maydis*	Yes
UM01840 (proton nucleoside cotransporter)	*U. maydis*	Yes
UM02739 (probable nik1 protein)	*U. maydis*	Yes
UM04478 (probable ITR2-myo-inositol transporter)	*U. maydis*	Yes
UM04773 (conserved hypothetical protein)	*U. maydis*	Yes
UM05361 (related to laccase I precursor)	*U. maydis*	Yes
UM05436 (conserved hypothetical Ustilago-specific protein)	*U. maydis*	Yes
UM05567 (probable ATG8 – essential for autophagy)	*U. maydis*	Yes
UM05732 (related to phytochrome)	*U. maydis*	Yes
UM06363 (related to APG1 – essential for autophagocytosis)	*U. maydis*	Yes
UM06414 (related to polyketide synthase)	*U. maydis*	Yes
UM10146 (probable CPC2 protein, guanine nucleotide-binding protein beta subunit-like)	*U. maydis*	Yes
UM11400 (probable thiamin biosynthetic enzyme)	*U. maydis*	Yes
UM11574 (related to carboxymuconolactone decarboxylase)	*U. maydis*	Yes
UM11922 (related to chitin deacetylase precursor)	*U. maydis*	Yes
UM11957 (related to histidine kinase)	*U. maydis*	Yes
UM12169 (related to sodium/nucleoside cotransporter 2)	*U. maydis*	Yes
UM10957 (probable adenylyl-cyclase-associated protein CAP)	*U. maydis*	Yes
UM00011 (amd2 probable AMD2 – amidase)	*U. maydis*	No
UM00027 (putative protein)	*U. maydis*	No
UM00336 (putatuve protein)	*U. maydis*	No
UM00455 (related to MIR1)	*U. maydis*	No
UM00466 (hypothetical protein)	*U. maydis*	No
UM02161 (conserved hypothetical) protein	*U. maydis*	No
UM02791 (related to GABA permease)	*U. maydis*	No
UM03522 (related to uga4 – GABA permease)	*U. maydis*	No

(continued)

Table 2 (continued)

Protein entry (similarity)	Organism	Gene deleted in target species
UM05439 (related to chitin-binding protein)	*U. maydis*	No
UM06013 (related to intersectin 1)	*U. maydis*	No
UM10726 (related to YPD1 – two-component phosphorelay intermediate)	*U. maydis*	No
VDAG_08656.1 (cell pattern formation-associated protein stuA)	*V. dahliae*	No
RSc3113 (*gspN* – part of Type II secretion system)	*R. solanacearum*	Yes
RSc3286 & RSc3287, (*solI* & *solR* – acylhomoserine lactone synthase and sensor (LuxIR orthologs))	*R. solanacearum*	Yes

Subsequent steps are unique to creating unmarked mutants by two-step allelic replacement. First, transformants are selected on agar medium supplemented with tetracycline to recover colonies where a single homologous cross-over event (e.g., in the upstream flank) has introduced the entire recombinant vector into the bacterial genome. Second, several colonies are then plated on a minimal medium supplemented with 5% sucrose as the sole carbon source to select for cells that have experienced a second homologous cross-over event and evicted the plasmid. If the second cross-over was in the downstream flank, this leaves the unmarked deletion allele behind in the genome. Desired deletion mutants are identified by PCR with primers 1 and 4 directly from colonies or from small pools of colonies, because the amplicon is substantially smaller than that from a wild-type genomic template. Digestion of this amplicon with I-*Sce*I also should result in two almost equal sized byproducts. DNA sequencing is not required for confirmation. We have used this approach to successfully delete genes in *Ralstonia solanacearum* (Table 2), and it should work for any Gram-negative bacterium.

4. Notes

1. While the rest of media are generally stored at 4°C, PDB tends to precipitate in cold and it is stored at RT.
2. This is a very dense solution, to prepare it just mix the PEG 4000, sorbitol and dH_2O and directly autoclave.

After autoclaving a clear viscous solution is obtained to which the rest of components are added.

3. If the 10× *taq*-polymerase reaction buffer used already contains the $MgCl_2$, do not add any extra to the reaction and increase volume of ddH_2O accordingly.

4. As with any PCR, the annealing temperature may have to be adjusted for your particular primer combination. However, 60°C generally works well for the primer lengths proposed and following general recommendations for primer design.

5. Although PCR products can be used directly for BP clonase reactions, if not eliminated from the sample, primers can also recombine with the Donor vector and greatly increase the background of colonies that do not harbour the desired construct. We therefore strongly recommend to perform this clean up step.

6. The $PEG/MgCl_2$ precipitation removes primer-dimers and DNA molecules smaller than 300 bp. Purification through QIAquick columns eliminates primers and DNA molecules smaller than 100 bp. We favour the use of $PEG/MgCl_2$ precipitation since it is an inexpensive method and produces very clean DNA.

7. It is important that both flanks are roughly at the same concentration during the BP clonase reaction, therefore if one of the flanks was amplified more efficiently it is advised to mix appropriate volumes of each PCR product to result in similar molar ratios in the combined purified sample.

8. Increasing the amount of PCR product usually result in more colonies, however, do not exceed 250 ng for a 10 μl reaction.

9. Donor vector DNA suitable for the BP clonase reaction can be purified using standard methods. This includes alkaline lysis, however this DNA cannot be quantified by A_{260} nm due to contamination with RNA and we therefore prefer to use the QIAprep miniprep method from Qiagen or any other "clean" plasmid purification method.

10. This incubation time can be reduced to as short as 1 h. However, as in our method a linear DNA molecule is generated after the BP clonase reaction, and the frequencies of transformation with linear DNA are low, we favor overnight incubations which increase the number of colonies by fivefold to tenfold.

11. Transformation can also be performed by electroporation, however, in our experience this method increases the background of colonies that do not have the right construct.

12. We have consistently found that selection for kanamycin resistance works much better on low Na LB than on standard LB. In standard LB sometimes colonies appear that seem to contain no plasmid.

13. Plasmids which are clearly smaller that the pDONR vector usually contain only one or neither of the flanks.

14. Any other method for purification of plasmid DNA compatible with digestion with restriction enzymes can be used.

15. These bands should be the size of the amplified flanks when using gene-specific primers combined with the SceI primers and larger when using vector primers instead of the gene-specific primers.

16. We start with 3–5 µg of plasmid DNA to make sure that after the precipitation step we have enough DNA for transformation. We regularly use ≥1 µg of linear DNA for each transformation.

17. The DelsGate method can be used with any organism with an efficient recombination system. We have used deletion constructs generated by DelsGate for protoplast-mediated transformation to produce deletion mutant in *U. maydis* and this is the methodology we describe here. For other fungi, aspects such as formation of protoplasts or amount of linear DNA that give efficient rates of fungal transformation may vary.

18. Since the sale of Novozyme, previously used for the formation of protoplasts, was discontinued we have tried several commercial enzyme mixtures; we have found that Vinoflow generates protoplasts very efficiently and that Lallzyme MMX, although requiring longer incubation times, also generates protoplasts with high transformation frequencies.

19. Be certain to use a variable speed centrifuge and that you have it set at the appropriate setting since centrifugation at higher speed at this stage will burst protoplasts.

20. This approach can only be used when the length of the ORF is different from that of the vector in the deletion construct. Therefore is not suitable for genes of about 4.4 kb when using pDONR-Cbx to generate the deletion construct, for genes of about 4.8 kb when using pDORN-Hyg, and for genes of about 3.5 kb when using pDONR-A-Hyg. However, when it can be applied we consider this approach desirable because whether there was homologous or ectopic integration a positive PCR result is expected, allowing these events to be distinguished by amplicon length. We therefore find this approach very accurate in testing for gene deletion by PCR. Its disadvantage is that larger bands need to be amplified, however, except for large ORFs the expected size of bands are still

within the range that can be amplified with standard *taq*-polymerase based PCR reactions.

21. Since larger bands need to be amplified with this approach longer extension times are used.

5. Note Added in Proof

DelsGate has been shown to work very effectively in ***Fusarium verticillioides*** (Personal communication, Dr. Anthony Glenn, USDA ARS, Athens, GA). Multiple genes have been confirmed to be deleted in this species with constructs generated by DelsGate methodology.

Additionally, we have modified the system for effective use in assembling deletion constructs for ***Agrobacterium tumefaciens***-mediated fungal transformation systems. We call this new method OSCAR for One Step Construction of Agrobacterium Recombination-ready plasmids. This system replaces multisite approaches. We have effectively tested this method for gene deletion in ***Verticillium dahliae***. A report on this system should be published in 2010 with Z. Paz as the primary author.

References

1. García-Pedrajas MD, Nadal M, Kapa LB, Perlin MH, Andrews DL, Gold SE (2008) DelsGate, a robust and rapid gene deletion construction method. Fungal Genet Biol 45:379–388
2. Walhout AJ, Temple GF, Brasch MA, Hartley JL, Lorson MA, van den Heuvel S, Vidal M (2000) GATEWAY recombination cloning: application to the cloning of large numbers of open reading frames or ORFeomes. Methods Enzymol 328:575–592
3. Suzuki Y, Kagawa N, Fujino T, Sumiya T, Andoh T, Ishikawa K, Kimura R, Kemmochi K, Ohta T, Tanaka S (2005) A novel high-throughput (HTP) cloning strategy for site-directed designed chimeragenesis and mutation using the Gateway cloning system. Nucleic Acids Res 33:e109
4. Liu H, Zhang S, Schell MA, Denny TP (2005) Pyramiding unmarked mutations in *Ralstonia solanacearum* shows that secreted proteins in addition to plant cell wall degrading enzymes contribute to virulence. Mol Plant-Microbe Interact 18:1296–1305
5. Muyrers JPP, Zhang Y, Stewart AF (2001) Techniques: recombinogenic engineering–new options for cloning and manipulating DNA. Trends Biochem Sci 26:325–331
6. Aslanidis C, de Jong PJ (1990) Ligation-independent cloning of PCR products (LIC-PCR). Nucleic Acids Res 18:6069–6074
7. Tsukuda T, Carleton S, Fotheringham S, Holloman WK (1988) Isolation and characterization of an autonomously replicating sequence from *Ustilago maydis*. Mol Cell Biol 8:3703–3709
8. Carroll AM, Sweigard JA, Valent B (1994) Improved vectors for selecting resistance to hygromycin. Fungal Genet Newsl 41:22

Chapter 6

Gene Silencing for Functional Analysis: Assessing RNAi as a Tool for Manipulation of Gene Expression

Carmit Ziv and Oded Yarden

Abstract

The availability of a large number of gene-disrupted mutants (either from natural mutants' collections or from knockout projects) is a great advantage for functional analysis studies. However, disfunction of many fungal genes, involved in key developmental processes, leads to dramatic and pleotropic changes in cell morphology, conferring a major difficulty in studying null mutants. Therefore, obtaining variable levels of reduction in gene expression, especially of essential genes or genes whose impaired expression confers a pleiotropic phenotype, is extremely beneficial for studying their function. Here, we describe the use of RNAi as a gene silencing mechanism, in a manner that might facilitate the functional analysis of such essential genes. Two alternative strategies for the construction of an RNAi-induced inverted-repeat construct are demonstrated and a third alternative is suggested. In addition, DNA-mediated transformation of conidia by electroporation, RNA extraction from fungal mycelium and northern blot analysis are described in detail.

The experimental results presented, demonstrate that RNAi can be employed as a gene silencing tool in *Neurospora crassa*, both for nonessential *(al-2)* and essential *(cot-1)* genes, resulting in a range of stable, partially silenced mutants, exhibiting different phenotypes.

Key words: RNA interference, Post transcriptional gene silencing, Essential genes, Functional analysis, dsRNA, Inverted-repeat construct, Cloning

1. Introduction

The ascomycete mold *Neurospora crassa* is an important model organism for the study of filamentous fungi. The magnitude and range of the biological understanding of *Neurospora*, from the molecular to the population level, make it an ideal reference organism for many other fungi (1–4). As the complete genome sequence of *N. crassa* (40 Mb) and its annotation have been published (5, 6), the primary structural features of every *N. crassa* gene are available and enable probing gene regulation on an organism-level scale (5–7). This is further facilitated by the availability of a growing number of knockout strains (8) in addition to a comprehensive collection of methods and protocols gathered in an online publicly available manual (www.fgsc.net/Neurospora/NeurosporaProtocolGuide.htm).

However, unsuccessful attempts to inactivate some of the genes and obtain viable mutants, for example of genes involved in hyphal elongation and morphogenesis, might be a result of these genes being essential or detrimental in a highly pleiotropic manner, when mutated. It is therefore difficult to functionally analyze fully inactivated (null) alleles of these genes in order to learn their role in the mentioned processes. Thus, it is highly advantageous to be able to actively manipulate their expression to various levels and not to be dependent on new or available collections of *N. crassa* mutants and gene disruption strains.

RNA interference (RNAi) is a posttranscriptional gene silencing (PTGS) process, in which the presence of double-stranded RNA (dsRNA) homologous to a gene of interest results in degradation of the corresponding message, thereby reducing the expression of the gene product (9). This mechanism is increasingly being recognized as a principal switch for controlling eukaryotic gene expression and has since been employed as a tool for elucidating gene function (10, 11).

In *N. crassa*, the introduction of transgenic copies blocks the expression of endogenous genes by mRNA degradation of the transgene homologue (12). This type of PTGS (termed quelling) involves accumulation of small interfering RNA (siRNA) (13, 14). Both mRNA degradation and siRNA accumulation have been found to be common traits of different PTGS phenomena (15–18). Furthermore, the proteins involved in quelling, cosuppression (in plants) and RNAi (in animals) were found to be evolutionarily conserved (14). The components of *N. crassa* quelling mechanism have been identified and analyzed and include the following proteins:

1. QDE1 – RNA depended RNA Polymerase (RdRP), NCU07534.3 (19)
2. QDE2 – Argonaute, part of RISC (RNA-induced silencing complex), involved in sequence-specific mRNA degradation, NCU04730.3 (13, 20)
3. QDE3 – RecQ DNA helicase, NCU08598.3 (21)
4. Two Dicers: NCU06766.3 and NCU08270.3 (22)
5. QIP – an exonuclease that interacts with QDE2, NCU00076.3 (23).

RNAi has been described in a variety of organisms (5). This includes several fungal species like the ascomycota *Schizosaccharomyces pombe* (17), *N. crassa* (24), *Sclerotinia sclerotiorum* (25), *Magnaporthe oryzae* (18), *Aspergillus nidulans* (26) and the basidiomycota *Cryptococcus neoformans* (27),

Coprinus cinereus (28), and *Phanerochaete chrysosporium* (29) (see also review by Nakayashiki and Nguyen (30) and Nakayashiki (31). However, accumulating genomic data indicate that although the majority of fungi possess the silencing machinery, it is not necessarily a general trait as the RNAi machinery is absent from the genomes of *Saccharomyces cerevisiae*, *Ustilago maydis* and a few *Candida* species (31–34).

RNAi is one of the most powerful tools to explore gene function (31). To ensure stably inherited silencing, the expression of a hairpin RNA form of the gene is induced as a source of dsRNA for gene silencing. The use of constructs based on direct inverted-repeats designed to produce hairpin structured dsRNA was shown to be effective for inducing RNA silencing in fungi (18, 25, 27, 35) as well as in plants (36, 37), nematodes (38) and trypanosomes (39).

In order to assess RNAi as a tool for partial gene silencing in *N. crassa*, we introduced the *albino-2* (*al-2*) hairpin-structured RNAi construct. This nonessential gene is involved in the carotenoid biosynthetic pathway, and when impaired, results in altered conidial color that can be visually detected. Next, in order to demonstrate the use of RNAi for the manipulation of essential genes, we silenced *cot-1*, a gene whose knockout strain can be maintained only as a heterokaryon, and its temperature sensitive impaired allele results in dramatic and pleiotropic changes in hyphal morphology (40–42). Using RNAi, we were able to dissect the pleiotropic consequences of COT1 disfunction by determining the outcome of quantitative changes in *cot-1* expression.

In this chapter we demonstrate two alternative strategies for the construction of an RNAi inverted-repeat construct and we suggest a third alternative (see Note 1). The methods described here also include DNA-mediated transformation of conidia by electroporation, RNA extraction from fungal mycelium and northern blot analysis.

The experimental evidence provided determine that RNAi can be employed as a gene silencing mechanism both for nonessential *(al-2)* and essential *(cot-1)* genes, resulting in a range of mutants, expressing different levels of the silenced genes and are stable over time.

2. Materials

2.1. Designing and Constructing the Inverted-Repeat RNAi Construct

1. Restriction enzymes and T4 DNA ligase, store at −20°C.
2. DNA Taq polymerase, store at −20°C.
3. Nuclease-free water: Ultra pure water, DNAse, and RNAse-free. Can be stored at room temperature. However, it is recommended to aliquot the water under sterile conditions and to store at 4°C or −20°C.

4. Primers are synthesized as lyophilized nucleic acids (stored at room temperature). For PCR: dissolve primers in nuclease-free water to a final concentration of 5×10^{-5} M, divide into aliquots and store at −20°C.

5. Commercial vectors:
    (a) pDrive cloning vector (QIAgen, Hilden, Germany) used for direct cloning of PCR products.
    (b) pBluescript SK vector (Stratagene, La Jolla, California).
    (c) pUC118 (ATCC).

2.2. Culture Media and Transformation

1. Vogel's minimal medium (2): can be obtained from the Fungal Genetic Stock Center (FGSC) as 50× solution. Store at room temperature.

2. VgS: either liquid or solid (supplemented with 1.5% agar) Vogel's minimal medium with 1.5% (w/v) sucrose. Autoclave (20 min at 121°C) and store at room temperature.

3. 1 M Sorbitol, in double-distilled, deionized water (DDW). Autoclave (20 min at 121°C) and store at room temperature.

4. Electroporator: Biorad Gene Pulsar II system, with 0.2 cm gap cuvette.

5. FGS (10×): 20% (w/v) sorbose, 0.5% (w/v) glucose, and 0.5% (w/v) fructose. Autoclave (20 min at 121°C) and store at room temperature.

6. Plating medium (PM): Vogel's minimal medium supplemented with 1.5% (w/v) agar in 90% of the volume. After autoclaving the medium, add 10% (v/v) of 10× FGS.

7. Hygromycin B: 100 mg/mL. Store at −20°C.

8. Race tubes: Glass tubes that are used for determining the growth rate of fungal culture. Can be obtained from the FGSC in the following specifications: made of Borosilicate glass, 40 cm in length and outside diameter of 16 mm with 2 cm 45 bends on both ends.

9. ImageJ 1.37 V software (Rasband, W.S., U.S. National Institutes of Health, Bethesda, MD, http://rsb.info.nih.gov/ij/, 1997–2006), free download is available.

2.3. Northern Blot Analysis

1. 0.5 mm zirconia/silica beads: washed, autoclaved, and dried at 60°C.

2. Tri-reagent. Toxic, store at 4°C.

3. Chloroform: ACS grade. Toxic, store at room temperature.

4. Isopropanol (2-propanol): ACS grade. Toxic, store at room temperature.

5. Ethanol 75%: ACS grade (with nuclease-free water), keep at −20°C.

6. Formaldehyde: ACS grade. Toxic, store at room temperature.
7. Formamide: ACS grade. Toxic, store at 4°C.
8. DEPC-treated water: prepare 0.1% diethylpyrocarbonate (DEPC) in DDW by adding 1 mL DEPC per 1 L DDW. Mix and incubate overnight at room temperature. Autoclave for 20 min at 121°C to destroy DEPC by causing its hydrolysis. Store at room temperature. *Important*: DEPC is highly toxic.
9. SSC (20×): 3.0 M NaCl, 0.3 M Na-citrate, and pH 7.0 (43) in DEPC-treated water. Adjust pH with concentrated HCl. Autoclave (20 min at 121°C) and store at room temperature. For lower concentrations, dilute with DEPC-treated water.
10. SDS 20% (w/v) in DDW (43). Toxic, store at room temperature.
11. EDTA 0.5 M, pH 8.0 in DDW (43). Adjust pH with NaOH (pellets). Filter sterilize, and store at room temperature.
12. MEN (10×): 0.2 M MOPS, 0.05 M NaAc, and 10 mM EDTA (see Table 1). Dissolve in 800 mL DEPC-treated water, bring to pH 7.2 with NaOH and make up to a final volume of 1 L. Filter sterilize, and store at 4°C in the dark.
13. Sample buffer: 50% (v/v) Formamide, 6% (v/v) Formaldehyde, and 1× MEN (see Table 2). Toxic, prepare fresh.

Table 1
Stock solutions and required amounts for the preparation of MEN (×10)

Final concentration	Stock	For 1 L (g)
0.2 M MOPS		41.8
0.05 M NaAc	NaAc·3H$_2$O	6.8
0.01 M EDTA	EDTA·2H$_2$O	3.72

Table 2
Stock solutions and required amounts for the preparation of sample buffer

Final concentration[a]	Stock	Volume for ~20 samples (μL)
50% Formamide[b]	100%	150
6% Formaldehyde[c]	35–37%	51.5
1× MEN	10×	30

[a]Final concentration is calculated for the sample buffer *including* the RNA sample
[b]Toxic, causes eye and skin irritation
[c]Toxic, causes burns. Very toxic by inhalation and through skin absorption

14. Dye marker: 50% (w/v) sucrose, 0.5% (w/v) bromophenol blue in DEPC-treated water. Autoclave (20 min at 121°C) and store aliquots at −20°C.
15. Ethidum Bromide (EtBr) 1:50. Dilute with nuclease-free water. Toxic, prepare fresh.
16. Acetic acid glacial 5% (v/v) in DEPC-treated water. Autoclave (20 min at 121°C) and store at room temperature.
17. Hybond-XL nylon membrane (Amersham Biosciences, UK): charged nylon membrane designed for nucleic acid transfer applications.
18. ULTRAhyb solution (Ambion, Austin, TX), store at 4°C. Before use, incubate in a hot water bath at 65°C until completely dissolved.
19. Labeled nucleotides (radiation hazards) and labeling kits:
    (a) For DNA probe: [α-^{32}P]dCTP, 50 µCi at 3,000 Ci/mmol is to be used with Prime-A-Gene labeling system (Promega, Madison, WI).
    (b) For RNA probe: [α-^{32}P]UTP, at about 800 Ci/mmol and 10 mCi/mL or greater, in aqueous solution, is to be used with MAXIscript kit (Ambion).

2.4. Stripping of Northern Blots

1. SSC (20×).
2. SDS 20% (w/v).
3. EDTA 0.5 M, pH 8.0.

For details see Subheading 2.3.

3. Methods

3.1. Use of RNAi for Partial Inactivation of the Non-essential al-2 Gene

N. crassa mycelia and conidia are characteristically orange due to the production of carotenoids by a biosynthetic pathway that includes three structural *albino* genes: *al-1*, *al-2*, and *al-3* (44). In *N. crassa* mycelia, the transcription of *al-1* and *al-2* is light-induced, and the absence or impairment of their transcripts can be easily detected visually (12, 45). To select for RNAi-induced impaired gene expression by visual inspection of conidial color, the nonessential *al-2* gene (encodes the enzyme phytoene synthase (46)) was chosen for RNAi.

3.1.1. Constructing an al-2 RNAi Inverted-Repeat Construct

In order to assess RNAi as a tool for partial gene silencing in *N. crassa*, we constructed an inverted-repeat structure of *al-2*, that is expected to be transcribed as a double stranded hairpin RNA. The cloning was based on directed ligation into a commercial vector using restriction sites that were introduced by PCR (see Note 1).

The plasmid vector pTJS542 containing the *al-2* gene (46) was obtained from the FGSC. Two partially overlapping fragments of the *al-2* coding region were amplified by PCR with pTJS542 plasmid as a template: a 3.3 kb sense fragment contained nucleotide positions 21 and 3,337 and a 2.99 kb antisense fragment corresponded to nucleotide positions 25 and 3,015. The primers used were designed to introduce restriction sites at the product end to facilitate directional cloning (see Table 3, underline indicates the introduced restriction site). Each of the two amplified products, containing the same gene, yet with different restriction sites, was cloned into a pDrive cloning vector. Each of the resulting plasmids (pDrive + one of the PCR products) was digested with the appropriate restriction enzymes to excise the *al-2* PCR product with sticky ends. Then, the two digested PCR products were ligated into a pBluescript SK vector in opposing orientation (Fig. 1) by two successive steps of ligation. The inverted-repeats construct of *al-2* was designated pCZ2. The expression of this hairpin construct was driven by the native promoter of *al-2*, present in the sense fragment of the construct.

Table 3
Primers used for the constructing of *al-2* RNAi induced construct

Name	Sequence
al2-*Xba*I	GCT<u>CTAGA</u>TGGTGTTGCGTTGCTGTTG
al2-*Sma*I	TAA<u>CCCGGG</u>AACTAGCCTAACACCACCAA
al2-*Hind*III	AT<u>AAGCTT</u>GTGCGTCAACCCTTCCTTCT
al2-*Apa*I	AA<u>GGGCCC</u>GATGGTGTTGCGTTGCTGT

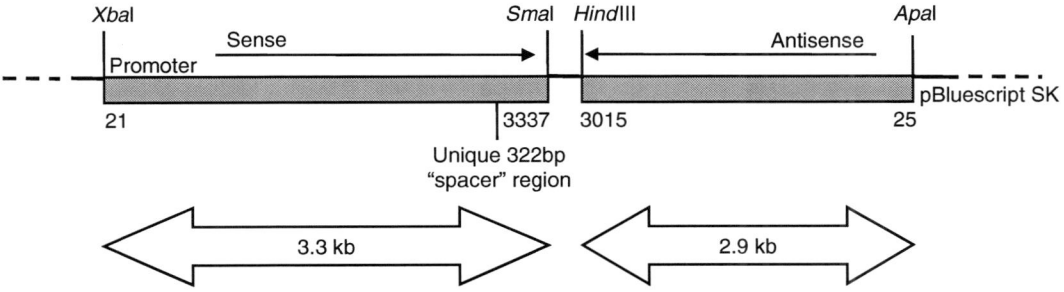

Fig. 1. Schematic illustration of pCZ2. This construct is designed to transcribe dsRNA hairpin homologues of *al-2*. Shaded bars represent *al-2* genomic DNA fragments. Numbers beneath the bars indicate the relevant nucleotide positions within the *al-2* gene.

3.1.2. Transformation with al-2 RNAi Cassettes

Plasmid pAB1004 [a Bluescript-based modification of plasmid pCB1004, harboring a hygromycin resistance cassette (47)) was used to facilitate transformant selection.

The construct pCZ2 was cotransformed with pAB1004 to *N. crassa* wild type (74-OR23-1 A, obtained from FGSC #987) by electroporation (48). The electroporation procedure includes the following steps: harvesting of the conidia, electroporation, plating and characterizing the resulting hygromycin-resistant transformants. Important: Conidia must be kept at 4°C at all stages from harvest to the end of the electroporation procedure.

1. Grow the cultures in a 250 mL Erlenmeyer flask plugged with a large cotton stopper, containing 50 mL solid VgS medium. For efficient transformation, it is advisable to use freshly harvested conidia. It is best to use cultures 9–11-day-old, efficiency appears to drop as cultures become older.
2. Once the culture has produced conidia, and the conidia have been given a few days to mature by exposing the culture to light, harvest conidia with 50 mL of ice- cold 1 M sorbitol and swirl to wet the conidia.
3. Filter the conidial suspension through sterile cheesecloth or muslin lined funnel to separate the conidia from the hyphae.
4. Pellet conidia by centrifugation for 5 min at $2,800 \times g$, 4°C and discard the supernatant.
5. Wash conidia three times in 50 mL of ice-cold 1 M sorbitol, pelleting by centrifugation for 5 min at $2,800 \times g$, 4°C between each wash. It is important to break up the clumps of conidia after discarding the supernatant by vigorously vortexing in a small amount of 1 M sorbitol between washes.
6. Count and dilute in ice-cold 1 M sorbitol to give a final concentration of 3×10^9 conidia per mL (see Note 2).
7. Add 30 μL of conidia to prechilled 1.5 mL microfuge tubes.
8. Add 10 μL of prechilled DNA (linearized DNA is preferred, ~500 ng/μL) to the conidial suspension and mix by pipetation. Lower DNA concentrations will work, but at lower efficiency.
9. Transfer the 40 μL conidia/DNA suspension to a prechilled cuvette (0.2 cm gap) and tap the cuvette on the bench to ensure that the suspension is evenly dispersed on the bottom of the cuvette.
10. Place the loaded cuvettes back on ice.
11. Set up the electroporator as follows: Resistance – low range – 200 ohms, Capacitance – 50 μF, Volts – 1.5 kV (time constant following electroporation is usually 8–9 ms).
12. Wipe the cuvette and place in the cuvette holder of the electroporator.

13. Measure the resistance of the cuvette plus conidia. Resistance should be similar or higher than the set value (200 Ω).
14. Electroporate the sample and place back on ice immediately.
15. Add 360 μL of VgS liquid medium to the cuvette following electroporation.
16. Transfer as much of the cell suspension as possible to a clean 1.5 mL microfuge tube at room temperature, and incubate in the 1.5 mL tube for 3 h at 25°C prior to plating with a medium containing the antibiotic.
17. Dissolve the PM and cool down to 55–60°C, then add 100 μg/mL hygromycin.
18. Transfer 1–200 μL of the electroporated conidia (suspended in VgS) to 15 mL medium (PM supplemented with hygromycin). Mix and plate, and incubate at the appropriate temperature (34°C). Repeat the plating procedure with serial dilutions.
19. Colonies usually appear within 3–5 days. In order to characterize the different transformants, transfer each of the colonies to a separate slant or plate containing VgS (supplemented with 1.5% agar and 100 μg/mL hygromycin) and incubate at the appropriate temperature (34°C) for a few days until enough biomass has accumulated and the culture has produced spores.
20. Incubate the cultures for 2–3 days in the sun to mature the spores. Then, the transformants' cultures can be stored at −20°C for few years.

3.1.3. Characterization of the al-2 RNAi-Induced Transformants

In order to determine the change in carotenoid production as a result of the RNAi-induced silencing of *al-2* in the different transformants, perform a methanol and acetone extraction and quantification procedure [according to (49) with minor modifications].

1. Inoculate 2×10^5 spores/mL of liquid VgS at the appropriate temperature in a rotary shaking water bath or incubator and incubate in the dark.
2. Filter a 1 day liquid culture through filter paper circles; expose the mycelium mat to light for 1 h, and then freeze in liquid nitrogen.
3. Dry the mycelium by lyophilization and determine the weight of the dried mycelium.
4. Grind the mycelium by adding 0.5-mm glass beads and subjecting the sample to a 10 s cycle with a bead-beater at max speed.
5. Add 200 μL of methanol per mg dried mycelium, mix vigorously, and incubate for 20 min at 60°C.

6. Recover the supernatants of the homogenates by centrifugation for 10 min at $10,000 \times g$, 4°C. Keep the recovered supernatant at 4°C in the dark.

7. Add 200 μL of acetone per mg dried mycelium to the pellet, mix vigorously and incubate again for 20 min at 60°C.

8. Repeat step 5 to recover the supernatant.

9. Mix the acetone extraction with the stored methanol extraction (keep at 4°C in the dark until measuring).

10. Determine the OD of the methanol: acetone extraction mixture at 470 nm. Use a mixture of methanol:acetone 1:1 as a blank.

N. crassa colonies were screened for hygromycin-resistant transformants that were also impaired in the production of carotenoids. Homokaryon derivatives of the pCZ2 transformants were obtained by multiple successive single-colony isolations, and by microconidial isolation (50) (see Note 3). 20% of the hygromycin-resistant transformants presented a fully silenced phenotype (white colonies), while 28% of the transformants presented a partially silenced phenotype (yellow/pink colonies) (Fig. 2).

3.1.4. RNA Extraction

Northern analyses (Figs. 3 and 4) were performed to determine the level of the light induced *al-2* mRNA in the different mutants, in order to verify the silencing mechanism and to ensure that the silenced phenotype was obtained as a result of an RNAi mechanism (see Note 4).

1. For total RNA isolation, inoculate 2×10^5 spores/mL of wild type and transformants in a 250 mL Erlenmeyer flask plugged with a large cotton stopper, containing 50 mL liquid VgS medium. Incubate at the appropriate temperature (34°C) in a rotary shaker (140–160 rpm) in the dark for 20–24 h.

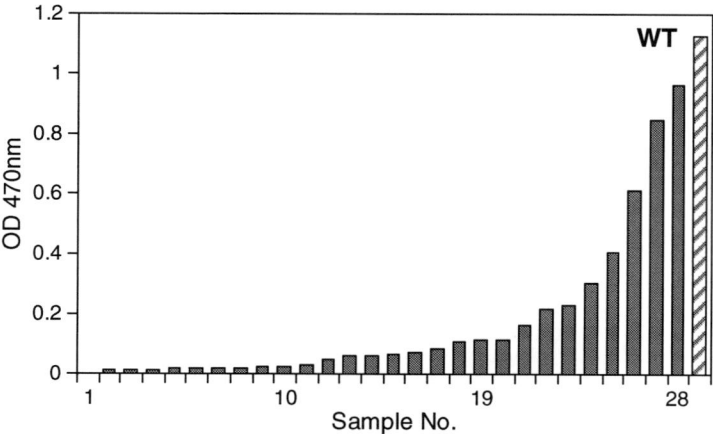

Fig. 2. RNAi results in a multi-range gene silencing mechanism. Colorimetric carotenoid quantification (as measured by OD at 470 nm) of 28 *N. crassa al-2* RNAi transformants.

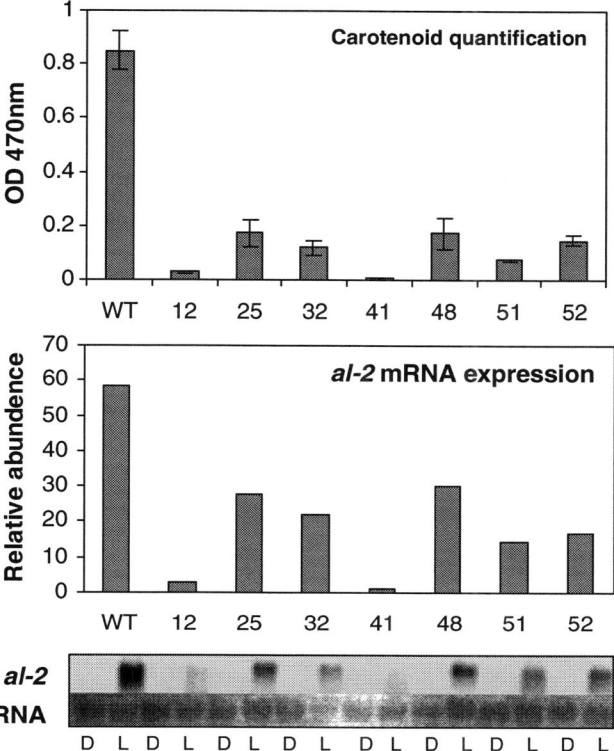

Fig. 3. RNAi silencing of *al-2* results in a reduced transcript abundance that is correlated to a reduced carotenoid biosynthesis. Upper chart: Carotenoid content (as measured by OD at 470 nm) in extracts of *N. crassa* wild-type and seven selected pCZ2 transformants. Bars indicate standard error of three repeats. Lower chart: northern blot and transcript abundance (as determined by Gel-Pro-Analyzer 3.0, Media Cybernetics, Silver Spring, MD) of *al-2* in extracts prepared from mycelia of wild-type and the transformants grown in either complete darkness (D) or after 1 h light induction (L). For normalization, ribosomal RNA was visualized by Methylene Blue.

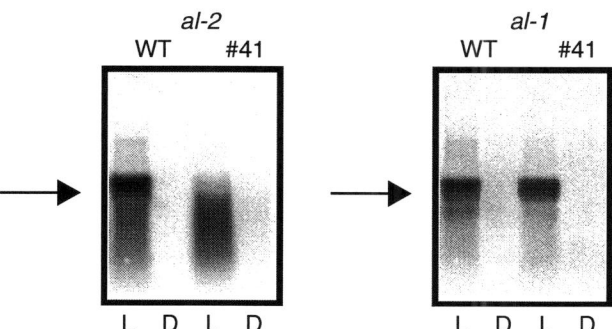

Fig. 4. Northern blot analysis of *N. crassa* wild-type (WT) and transformant #41. RNA samples were extracted from cultures grown in either constant darkness (D) or after 1 h light induction (L). Hybridization was carried out using specific *al-2* and *al-1* probes. Arrows mark the intact transcript.

2. Filter a 1 day liquid culture through filter paper discs in the dark. Cut the filter paper with the mycelium mat into two equal halves, expose one half to light for 1 h and keep the other half in the dark.

3. Freeze the mycelium mat in liquid nitrogen. Frozen culture samples can be stored at −80°C for a few weeks until extraction of RNA (see Note 5).

4. Keep the frozen culture samples on ice. Add 1.25 mL of Tri-reagent to 0.5 g of frozen mycelium in addition to 0.5-mm zirconia/silica beads.

5. Grind the samples with a bead beater for 30 s at maximum speed.

6. Repeat step 3.

7. Keep the samples on ice for 5 min, and then add 250 μL of chloroform to the extracts and grind the samples again for 15 s at maximum speed.

8. Keep the samples on ice for 5–10 min, and then centrifuge the homogenates for 10 min at $12,000 \times g$, 4°C.

9. Collect the aqueous layer (~600 μL) into a new microfuge tube and gently mix with 600 μL of isopropanol.

10. Incubate the samples for 10 min at room temperature to precipitate the RNA.

11. Centrifuge the samples for 10 min at $12,000 \times g$, 4°C.

12. Carefully remove the supernatant without disposing the pellet.

13. Wash the pellet by adding 1–1.5 mL cold ethanol (75%, v/v) and following by a short vortex (only until the pellet moves) and recentrifuge for 5 min at $9,000 \times g$, 4°C.

14. After the centrifugation, remove the ethanol entirely, by vacuum. Let the tubes dry by placing, upside-down on a filter paper for 10–15 min. Make sure not to over-dry the pellet, as it will be difficult to dissolve it in water.

15. Resuspend the pellet in 100 μL of nuclease-free water. Incubate the tubes at 65°C for 10 min until the pellet is completely dissolved. Sometimes, incubation at 37°C is enough for dissolving the pellet completely.

16. RNA samples may be stored at −20°C for 1–2 days before use. For longer storage, it is recommended to aliquot the RNA and to maintain the samples at −80°C. (see Note 6 for important guidelines of working with RNA).

3.1.5. Northern Blot Analysis

For northern blot analysis, total RNA samples (15 μg) are transferred to Hybond-XL nylon membranes.

1. Add 10.5 μL of sample buffer to 15 μg of RNA (3–5 μL).

2. Denature for 10 min at 65°C and place on ice.

Table 4
Stock solutions and required amounts for the preparation of 6% formaldehyde – 1% agarose gel

Ingredient	Mini gel	Midi gel	Maxi gel
MEN (10×) (mL)	5	10	15
DEPC-treated water (mL)	37	74	111
Agarose (g)	0.5	1.0	1.5
ªformaldehyde (mL)	8.3	16.6	24.9

ªAdd when gel is at 65°C (see Note 7)

3. Add 3.5 μL of dye marker and 0.5 μL of EtBr (1:50), keep on ice.
4. Prepare 6% formaldehyde – 1% agarose gel according to Table 4 (in the chemical fume hood).
5. Add 1× MEN (dilute the 10× MEN with DEPC-treated water) to gel tank (300 mL for mini gel, 800 mL for midi gel, and 1,300 mL for maxi gel).
6. Perform the electrophoresis (for midi gel apparatus, 100 V–110 V for 1–1.5 h).
7. After resolving the nucleic acids, take a photo with a ruler under UV light for orienting the location of the ribosomal RNA.
8. Shake the gel gently in 20× SSC for 30 min to remove formaldehyde residues.
9. Wet the membrane with 2× SSC.
10. Transfer the RNA to the membrane by upward capillary transfer (43) with 10× SSC, overnight. Do not forget to remove air bubbles.
11. Next morning, mark the wells and the side of RNA and cut a membrane corner for later orientation.
12. Wash the membrane in 2× SSC for a few minutes and then crosslink by exposure to UV (perform "optimal crosslink," twice in a UV oven).
13. Shake the membrane for 15 min at room temperature in 5% acetic acid.
14. Discard the acetic acid and shake with DEPC-treated water with a few grains of methylene blue until the ribosomal bands are visible (15–30 min). Wash the background with DEPC-treated water for visualization and take a photo with a ruler.
15. Wrap the wet membrane with Saran wrap and keep at −20°C.

Table 5
Primers used for the synthesis of probes for the northern blot analysis of *al-1* and *al-2* transcripts

Gene detected by the probe	Primers used for PCR	PCR product (bp)
al-1 (NCU00552.3)	al1-801: AAAGAAGCCAAAGGCTACGG al1-1346: TTGGTCTCCCAGGTGTAAGG	540
al-2 (NCU00585.3)	al2-1514: CAGGATACATCTCATCCAAACAG al2-1996: GATAGAAGGTGGGTAGGAACATC	483

3.1.6. Hybridization

RNA samples blots were probed with a [α-^{32}P]dCTP hexamer-labeled DNA probe prepared from the 480 bp *al-2* PCR product and from the 540 bp *al-1* RT-PCR product (using *N. crassa* cDNA of culture exposed to 1 h light as a template). See Table 5 for the primers used to synthesize the PCR products that were used as templates for the probes.

1. Place the membrane(s)/blot(s) in a hybridization tube with the RNA side facing up (toward the inner part of the tube).
2. Add 10–15 mL of ULTRAhyb solution (see Note 8) and prehybridize at 42°C for at least 1 h in a rotary hybridization oven.
3. Synthesize the [α-^{32}P]dCTP hexamer-labeled DNA probe using Prime-A-Gene labeling system according to the manufacturer's instructions. Alternatively, if you are reusing a probe (see Note 9), go directly to step 6.
4. Denature the probe by boiling for 2–5 min and quickly place on ice. Add EDTA to a final concentration of 20 mM and keep on ice.
5. Transfer 5–8 mL of the hybridization buffer used for prehybridization from the hybridization tube into a 15 mL tube. Add the probe to the prehybridization ULTRAhyb solution, mix well and transfer back to the hybridization tube.
6. Hybridize overnight at 42°C in a rotary hybridization oven.
7. Wash the blot briefly with approximately 100 mL of 2× SSC, 0.1% SDS solution while in the hybridization tube.
8. Low stringent washes: wash the blot for 5 min at 42°C with 100 mL of prewarmed solution of 2× SSC, 0.1% (w/v) SDS. Discard the wash solution and repeat the wash.
9. High stringent washes: Wash the blot for 15 min at 42°C with 100 mL of prewarmed solution of 0.1× SSC, 0.1% (w/v) SDS. Discard the wash solution and repeat the wash.
10. Check the signal with a monitor. If the radioactive background signal is too high, repeat the high stringent washes.
11. Seal membrane in a plastic bag or wrap in Saran wrap and expose to film or PhosphorImager.

Carotenoid extraction and analysis was performed using methanol and acetone extraction as described above in Subheading 3.1.3. The results indicate that there is a clear correlation between *al-2* transcript abundance and transformant color intensity (Fig. 3).

3.1.7. Stripping the Northern Blot and Re-probing it for Control

Transformant #41, which exhibited a complete silenced *al-2* phenotype, was further examined in order to verify the nature of *al-2* degradation. Following a 1 h light induction of cultures that were pregrown in the dark, *al-1* and *al-2* mRNA abundance was determined by northern blot analysis. The northern blot analysis of *al-2* expression was performed as described above (in Subheading 3.1.5). Then the blot was stripped (see Note 10) and reprobed with *al-1* probe.

For Stripping:

1. Place the membrane(s)/blot(s) in the hybridization tube with the RNA side facing up (toward the inner part of the tube).
2. Boil 200 mL of 0.05× SSC, 0.01 M EDTA.
3. Add 20 mL of 20% (w/v) SDS to the 0.05× SSC, 0.01 M EDTA solution (to a final concentration of 1%) and transfer into the hybridization tube.
4. Incubate the blots at 95°C for 15 min in a rotary hybridization oven.
5. Repeat steps 1–3 with fresh solutions.
6. Rinse blots in 0.01× SSC at room temperature.
7. Check the signal with a monitor to verify that the radioactive signal was entirely removed.

Our results indicate that the *al-2* transcript was subjected to degradation in transformant #41, as evident by the size range and signal intensity, compared to the wild-type strain (Fig. 4). However, there was no apparent difference in the pattern of the *al-1* transcript between these strains, demonstrating the targeted nature of the silencing mechanism.

To conclude, this experiment demonstrates a clear correlation between the expression levels (mRNA) of a gene encoding a biosynthetic enzyme and its enzymatic reaction end product, i.e., carotenoid levels in *al-2* RNAi induced strains.

3.2. Use of RNAi for Partial Inactivation of cot-1, a Gene Encoding a Kinase Required for Hyphal Growth in N. crassa

The *N. crassa cot-1* gene encodes a Ser/Thr protein kinase involved in apical hyphal cell elongation and polarity. The *cot-1* temperature-sensitive (ts) mutant is defective in hyphal extension that results in colonial growth at the restrictive temperature (32°C and above). The *cot-1* ts mutant ceases hyphal elongation after being shifted to restrictive temperature and produces new hyphal tips along the entire cell, indicating that COT1 is required for normal hyphal elongation (40). However, since defects in COT1 confer pleiotropic morphological effects, it is clear that COT1 may be involved in the regulation of several different cellular processes that affect hyphal morphogenesis. Nevertheless, the actual role of COT1 kinase in the fungal cell has yet to be determined (51).

We adapted RNAi as a silencing mechanism to reduce *cot-1* gene expression to learn its role in the hyphal elongation process.

3.2.1. Constructing a cot-1 RNAi Inverted-Repeat Construct

Constructing an inverted-repeat structure of *cot-1* that will be transcribed as a double stranded hairpin RNA was based on directed ligation using restriction sites that were available in the original plasmids/ORF (see Note 1). A hairpin construct of the *cot-1* gene was assembled using the first 2,000 bp of the genomic fragment that includes the *cot-1* promoter and the first two exons. Plasmids pOY18 containing the *cot-1* wild-type allele (4 kb), capable of complementing the *cot-1* ts mutant (40) and pCZ13 (which is the 4 kb genomic *Sma*I/*Eco*RI *cot-1* fragment from pOY18 ligated into pUC118), were used for the constructing of *cot-1* inverted-repeats construct. Since the *Eco*RV restriction site appears twice at the genomic *Sma*I/*Eco*RI *cot-1* fragment (at positions 2,027 and 4,031), digesting this fragment with *Eco*RV excludes the 3′ half of this fragment. To do so, both pOY18 and pCZ13 were digested with *Eco*RV and religated to create pCZ31 and pCZ32, respectively. Then, the *Xba*I/*Nru*I 1,800 bp fragment from pCZ31 was ligated into pCZ32 (that was previously digested with *Eco*RV and *Spe*I) to create pCZ33 that includes *cot-1* promoter in addition to part of the ORF (the first two exons and introns) in inverted-repeats in a manner that the nonoverlapping 240 bp of the long fragment create the "spacer" that forms the loop of the RNA hairpin structure (for schematic illustration, see results Fig. 5). The expression of a hairpin construct was driven by the native promoter of *cot-1*, present in the both orientations of the construct.

3.2.2. Transformation and Screening for cot-1 RNAi Mutants

pCZ33 was cotransformed with pMP6 (containing a hygromycin-resistance cassette, kindly provided by M. Plamann) to *N. crassa* conidia by electroporation (see Subheading 3.1.2). Hygromycin-resistant transformants were screened for altered growth rate and/or

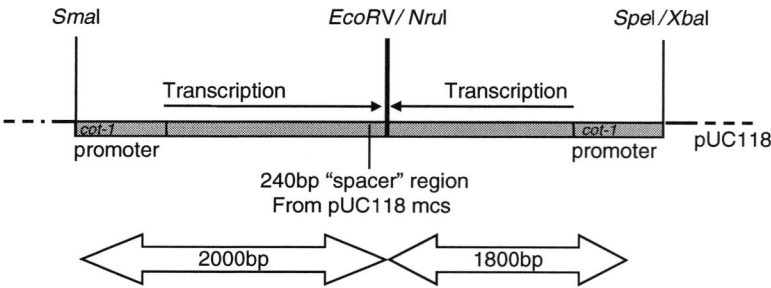

Fig. 5. Schematic illustration of pCZ33. Schematic illustration of the construct, designed to transcribe dsRNA hairpin homologs of *cot-1*. *Gray bars* represent *cot-1* genomic DNA fragments.

hyphal morphology and subsequently subjected to homokaryon purification (50) (see Note 3). Homokaryon mutants harboring the *cot-1* inverted-repeats construct (verified by Southern analysis, see Note 4) were subjected to morphological (Figs. 6 and 7) and molecular characterization (Fig. 8).

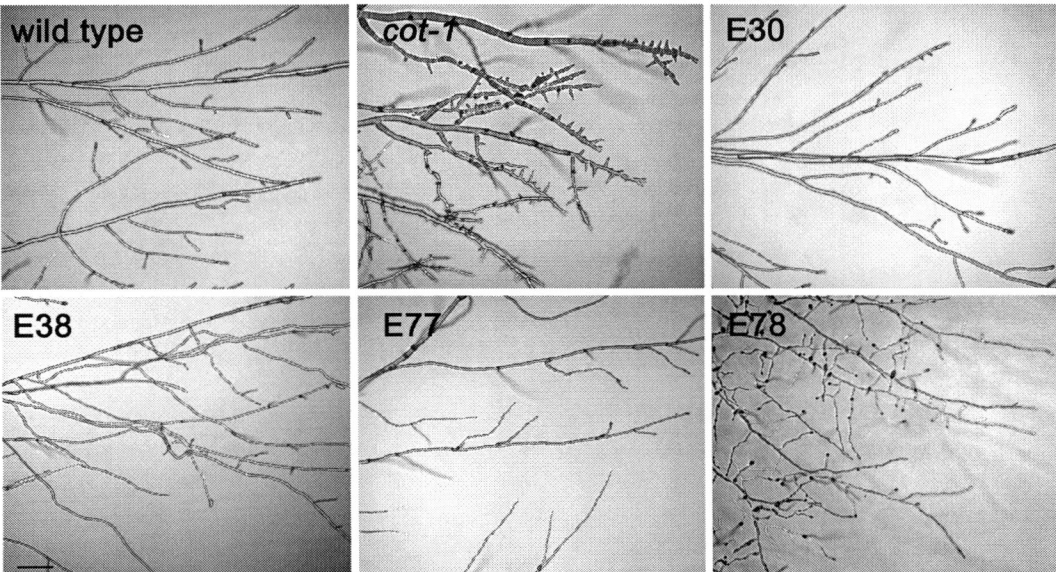

Fig. 6. RNAi silencing of *cot-1*. Phenotype of several *N. crassa cot-1* RNAi transformants. Cultures were documented 4 h after a shift from 25°C to 36°C. Bar indicates 100 μm.

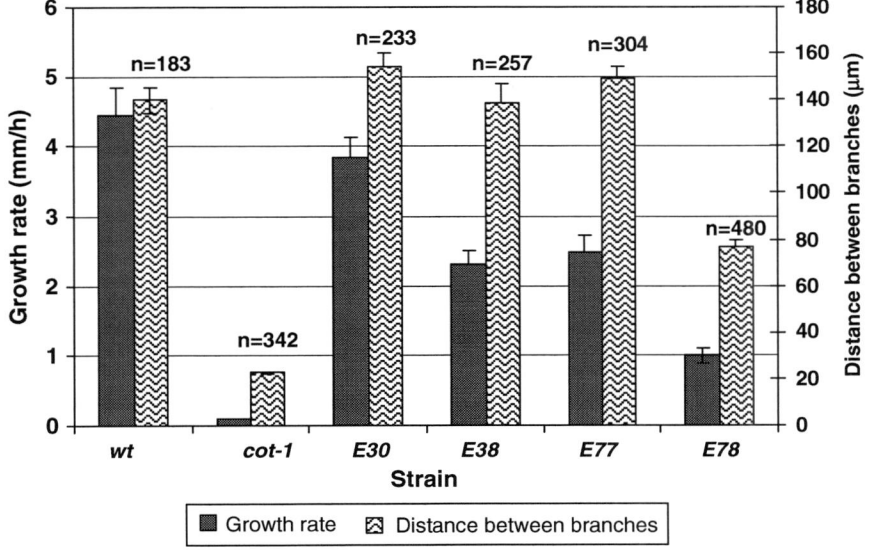

Fig. 7. Growth rate versus distance between branches. Cultures were grown on VgS at 34°C. Standard error are presented. Numbers above columns (n) indicate the number of branches analyzed.

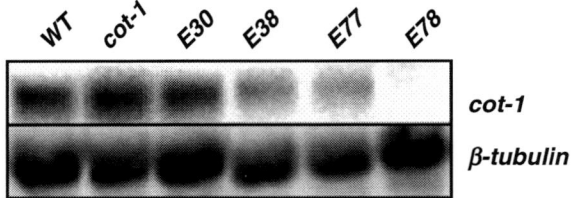

Fig. 8. Northern blot analysis of *cot-1* transcription in different *cot-1* RNAi mutants. Cultures were harvested 4 h after a shift from 25°C to 36°C.

Light microscopy was performed with a microscope equipped with a digital camera. Conidia of the different strains were inoculated on glass slides covered with 1 mL of solid VgS medium and incubated at 25°C at high humidity overnight, and then transferred to 34°C for 4–6 h. The edge of the growing colonies were observed directly and recorded. ImageJ software was used to analyze the microscopic documentation in order to determine the distance between branches. Growth rate of the different strains was determined by race tubes (2).

The most severe phenotype was observed in mutant E78 that presents a very slow growth rate and increased branching pattern. The other mutants showed intermediate phenotypes with reduced growth rate but with minor effect on branching pattern (Fig. 7).

3.2.3. Northern Blot Analysis of the cot-1 RNAi Mutants

Northern analyses were performed to determine the level of *cot-1* mRNA in the different mutants as described in Subheading 3.1.4 with few modifications (see Note 4).

1. Cultures of wild type, *cot-1* (ts) mutant and the transformants were grown by inculcating 2×10^5 spores/mL of in 50 mL liquid VgS and incubating at the permissive temperature (25°C) in a rotary shaker (140–160 rpm) for 16–20 h.

2. When enough biomass have accumulated, the incubating temperature was elevated to the restrictive temperature (36°C) for 4–6 h, until the expected morphological changes of *cot-1* (ts) mutant were visible by microscopic examination.

3. The cultures were filtered through filter paper circles and immediately frozen in liquid nitrogen (see Note 5).

4. [α-^{32}P]UTP–RNA probe was synthesized using MAXIscript kit according to the manufacturer's instructions. For templates, a fragment of the targeted gene was PCR amplified using the primers which are listed in Table 6 and then ligated into pDrive cloning vector. The resulting plasmids were linearized and were used for the synthesis of the RNA probe with SP6 RNA polymerase.

5. RNA radioactively labeled probes were used for hybridization at 68°C in the presence of ULTRAhyb solution.

Table 6
Primers and plasmids used for synthesizing the probes used for northern blot analyses of *cot-1* expression

Gene detected by the probe	Primers used for PCR	PCR product (bp)	Enzyme used for linearization/transcription of the probe
cot-1 (NCU07296.3)	cot1-1734F: ACCCTTTTCAGACAGAGCGA	1,158	*Bam*HI/SP6 RNA polymerase
	cot1-2891R: CTTGATTTCGTGAGCACCAC		
tub-2 (β-tubulin, NCU04054.3)	btub-1304F: CCGTCTCCACTTCTTCATGG	559	*Sna*BI/SP6 RNA polymerase
	btub-1862R: AGCATCCTGGTACTGCTGGT		

6. Washes were performed at 68°C with prewarmed buffers. The most stringent washes were carried out at 68°C with 0.1× SSC and 0.1% (w/v) SDS.

7. The blot of *cot-1* expression was stripped as described in Subheading 3.1.7 and reprobed with *tub-2* probe to detect the β-tubulin mRNA as a control.

Northern analyses of the different mutants indicated that in the resulting mutants, *cot-1* mRNA levels (Fig. 8) were correlated with the phenotype of the RNAi-silenced mutants. Interestingly, although in all the mutants tested, changes in growth rates were correlated with the level of silencing (Figs. 7 and 8), in some mutants (E38 and E77), the effect on branching density was not coupled with low growth rates. These results demonstrate that COT1 may regulate elongation and branching via alternative routes or that COT1 may be subjected to different regulatory thresholds for its proper function in determining elongation rate vs. branching pattern. Thus, this broad-range silencing system enabled us to discriminate between two different functions of *cot-1* (i.e., regulation of elongation and regulation of branching) by disassociating the two phenotypic defects from one another.

3.3. RNAi Can Be Adopted as a Partial Multi-Range Gene Silencing Tool

We adapted RNAi as a silencing mechanism to reduce genes' expression for their functional analysis. Our results demonstrate that RNAi can be employed as a gene silencing mechanism both for nonessential (*al-2*) and essential (*cot-1*) genes, resulting in a range of different mutants. In this study we determined that, following homokaryon purification, the silenced phenotype was stable, indicating that RNAi is stable in *N. crassa* through vegetative cycles. This is in agreement with (24) which determined that expressing of an hairpin construct to induce RNAi results in a more stable silencing compared with quelling. A-sexual stability of RNAi was also observed in *S. sclerotiorum* (25). However, in *Trichoderma asperellum*, the RNAi construct was found to be unstable as a high incidence of revertants among the transformants was observed (Ada Viterbo, personal communication). Lack of stability of RNAi transformants has also been reported to occur in *Aspergillus fumigatus* (52).

To conclude, the ability to use RNAi as a mechanism to silence specific genes and to create stable partially silenced phenotypes is a valuable tool for the manipulation of essential genes (12) and can also contribute to functional analysis and dissection of genes conferring pleiotropic phenotypes, when reduced expression levels result in partial loss of function.

4. Notes

1. Cloning of the inverted-repeat construct that is transcribed as a hairpin RNA structure is demonstrated here by the use of specific restriction enzymes for directed ligation. The restriction sites used can be present at the original sequence of the template plasmid (as in the case of *cot-1*) or can be introduced by designing PCR primers that include the restriction sites at the 5′ and 3′ (as in the case of *al-2*). The two fragments that constitute the RNAi-inducing construct should be 500–1,000 bp in length and partially overlapping and ligated as inverted-repeats. The nonoverlapping fragment (the "spacer") is used for the loop of the hairpin. One of the fragments should include a promoter (the native promoter of the gene or an alternative one, like the *A. nidulans TrpC* or *N. crassa cpc-1* promoter). The two digested fragments can be ligated into a commercial cloning vector possessing the appropriate multiple cloning site or into pSilent (35) – a 6.9 kb plasmid designed specifically for the constructing of RNAi constructs for ascomycetes. The pSilent plasmid contains Amp^R and Hyg^R genes and *TrpC* promoter and terminator for the hairpin construct (the use of pSilent was demonstrated by (25)). pSilent can be obtained from the FGSC.

2. The working concentration of 3×10^9 conidia per mL assumes you will be adding 10 μL of DNA to 30 μL of conidia, to give a total of 9×10^7 conidia in the electroporation procedure. If you deviate from 10 μL of transforming DNA, adjust accordingly.

3. Since RNAi is a cytoplasmatic silencing mechanism (53), the phenotype can be apparent in primary, heterokaryon state,

transformants (31). However, for long term stability and performing sexual crosses, it is advisable to work with homokaryons.

4. Verification of RNAi: after transformation of the constructs, there are several levels of molecular validations:

   (a) To prove the integration of the construct, we recommend Southern blot analysis. Proof by amplifying the construct by PCR can be problematic when an inverted-repeat is a template (either in the original plasmid or in the DNA).

   (b) Reduction in gene expression can be detected by northern blot (as demonstrated here, we recommend with RNA radio-labeled probe) or by real-time PCR (as was demonstrated in (25)). Please note to use polydT primers for the reverse transcription reaction and NOT random hexamers as primers that can detect degraded mRNA.

   (c) To prove that RNAi is taking place: one can detect the presence of siRNA which are indicative of RNAi (13, 14).

5. We found that it is more convenient to freeze the 0.5 g culture samples for RNA extraction in liquid nitrogen in 2 mL tubes (with a screwcap) that already contain the beads. These tubes can be kept at −80°C until the extraction of RNA. However, it is important to work very quickly with the mycelium (cutting and weighing) before freezing, to reduce changes in RNA. Do not be tempted to use larger amounts of tissue for RNA extraction. If the resulting RNA is degraded or not clean, *reduce* the amount of tissue used for the RNA extraction.

6. Guidelines for working with RNA: Wear gloves. Keep RNA samples on ice. Use only nuclease-free water or DEPC-treated water. Use sterile or autoclaved tubes and tips. Treat the gel apparatus and glassware with 1 M NaOH + 0.2% (w/v) SDS for 1 h and then wash thoroughly (to remove all traces of NaOH) with DEPC-treated water. Alternatively, treat with RNAseZap (Ambion, Austin, TX) and rinse with DEPC-treated water.

7. Boil to dissolve the agarose in 1× MEN in an Erlenmeyer flask (in a microwave oven). Add a magnetic stirrer and a thermometer and place in the chemical fume hood with gentle mixing. When the agarose solution reaches 65°C, gently pour the formaldehyde by "sliding" it along the thermometer to avoid the formation of bubbles. When the gel solution is completely homogeneous, cast the gel. Be aware that a formaldehyde–agarose gel is less flexible than a regular agarose gel and tends to break more easily.

8. The hybridization buffer should cover the membrane but should not be in excess, as too large volume will dilute the probe and reduce the hybridization efficiency.

9. It is possible and recommended to reuse a probe for hybridization of northern blot. After the first use of the probe, keep the hybridization buffer with the probe at −20°C up to 7–10 days (depending on the labeling efficiency). For reuse of the probe, incubate the probe at 65°C till completely dissolved. After prehybridization, discard the "new" buffer and add the "old" hybridization buffer with the probe. Probe can be used 2–4 times.

10. Stripping a blot for reprobing is not always necessary. The radioactive signal will decay almost completely after a few weeks, and the blot can be reprobed with a different probe without stripping. The same blot can be stripped for 2–3 times without major reduction in its quality.

References

1. Davis RH, Perkins DD (2002) Timeline: *Neurospora*: a model of model microbes. Nat Rev Genet 3:397–403
2. Davis RH (2000) *Neurospora*: contributions of a model organism. Oxford University Press, Oxford, UK
3. Divon HH, Ziv C, Davydov O, Yarden O, Fluhr R (2006) The global nitrogen regulator, FNR1, regulates fungal nutrition-genes and fitness during *Fusarium oxysporum* pathogenesis. Mol Plant Pathol 7:485–497
4. Scheffer J, Ziv C, Yarden O, Tudzynski P (2005) The COT1 homologue CPCOT1 regulates polar growth and branching and is essential for pathogenicity in *Claviceps purpurea*. Fungal Genet Biol 42:107–118
5. Galagan JE, Calvo SE, Borkovich KA, Selker EU, Read ND, Jaffe D et al (2003) The genome sequence of the filamentous fungus Neurospora crassa. Nature 422:859–868
6. Borkovich KA, Alex LA, Yarden O, Freitag M, Turner GE, Read N et al (2004) Lessons from the genome sequence of *Neurospora crassa*: tracing the path from genomic blueprint to multicellular organism. Microbiol Mol Biol Rev 68:1–108
7. Kasuga T, Townsend JP, Tian C, Gilbert LB, Mannhaupt G, Taylor J et al (2005) Long-oligomer microarray profiling in *Neurospora crassa* reveals the transcriptional program underlying biochemical and physiological events of conidial germination. Nucleic Acids Res 33:6469–6485
8. Colot HV, Park G, Turner GE, Ringelberg C, Crew CM, Litvinkova L et al (2006) A high-throughput gene knockout procedure for *Neurospora* reveals functions for multiple transcription factors. Proc Natl Acad Sci U S A 103:10352–10357
9. Tomari Y, Zamore PD (2005) Perspective: machines for RNAi. Genes Dev 19:517–529
10. Cottrell TR, Doering TL (2003) Silence of the strands: RNA interference in eukaryotic pathogens. Trends Microbiol 11:37–43
11. Hutvagner G, Zamore PD (2002) RNAi: nature abhors a double-strand. Curr Opin Genet Dev 12:225–232
12. Romano N, Macino G (1992) Quelling: transient inactivation of gene expression in *Neurospora crassa* by transformation with homologous sequences. Mol Microbiol 6: 3343–3353
13. Catalanotto C, Azzalin G, Macino G, Cogoni C (2002) Involvement of small RNAs and role of the qde genes in the gene silencing pathway in Neurospora. Genes Dev 16:790–795
14. Fulci V, Macino G (2007) Quelling: post-transcriptional gene silencing guided by small RNAs in *Neurospora crassa*. Curr Opin Microbiol 10:199–203
15. Hamilton A, Voinnet O, Chappell L, Baulcombe D (2002) Two classes of short interfering RNA in RNA silencing. EMBO J 21:4671–4679
16. Zamore PD, Tuschl T, Sharp PA, Bartel DP (2000) RNAi: Double-stranded RNA directs

the ATP-dependent cleavage of mRNA at 21 to 23 nucleotide intervals. Cell 101:25–33
17. Raponi M, Arndt GM (2003) Double-stranded RNA-mediated gene silencing in fission yeast. Nucleic Acids Res 31:4481–4489
18. Kadotani N, Nakayashiki H, Tosa Y, Mayama S (2003) RNA silencing in the phytopathogenic fungus *Magnaporthe oryzae*. Mol Plant Microbe Interact 16:769–776
19. Cogoni C, Macino G (1999) Gene silencing in *Neurospora crassa* requires a protein homologous to RNA-dependent RNA polymerase. Nature 399:166–169
20. Catalanotto C, Azzalin G, Macino G, Cogoni C (2000) Gene silencing in worms and fungi. Nature 404:245
21. Cogoni C, Macino G (1999) Posttranscriptional gene silencing in Neurospora by a RecQ DNA helicase. Science 17:2342–2344
22. Catalanotto C, Pallotta M, ReFalo P, Sachs MS, Vayssie L, Macino G, Cogoni C (2004) Redundancy of the two Dicer genes in transgene-induced posttranscriptional gene silencing in *Neurospora crassa*. Mol Cell Biol 24:2536–2545
23. Maiti M, Lee HC, Liu Y (2007) QIP, a putative exonuclease, interacts with the Neurospora Argonaute protein and facilitates conversion of duplex siRNA into single strands. Genes Dev 21:590–600
24. Goldoni M, Azzalin G, Macino G, Cogoni C (2004) Efficient gene silencing by expression of double stranded RNA in *Neurospora crassa*. Fungal Genet Biol 41:1016–1024
25. Erental A, Harel A, Yarden O (2007) Type 2A phosphoprotein phosphatase is required for asexual development and pathogenesis of *Sclerotinia sclerotiorum*. Mol Plant Microbe Interact 20:944–954
26. Khatri M, Rajam MV (2007) Targeting polyamines of *Aspergillus nidulans* by siRNA specific to fungal ornithine decarboxylase gene. Med Mycol 45:211–220
27. Liu H, Cottrell TR, Pierini LM, Goldman WE, Doering TL (2002) RNA interference in the pathogenic fungus *Cryptococcus neoformans*. Genetics 160:463–470
28. Namekawa SH, Iwabata K, Sugawara H, Hamada FN, Koshiyama A, Chiku H et al (2005) Knockdown of LIM15/DMC1 in the mushroom *Coprinus cinereus* by double-stranded RNA-mediated gene silencing. Microbiology 151:3669–3678
29. Matityahu A, Hadar Y, Dosoretz CG, Belinky PA (2008) Gene silencing by RNA Interference in the white rot fungus *Phanerochaete chrysosporium*. Appl Environ Microbiol 74:5359–5365
30. Nakayashiki H, Nguyen QB (2008) RNA interference: roles in fungal biology. Curr Opin Microbiol 11:494–502
31. Nakayashiki H (2005) RNA silencing in fungi: mechanisms and applications. FEBS Lett 579:5950–5957
32. Aravind L, Watanabe H, Lipman DJ, Koonin EV (2000) Lineage-specific loss and divergence of functionally linked genes in eukaryotes. Proc Natl Acad Sci U S A 97:11319–11324
33. Kamper J, Kahmann R, Bolker M, Ma L-J, Brefort T, Saville BJ et al (2006) Insights from the genome of the biotrophic fungal plant pathogen Ustilago maydis. Nature 444:97–101
34. Feldbrugge M, Zarnack K, Vollmeister E, Baumann S, Koepke J, Konig J et al (2008) The posttranscriptional machinery of *Ustilago maydis*. Fungal Genet Biol 45:S40–S46
35. Nakayashiki H, Hanada S, Nguyen BQ, Kadotani N, Tosa Y, Mayama S (2005) RNA silencing as a tool for exploring gene function in ascomycete fungi. Fungal Genet Biol 42:275–283
36. Stoutjesdijk PA, Singh SP, Liu Q, Hurlstone CJ, Waterhouse PA, Green AG (2002) hpRNA-mediated targeting of the Arabidopsis *FAD2* gene gives highly efficient and stable silencing. Plant Physiol 129:1723–31
37. Han Y, Grierson D (2002) The influence of inverted repeats on the production of small antisense RNAs involved in gene silencing. Mol Genet Genomics 267:629–635
38. Tavernarakis N, Wang SL, Dorovkov M, Ryazanov A, Driscoll M (2000) Heritable and inducible genetic interference by double-stranded RNA encoded by transgenes. Nat Genet 24:180–183
39. Ngo H, Tschudi C, Gull K, Ullu E (1998) Double-stranded RNA induces mRNA degradation in *Trypanosoma brucei*. Proc Natl Acad Sci U S A 95:14687–14692
40. Yarden O, Plamann M, Ebbole DJ, Yanofsky C (1992) *cot-1*, a gene required for hyphal elongation in *Neurospora crassa*, encodes a protein kinase. EMBO J 11:2159–2166
41. Gorovits R, Sjollema KA, Sietsma JH, Yarden O (2000) Cellular distribution of COT1 kinase in *Neurospora crassa*. Fungal Genet Biol 30:63–70
42. Steele GC, Trinci AP (1977) Effect of temperature and temperature shifts on growth and branching of a wild type and a temperature sensitive colonial mutant (*Cot 1*) of *Neurospora crassa*. Arch Microbiol 113:43–48

43. Sambrook J, Fritsch EF, Maniatis T (1989) Molecular cloning: a laboratory manual. CSH Laboratory Press, Cold Spring Harbor, NY
44. Perkins DD, Radford A, Sachs MS (2001) The *Neurospora* compendium. Academic, San Diego, California
45. Li C, Schmidhauser TJ (1995) Developmental and photoregulation of *al-1* and *al-2*, structural genes for two enzymes essential for carotenoid biosynthesis in *Neurospora*. Dev Biol 169:90–95
46. Schmidhauser TJ, Lauter FR, Schumacher M, Zhou W, Russo VE, Yanofsky C (1994) Characterization of *al-2*, the phytoene synthase gene of *Neurospora crassa*. Cloning, sequence analysis, and photoregulation. J Biol Chem 269:12060–12066
47. Carroll AM, Sweigard JA, Valent B (1994) Improved vectors for selecting resistance to hygromycin. Fungal Genet Newsl 41:22
48. Margolin BS, Freitag M, Selker EU (1997) Improved plasmids for gene targeting at the his-3 locus of *Neurospora crassa* by electroporation. Fungal Genet Newsl 44:34–36
49. Linden H, Rodriguez Franco M, Macino G (1997) Mutants of *Neurospora crassa* defective in regulation of blue light perception. Mol Gen Genet 254:111–118
50. Ebbole DJ, Sachs MS (1990) A rapid and simple method for isolation of *Neurospora crassa* homokaryons using microconidia. Fungal Genet Newsl 37:17–18
51. Gorovits R, Yarden O (2003) Environmental suppression of *Neurospora crassa cot-1* hyperbranching: a link between COT1 kinase and stress-sensing. Eukaryot Cell 2:699–707
52. Henry C, Mouyna I, Latge JP (2007) Testing the efficacy of RNA interference constructs in *Aspergillus fumigatus*. Curr Genet 51:277–284
53. Agrawal N, Dasaradhi PVN, Mohmmed A, Malhotra P, Bhatnagar RK, Mukherjee SK (2003) RNA interference: biology, mechanism, and applications. Microbiol Mol Biol Rev 67:657–685

Part II

Detection and Quantification of Fungi

Chapter 7

Analysis of Fungal Gene Expression by Real Time Quantitative PCR

Shahar Ish-Shalom and Amnon Lichter

Abstract

The Real-Time quantitative PCR (qPCR) method has become central for the quantification of gene expression as well as other applications. The major advantages of qPCR are the utilization of small amount of template, high sensitivity and the ability to detect products during the reaction. After selecting qPCR among other options (northern blot, semi-quantitative PCR), one should consider several factors. The first and critical step in qPCR of fungi is the selection of an appropriate growth medium and RNA extraction method, which will avoid accumulation of inhibitors. In this chapter, we focus on detection of the accumulating product with the Syber Green dye, but other detection technologies, such as hybridization probes, might be considered as well. Accurate qPCR analysis with Syber Green depends mainly on optimal PCR reaction, and therefore it is important to design primers that will avoid formation of interfering structures. It is possible to use absolute quantification of the template in the sample, or to conduct a relative analysis, as described in this protocol. In the relative analysis method, expression of the gene of interest is compared with expression of a reference gene. According to our experience as well as according to the literature, it is recommended to use at least three reference genes in order to obtain reliable results.

Key words: qPCR, Real time, *Botrytis cinerea*, Fungi, Gene expression, Syber Green, Primer design, ddCT analysis, Relative analysis

1. Introduction

Analysis of gene expression is fundamental to biological sciences. Traditionally, northern analysis and RNAase protection assays have been used as the methods of choice, mainly in a qualitative or "semi quantitative" manner. These methods are very reliable but required large amounts of RNA, suffered from low throughput and the use of radio-labeled probes. Polymerase chain reaction (PCR) revolutionized experimental biology (1), and soon after its introduction it was coupled to analysis of gene expression (2). In reverse transcription (RT) PCR, cDNA is used to amplify specific products and compare their abundance with respect to reference genes to normalize the expression. This technology is robust but because of the exponential nature of the PCR reaction, end-point analysis of the amount of product, typically analyzed by gel electrophoresis, can introduce significant errors. Therefore, the development of technologies for "Real Time" detection of the amount of product that accumulates during the PCR reaction had a significant contribution to the ability to quantify the amount of nucleic acid in general, and the analysis of gene expression in particular. There are different methods, with various degree of specificity for the detection of PCR products by qPCR. The nonspecific dye Syber Green binds to double stranded DNA and is widely used, but specific hybridization probes such as the TaqMan® probes (Applied Biosystems, Carlsbad, USA) and LUX fluorescent hairpins (Invitrogen Corporation, Carlsbad, USA) can solve background problems. Central to qPCR is the question of absolute or relative analysis of gene expression. Absolute quantification is based on determination of the number of molecules in the sample, while relative quantification depends on comparison to normalize genes (3). In this chapter, we describe the methods we use for the analysis of gene expression in *Botrytis cinerea*. The methodology is based on Syber Green detection and relative gene expression analysis with three reference genes.

2. Materials

2.1. Preparation of the Fungus

1. Conidia of *B. cinerea* at a final concentration of 100/mL from 7 to 10 days old culture of the fungus grown on potato dextrose agar (PDA).
2. Four layers of sterile cheesecloth for separation of conidia from hyphae, using a spreader stick.
3. Harvest solution: 1 L of sterile double distilled water (sDDH$_2$O) 0.001% Triton X-100.

2.2. Growth Conditions

1. Gamborg B5: Prepare 800 mL of 3.16 g/L Gamborg B5 including vitamins (Duchefa Biochemie, Haarlem, the Netherlands), 8 mM NaNO$_3$, 3% (w/v) D (+)-glucose. Autoclaved for 18 min at 121°C, adjust pH to 5.7 with 1 M HCl and complete to 1 L with sDDH$_2$O.
2. Harvest of conidia: a ceramics funnel (55 mm) covered with a sterile 3 MM paper disk wetted by sterile water. Collect hyphae with a straight edge metal spoon and immediately freeze in liquid nitrogen.

2.3. Preparation of RNA and cDNA

2.3.1. Protocol I

1. Liquid nitrogen, mortar, and pestle. Clean with 70% ethanol between samples.
2. Denaturation buffer for RNA extraction: 25 mM NaCitrate, pH 7.0, 0.5% (w/v) 10% Sarkosyl, 4 M guanidine thiocyanate, 0.1 M β-mercaptoethanol. Dissolve in diethylpyrocarbonate (DEPC)-treated sDDH$_2$O to a final volume of 528 mL. Store at 4°C.
3. PCA stock solution: phenol saturated with Tris–HCl, pH 8.0, chloroform, iso-amyl alcohol (25:24:1, v/v/v). Store at 4°C in a light protected glass bottle.
4. Phenol, pH 6.7. Store at 4°C.
5. LiAcetate and LiCl stock solution: 2 M NaAcetate, pH 4.0, 6 M LiCl. Autoclaved and store at RT.
6. 2-propanol and 70% ethanol. Stored at −20°C.

2.3.2. Protocol II: Rapid Extraction

1. Use a Plant/fungi total RNA purification kit (Norgen Biotek, Thorold, Canada). The kit contains lysis solution, wash solution, elution buffer as well as mini spin columns, 2 mL collection tubes, 1.7 mL elution tubes.
2. Prepare 70 and 96% ethanol and β-mercaptoethanol to be opened in a chemical hood.

2.3.3. RNA Gel Electrophoresis

1. Agarose (e.g., SeaKem LE, Lonza, Minsk, Belarus).
2. 40 mM Tris–acetate.
3. TAE: 1 mM ethylenediaminetetraacetate (EDTA) buffer prepared as a 50× stock solution (4).
4. Formaldehyde loading dye (e.g., Ambion, Applied Biosystems, Austin, USA).
5. Safeview (Applied Biological Materials, Richmond, Canada) solution.

2.3.4. cDNA Preparation and Analysis

1. DNAse treatment: RQ1 RNase- free DNase kit (Promega, Madison, USA). Contains 10× reaction buffer, RQ1 RNase-free DNase, stop solution and DEPC-treated water.
2. cDNA synthesis: First strand cDNA synthesis kit (e.g., EZ-First strand cDNA synthesis kit, Biological industries, Bet Haemek, Israel). Contains oligo (dT) primers, random hexamer primers, 10× reaction mix, 100 mM dithiothreitol (DTT) and water.
3. Components for standard PCR amplification for cDNA analysis: specific primers, template (genomic DNA and cDNA), sterile water and components for PCR reaction (e.g., ReadyMix, ABgene, Epsom, United Kingdom). Contains 0.25 U/μl thermoprime plus DNA polymerase, 75 mM Tris–HCl, pH 8.8, 20 mM $(NH_4)_2SO_4$, 1.5 mM $MgCl_2$, 0.01% (v/v) Tween 20, 0.2 mM each of the dNTP's, precipitant and red dye for electrophoresis.

4. Materials for nucleic acid analysis (see Subheading 2.3.3) with DNA size marker (e.g., GeneRuler 100 bp Plus DNA ladder, Fermentas, Burlington, Ontario).

2.4. Real Time qPCR

1. High quality cDNA.
2. Primers (see Table 1).
3. Real time qPCR reaction mix (e.g., 2×QPCR SYBR® Green Mix, ABgene, Epsom, United Kingdom). Contains SYBR® Green I dye, Thermo-Start® DNA Polymerase supplied in a proprietary reaction buffer).
4. Tubes: standard 0.2 mL tubes for qPCR or 0.1 mL tubes for qPCR (Corbett Life Science, Concorde, Australia).

Table 1
Type of primers described in different protocols

	Name	Accession	Sequence
B. cinerea	eif4A-12-F eif4A-12 R	BC1G_07971	TATTCATCGCATTGGTCGAA CAACATTCATTGGCATCTCG
	18 S rRNA-F 18 S rRNA-R	BC1G_06392.1	TTGGTTTCTAGGACCGCCG GGCAAATGCTTTCGCAGTAGT
	Actin-F Actin-R	BC1G_08198.1	CCCAATCAACCCAAAGTCCAACAG CAAATCACGACCAGCCATGTC
	β-tubulin-F β-tubulin-R	BC1G_00122.1	TTGGATTTGCTCCTTTGACCAG AGCGGCCATCATGTTCTTAGG
	HSP104-F HSP104-R	BC1G_15409.1	AAGGCTACGGAGAAGGATAAGTTG TGGTGGCGAGTTTGGGTTTG
N. crassa	L6_rRNA-F L6_rRNA-R	NCU02707.3	CAGAAATGGTACCCTGCTGAGG GCGGATGGTCTTGCGG
	ActF ActR	NCU04173.3	TCCATCATGAAGTGCGATGTC TTCTGCATACGGTCGGAGAGA
	Tub-2-F Tub-2-R	NCU04054.3	CCCGCGGTCTCAAGATGT CGCTTGAAGAGCTCCTGGAT
A. fumigatus	28 S rRNA-F 28 S rRNA-R	AB008401	GGCCCTTAAATAGCCCGGT TGAGCCGATAGTCCCCCTAA
	ActinRT1 ActinRT2	XM_746399.1	ATCGGCGGTGGTATCCTC TCTTCGTGCCATTCGTCTG
	α-tubulin-F α-tubulin-R	Afu2g14990	CGG CTAATGGAAAATACATGGC GTCTGGCCTTGAGAGATGCAA

Primer sequences are according to the following references. For the fungus *Neurospora crassa*: L6_ribosomal, (9); actin, (10), and tubulin (11). For the fungus *Aspergillus fumigatus*: 28 S rRNA, (12); actin and tubulin, (13)

3. Methods

The first stage in successful qPCR is the selection of growth conditions that will avoid subsequent interference in downstream RNA extraction such as accumulation of polysaccharides. That is why Gamborg B5, pH 5.7 medium was selected over other media for the growth of *B. cinerea* and the period of cultivation was set to 3–4 days at 22°C. The next important step is the selection of RNA extraction method. If the fungus of choice has a well established method for RNA extraction, it should be used, otherwise, selection of a reliable RNA extraction kit or procedure is a critical step and should be carefully evaluated. After considering whether qPCR is indeed the method which you should invest in (1), it is essential to figure out the scope of the work and the experimental setup. If, for example you need to analyze few genes of interest and over a long period of research, selection of hybridization probes (such as the TaqMan® probes), is likely to yield faster and more reliable results. If on the other hand, you are analyzing a gene family with high homology between member or a larger number of genes, Syber® Green technology is likely to be the method of choice. One should be aware that by choosing Syber® Green quantification, you commit yourself to rigorous optimization of the PCR reaction because if false products will be formed during the PCR reaction, they will yield erroneous results and nonoptimal efficiency will skew the mathematical models that allow correct quantification (3). There are many free or commerciality available softwares for primer design, some of which are listed below (see Note 6), but no available software is likely to solve all problems and in order to get high efficiency score, one should be prepared to design and order more than a single primer set. Next, it is important to select between relative and absolute quantification (3): in absolute quantification, one should have a plasmid containing the gene of interest (GOI) and accurate calculation of the number of plasmid molecules in the stock solution. This methodology also involves setting a calibration curve in each qPCR run, which often limits the number of experimental samples that can be accommodated in each run. Alternatively, the use of reference genes and relative quantification is sufficient, depending on the biological question. Reference genes must be carefully selected; stability of expression needs to be determined in each individual organism and experimental setup. It is strongly advised to use more than one reference gene and according to some reports, at least three reference genes are necessary. We have selected the widely used 18 S rRNA, actin, and β-tubulin genes as reference genes (Table 1), but many choices are available in the literature and the debate goes on (2, 5, 6). Last, and most importantly, the calculation method should be carefully considered.

Mathematical models suggest that the delta-delta CT method (7), which we used can yield reproducible and sufficient results. Others claim however that correction for efficiency and taking into account the presence of inhibitors in the cDNA synthesis step or the PCR steps is essential for accurate quantification by qPCR (8).

3.1. Growth Conditions

1. Harvest conidia of *B. cinerea* from a 7 to 10 days old PDA culture plate by adding 5 mL 0.001% Triton X-100 solution and scraping the spores gently with a spreader stick. Purify the conidia from hyphae by filtering it through four layers of cheesecloth. Centrifuge the conidia at $3,000 \times g$ wash the pellet twice with distilled water and add 1 mL of sterile water. Count the number of conidia using a hemocytometer. Adjust to 10^7 conidia/mL.

2. Lyophilization of the hyphae was shown to be one of the most important factors in deterioration and inconsistency in the quality of RNA.

3. Prepare 60 mL of Gamborg B5 medium (see Note 1) in 250 mL Erlenmayer flasks and add conidia to a final density of 100 conidia/mL. Incubate the culture at 22°C on an orbital shaker at 120 rpm for 4 days.

4. Harvest the biomass by filtering the hyphae over a ceramic funnel with a 3 MM paper disk. Scrape hyphae from the filter paper, add to a 15 mL polypropylene tube, freeze in liquid nitrogen and store at −80°C until preparation of RNA (Note 2).

3.2. Preparation of RNA and cDNA

3.2.1. Protocol I

1. Prepare fine powder from 80 mg of dry biomass by grinding with a mortar and pestle in liquid nitrogen.

2. Immediately add the powder to 1 mL denaturation solution to prevent RNAse activity.

3. Add 1 mL PCA solution to the denaturation solution in a 2 mL tube, vortex and centrifuge at $11,000 \times g$ at ambient temperature.

4. Remove the aqueous phase, add an equal volume of 2-propanol and incubated at −20°C for 1 h.

5. Centrifuge samples at $8,000 \times g$ for 10 min at ambient temperature, discard the supernatant, and suspend pellet in 0.8 mL DEPC-treated water.

6. Add the following solutions sequentially, vortex samples after adding each component: 0.08 mL of 2 M $(NH_4)_2$-acetate, 0.8 mL of phenol, pH 6.7, and 0.16 mL chloroform:isoamyl alcohol (49:1, v/v).

7. Incubate samples on ice for 10 min and then centrifuge at $11,000 \times g$ for 10 min. Remove the aqueous phase into a new 2 mL tube, add an equal amount of 6 M LiCl, incubate overnight at −20°C.

8. Centrifuge samples at 11,000×*g* for 10 min at 4°C and discard the supernatant. Add 1.5 mL of 70% EtOH, vortexe and incubated for 10 min at ambient temperature. Centrifuge at 8,000×*g* for at least 5 min at 4°C and discard the supernatant.

9. Dry pellets in a laminar hood for at least 30 min, dissolve in 100 μL of DEPC-treated water, and store at −80°C until further analysis.

3.2.2. Protocol II

1. Prepare fine powder from 50 mg of dry biomass by grinding with a mortar and pestle in liquid nitrogen.

2. Process samples according to manufacturer's instructions (Norgen Biotek).

3. Determine RNA quantity and quality (1.5 μL samples from protocol I or II) using a NanoDrop 1000 instrument (see Note 3).

4. RNA stability assay: divide a 10 μl sample into two tubes: incubate one tube at 37°C and the other tube on ice for 30 min.

5. Add 2.3 μl of formaldehyde loading dye to each sample and incubate for 10 min at 65°C. Place on ice until analysis.

6. Gel electrophoresis: prepare 1% of agarose gel with 0.5×TAE and 5 μl/100 mL SafeView dye. Load samples on the gel separate by electrophorsis. Visualize RNA for the appearance of 18 S and 28 S rRNA bands. Make sure that the bands are sharp and that there are no signs of degradation before or after the stability assay.

7. Removal of DNA traces: incubate 1 μg of RNA with RQ1 DNAse according to instructions (RQ1 kit).

8. Preparation of cDNA: Convert the DNA-free RNA samples into cDNA using a mixture of olido (dT) and random hexamer primers, according to the manufacturer's instructions (see Notes 4 and 5).

9. Preliminary PCR analysis of the cDNA: this step is carried out in order to verify amplification of specific products from the obtained cDNA samples. Use 2 μL of cDNA as template and 12.5 μM of the specific primers for a PCR. Add 10 μL of reaction mixture and sDDH$_2$O to a final volume of 20 μL. Use 5–10 ng genomic DNA as a positive control and "no template control" as a negative control. Analyze PCR products on a 2% agarose gel (see Subheading 2.3.3).

3.3. Real Time qPCR

1. Primer design: Set amplicon size for the GOI between 100 and 200 bp (see Note 6). When multiple genes are analyzed, prefer a uniform melting temperature (Tm) for the different primer sets in order to be able to compare the expression of different genes in the same run. Otherwise, select primer sets by optimal parameters. Free software for primer design are Primer 3

(http://frodo.wi.mit.edu/), IDT Scitools (http://www.idtdna.com/SciTools/SciTools.aspx), or other commercially available software (see Note 7). Parameters to be considered during primer design are: the free energy for the formation of hairpins, self dimerization or heterodimer formation should be below $\Delta G \leq -9$ kcal/mol. Verify that the amplicon will not have problematic secondary structures with values of $\Delta G \leq -9$ kcal/mol using the DINAMelt Server (see Note 8).

2. Selection of reference genes: it is recommended to use three reference genes for accurate quantification of the GOI. 18 S rRNA, actin, and tubulin are commonly used reference genes for many organisms and examples for primers for *B. cinerea*, *Neurospora crassa* and *Aspergillus fumigates* are presented (Table 1) (see Note 9).

3. Preparation of the qPCR reaction: qPCR is sensitive to experimental errors. Pipettors should be calibrated routinely and accurate pipetting should be practiced. Use pipettors and tips of 10 μL volumes. The tubes should be arranged on a cooled tray. Calculate the amount of reaction mixture to be used for each primer set and add to it water and primers. Wet the pipette tips before dispensing the aliquots. Dispense the mixture to each tube and add template by applying the droplet to the wall of the tube without touching it with the pipette tip.

4. Operation of the qPCR instrument: The instructions are from the RotorGene 6000 qPCR machine (Corbett Life Science, Concorde, Australia) and should be modified according to the specific instrument or reaction mixture properties as recommended (see Note 10). Load samples into the machine and select the appropriate program (e.g., Syber Green). Define number of samples and the type of tubes. Set the acquisition point of the fluorescence signal to 74°C.

5. Reaction conditions: preactivation of the enzyme for 15 min at 95°C, strand denaturation for 10 s at 95°C, annealing of primers for 20 s at the optimal melting temperature (Tm), elongation for 20 s at 72°C, the melting stage (72–99°C) at the end of the of the program (after 40 cycles) for calculation of the Tm of the PCR product. Define gain value to a range of 0.5–2.0 (for RotorGene 3000). Sample name and type can be defined during operation or later.

6. Analysis of the quality of the results: determine the reproducibility of the triplicates. If triplicates deviate from the average by more than one cycle it indicates pipetting error and one should consider repeating the qPCR (see Subheading 3.3). The next stage is checking the quality of the product by melt curve analysis. The melt curve should appear as a single and narrow peak without shoulders. If another peak appears before the real peak, it is possible to set another acquisition

temperature (B point) between the curves, for the next run. Consider that this situation will result in nonoptimal amplification of the specific product because the nonspecific product will reduce the efficiency of the reaction.

7. Efficiency of the reaction: in the first analysis of the primers and template, run a dilution curve with genomic DNA at initial concentration of 5–10 ng, consisting of four fivefold dilutions to check the efficiency of the reaction. The optimal slope of the curve should be 3.322, and the corresponding M-value should be 1.00 ± 0.02. In case of deviation from these values, consider optimizing primer concentration, the amount of template, the annealing temperature, or ordering a new primer set.

8. Data analysis: The threshold which determines at what fluorescence level and cycle time (CT), the data will be analyzed can be set automatically or manually. The automatic threshold setup determines the point on the curve of all samples, in which deviation from technical replication (triplicates) is minimal. In a manual threshold setup, the threshold is set to a point that according to experimental experience gives reliable results (see Note 11). The qPCR data can be analyzed by the Rotor-Gene 6 software, which is part of the qPCR machine or exported to excel (Table 2). The formula which should be copied into the ddCT column is:

$$ddCT = 2\char`^ - ((GOI\ CT_{tn} - REF\ CT_{tn}) - (\$GOI\ CT_{t0} - \$REF\ CT_{t0})).$$

9. Comparative analysis gene expression: data from qPCR analysis of different reference genes can be compiled in excel and reproduced as figures (see Table 3).

4. Notes

1. Acidic medium with low nitrogen content is chosen in order to minimize accumulation of polysaccharides in the growth medium and to allow efficient separation of hyphae from the medium, resulting in better RNA quality.

2. Lyophilization of the hyphae was shown to be one of the most important factors in deterioration and inconsistency in the quality of RNA.

3. Pure RNA should yield a ratio of A_{260}/A_{280} above 1.8 and the ratio of A_{260}/A_{230} should be approximately two.

4. It is recommended to use a combination of poly (dT) and random hexamer primers for first strand cDNA synthesis in

Table 2
Example of analysis of a gene of interest (GOI) and a reference gene (REF) in calculation of relative expression by the ddCT method. The specific example shows the expression of HSP104 and the 18 S rRNA after transition of *B. cinerea* strain B05.10 from 22°C to 0°C at different time points (0, 1, 4, 10, and 24 h)

Sample name	CT values		Average CT		ddCT
	GOI	REF	GOI	REF	
0	28.89	22.92			
0	28.25	24.25	27.99	23.39	1.00
0	26.84	23.00			
1	25.47	18.97			
1	26.07	21.53	25.89	20.25	0.49
1	26.14	20.24			
4	24.59	20.02			
4	24.84	21.36	24.57	20.90	1.92
4	24.27	21.33			
10	23.67	23.86			
10	23.73	22.65	23.58	22.28	9.87
10	23.34	20.33			
24	23.65	22.11			
24	23.01	21.22	23.31	21.47	6.79
24	23.26	21.07			

Table 3
Example of three genes analyzed by qPCR with three different reference genes at different time points after the beginning of the treatment. Relative expression of each of the three genes toward each of the reference genes shows a similar trend, but note that differences in expression levels may be significant

Time	Induction and repression			High induction			Repression		
h	rRNA	Actin	Tubulin	rRNA	Actin	Tubulin	rRNA	Actin	Tubulin
0	1.0	1.0	1.0	1.0	1.0	1.0	1.0	1.0	1.0
1	14.7	31.5	3.5	384.1	1,236.9	26.7	0.1	0.1	0.1
4	2.6	3.0	1.1	5,199.8	2,460.9	532.9	0.1	0.1	0.1
10	0.6	0.1	0.1	2,093.8	536.9	436.0	0.3	0.2	0.1
24	0.1	0.4	0.1	76.6	1,755.4	326.1	0.1	0.0	0.0

order to increase the yield and types of RNA molecules that will be used as templates.

5. The freshly made cDNA should be divided into aliquots and stored at −20°C until analysis.

6. Amplicon size can be larger than 200 bp if it is required because of homology and primer specificity considerations or in order to facilitate the design of better primers. For further details, consult http://www.genequantification.info/.

7. Other software packages are available for primer design, e.g., *Vector NTI* (Invitrogen), *Primer Express* (Applied Biosystems), and *CLC DNA workbench* (CLC bio USA, Massachusetts, USA).

8. The DINAMelt Server application enables to check the secondary structure of the amplicon (http://www.bioinfo.rpi.edu/applications/hybrid/twostate-fold.php).

9. Expression of reference genes should be stable and correct for experimental errors. Test the stability of expression among treatments and replications and avoid the use of reference genes with inconsistent expression pattern.

10. There are many commercially available reaction mixtures. Always make sure to follow the recommended time for preheating and activation of the enzyme. For example, the ABgene Syber Green I mixture requires 15 min at 95°C, while the Takara SYBR®Premix ExTaq™ (Company details) does not require a preactivation step.

11. Threshold acquisition of the experiment can be fixed automatically or manually. In cases of one experiment analysis, the instrument automated threshold is enough, but in cases of multiple experiment analyses, we use a manually fixed threshold value of 0.05 fitted to the Rotor Gene 3,000 machine.

References

1. Kubista M, Andrade J, Bengtsson M, Forootan A, Jona'k J et al (2006) The real-time polymerase chain reaction. Mol Aspects Med 27:95–125
2. VanGuilder H, Vrana K, Freeman W (2008) Twenty-five years of quantitative PCR for gene expression analysis. Biotechniques 44:619–626
3. Chini V, Foka A, Dimitracopoulos G, Spiliopoulou I (2007) Absolute and relative real-time PCR in the quantification of *tst* gene expression among methicillin-resistant *Staphylococcus aureus*: evaluation by two mathematical models. Lett Appl Microbiol 45:479–484
4. Sambrook F, Maniatis T (1989) Molecular cloning: a laboratory manual. Cold Spring Harbor Laboratory Press, New York
5. Vandesompele J, De Preter K, Pattyn F, Poppe B, Van Roy N et al (2002) Accurate normalization of real-time quantitative RT-PCR data by geometric averaging of multiple internal control genes. Genome Biol 3:1–12
6. Liu ZL, Slininger PJ (2007) Universal external RNA controls for microbial gene expression analysis using microarray and qRT-PCR. J Microbiol Methods 68:486–496
7. Schmittgen TD, Livak KJ (2001) Analysis of relative gene expression data using

real-time quantitative PCR and the 2(-delta delta C(T)) Method. Methods 25:402–408
8. Pfaffl M (2001) A new mathematical model for relative quantification in real-time RT-PCR. Nucleic Acids Res 29:e45
9. Froehlich A, Noh B, Vierstra R, Loros J, Dunlap J (2005) Genetic and molecular analysis of phytochromes from the filamentous fungus *Neurospora crassa*. Eukaryotic Cell 4:2140–2152
10. Görl M, Merrow M, Huttner B, Johnson J, Roenneberg T et al (2001) A PEST-like element in FREQUENCY determines the length of the circadian period in *Neurospora crassa*. EMBO J 20:7074–7084
11. Navarro-Sampedro L, Yanofsky C, Corrochano L (2008) A genetic selection for *Neurospora crassa* mutants altered in their light regulation of transcription. Genetics 178:171–183
12. Francesconi A, Kasai M, Harrington S, Beveridge M, Petraitiene R et al (2008) Automated and manual methods of DNA extraction for *Aspergillus fumigatus* and *Rhizopus oryzae* analyzed by quantitative real-time PCR. J Clin Microbiol 46:1978–1984
13. Semighini C, Marins M, Goldman M, Goldman G (2002) Quantitative analysis of the relative transcript levels of ABC transporter Atr genes in *Aspergillus nidulans* by real-time reverse transcription-PCR assay. Appl Environ Microbiol 68:1351–1357

Chapter 8

Identification of Differentially Expressed Fungal Genes *In Planta* by Suppression Subtraction Hybridization

Benjamin A. Horwitz and Sophie Lev

Abstract

In host–pathogen interactions, identification of pathogen genes expressed during plant infection poses a challenge, even though these genes may be strongly induced by signals from the host. Here, we describe the application of a PCR-based differential screening method to plant–fungal interactions. Suppression subtraction hybridization (SSH) provides a sensitive method to isolate fungal genes expressed in planta. Total RNA is isolated from infected plants for comparison with the pathogen in axenic culture, or, in the application described here, plants infected with a wild type isolate are compared with plants infected with a mutant. Following library construction, clones are sequenced and screened for differential expression in the two starting populations.

Key words: Differential gene expression, cDNA library, SSH, Host–pathogen

1. Introduction

Gene expression of fungal pathogens is reprogrammed as a result of interaction with the host. Identification of the genes induced at early stages of infection, however, poses a challenge because the fungal biomass is typically a small fraction of that obtained by harvesting infected tissue. Once the infection spreads, the amount of fungal biomass reaches levels significant enough to assess gene expression. Furthermore, it is often of interest to compare genes expressed in a wild type fungal isolate with those expressed in a mutant strain during host colonization. We describe a relatively simple, low-cost method to do such comparisons, providing a first set of differentially expressed genes.

Genome-wide profiling using oligonucleotide arrays has provided the most complete picture of the transcriptional profile. Genes regulated by the Kss1/Fus3 MAPK signaling pathways in yeast pheromone response and dimorphic development were followed by microarray hybridization (1, 2). Digital gene expression (see ref. (3)) based on massively parallel sequencing may soon replace the demand for custom-made microarrays. These genome-wide strategies, however powerful, are not always appropriate. Suppression subtraction hybridization (SSH) provides a relatively fast low budget alternative when there is a need for identification of a small representative subset of genes differentially expressed in two RNA populations This PCR-based differential cDNA screening procedure (4, 5) has been conveniently adapted as a commercial kit (Clontech). SSH was used to identify genes underexpressed in mutants lacking the pathogenicity-related MAP kinase genes of *Magnaporthe oryzae* and *Cochliobolus heterostrophus* (6, 7), and a Gα subunit of *Botrytis cinerea* (8). SSH can also identify fungal genes against a background of the uninfected host, for example: *Alternaria brassicicola* on *Arabidopsis* (9), genes of the white-rot fungus *Heterobasidion parviporum* expressed during colonization of Norway spruce stems (10), and in symbiosis, for arbuscular mycorrhizae (11). It is even possible to isolate differentially expressed mycoparasite genes in antagonistic fungal–fungal interactions (12, 13).

We describe here our protocol applying the SSH method to in planta differential screening. Infected plant tissue is harvested at an appropriate stage, total RNA isolated, followed by enrichment for polyA+RNA, differential cDNA library construction, and analysis of the libraries.

2. Materials

2.1. Plant Material and Fungal Cultures

1. Plants grown and infected under controlled conditions. Plant infection needs to conform to any local regulations to prevent release of imported or genetically engineered pathogens to the environment.

2. Fungal cultures. For *Cochliobolus*, complete medium (14) is used to grow cultures for inoculation of plants (see Note 1).

3. Polytron homogenizer.

4. Surfactant: 0.05% (v/v) Tween 80 or Tween 20 water solution is used for preparing fungal inoculum.

5. Complete medium (CM). Per liter: 10 mL solution A, 10 mL solution B, 0.5 mL Srb's micronutrients, 10 g glucose, 1 g yeast extract, 1 g enzymatically hydrolyzed casein, 20 g agar (Difco), deionized water to 1 L.

6. Solution A. Per 100 mL: 10 g $Ca(NO_3)_2 \cdot 4H_2O$, water to 100 mL.

7. Solution B. Per 100 mL: 2 g KH_2PO_4, 2.5 g $MgSO_4 \cdot 7H_2O$, 1.5 g NaCl, water to 100 mL.
8. Srb's micronutrients. Per 1 L: 57.2 mg H_3BO_3, 393 mg $CuSO_4 \cdot 5H_2O$, 13.1 mg KI, 60.4 mg $MnSO_4 \cdot H_2O$, 36.8 mg $(NH_4)_6Mo_7O_{24} \cdot 4H_2O$, 5,490 mg $ZnSO_4 \cdot H_2O$, 948.2 mg $FeCl_3 \cdot 6H_2O$. Complete with water to 1 L, store in a light protected bottle.

2.2. RNA Isolation and Library Construction

1. Diethylpyrocarbonate (DEPC). (see Note 2).
2. Phenol reagent. Phenol, chloroform, isoamyl alcohol (25:24:1).
3. Sodium acetate. 2 M NaAc, pH 5.2 (DEPC-treated).
4. Lithium acetate. 4 M LiAc (DEPC-treated).
5. NTES: 0.1 M NaCl, 0.01 M Tris–HCl pH 7.5, 1 mM EDTA, 1% SDS.
6. DEPC-treated water. Add 20 µL of diethylpolycarbonate per 100 mL water, working in fume hood mix well, and let stand overnight. Autoclave and store at RT.
7. PolyATract kit (Promega) or equivalent. Use kit for preparation of polyA + RNA.
8. SSH. Clontech PCR-Select kit.
9. Taq polymerase. Advantage Taq or equivalent enzyme for PCR amplification steps.

2.3. Library Analysis

1. Hybridization membrane. NYTRAN SuPerCharge nylon membranes (Schleicher & Schuell).
2. oligo-dT primer ($T_{17}V$).
3. dNTPs. 10 mM each of dATP, dCTP, dGTP, and dTTP.
4. $\alpha^{32}PdCTP$. Use caution, particularly with the undiluted source. Work in a radiation approved area appropriate for ^{32}P, behind shielding adequate for beta sources.
5. MMLV reverse transcriptase (Stratagene).
6. 20× SSC. 3 M NaCl, 0.3 M sodium citrate.

3. Methods

3.1. Plant Material and Fungal Cultures

1. Grow maize plants for 15–16 days at 22–24°C.
2. Inoculate liquid CM medium with mycelial plugs and incubate at room temperature with shaking for 3 days (see Note 1).
3. Homogenize mycelium briefly while still in growth medium (see Note 3) using Polytron homogenizer. Add 0.05% Tween 80 to the suspension as a surfactant.

4. Cover the soil and invert the plant pots to dip the leaves in the mycelial suspension until uniformLy coated.

5. Place the inoculated plants in sealed, transparent plastic bags presprayed with water (to maintain high humudity) and keep under continuous illumination.

6. After 24 h, wild type mycelia cause complete necrotic destruction of plant tissue. When comparing wild type and mutant strains, allow an appropriate time for the mutant to reach, if possible, a similar infection stage to that seen with the wild type.

7. Collect the infected leaves for RNA isolation, flash-freeze in liquid N_2, and store at –70°C until use.

3.2. RNA Isolation

1. Grind harvested material to a fine powder using RNAase-free (oven-baked) mortar and pestle in liquid nitrogen. Do not let the powder thaw during the procedure.

2. Transfer 1 g of powder to 15 mL propylene Falcon tubes, add 1.5 mL NTES and 1 mL phenol reagent to the powder, and shake the tubes vigorously for 10 min.

3. Transfer the mixture to 2 mL microfuge tubes and centrifuge for 10 min at top speed. Transfer the aqueous phase to new tubes without disturbing the interface.

4. Precipitate RNA by adding of 1/10 volume sodium acetate (2 M, pH 5.2) and 2 volumes of ethanol. Mix thoroughly and incubate at –20°C for at least 20 min. Centrifuge the tubes at top speed for 10 min.

5. Dissolve the pellet in 800 μL DEPC-treated water and centrifuge again for 10 min at top speed to get rid of polysaccharides (large gelatinous pellet).

6. Add 800 μL lithium acetate (4 M) to the supernatant. Incubate the tubes on ice overnight and then centrifuge 10 min at top speed.

7. Dissolve the pellet in 300 μL DEPC-treated water, and precipitate the RNA again with 1/10 volume sodium acetate 2 M and two volumes ethanol at –20°C for at least 2 h.

8. Centrifuged the tubes at top speed for 10 min, and dissolve the RNA pellet in 100 μL DEPC-treated water.

3.3. Equalizing of Fungal to Plant RNA Ratio in Tester and Driver RNA Pools

1. Northern blot. Total RNA (10 μg) from the tester and driver populations are separated on denaturing agarose gels, blotted to Hybond N+nylon membranes (Amersham), and hybridized with a radioactive fungal probe, following standard nucleic acid techniques (15). For *C. heterostrophus*, a *GAPDH* gene fragment generated by PCR with primers 5′-CCCTCGC-CTGACGCCCCAT-3′ and 5′-CGAGGACACGGCGGGA-GTAA-3′ is used to quantify fungal RNA (see Note 4).

2. The fungal *GAPDH* signal is quantified by densitometry of the blot image, and the fungal RNA proportion in tester and driver RNA pools is estimated.

3. Equalizing of fungal RNA ratio by adding RNA of uninfected plant tissue to the infected (tester) sample.

3.4. Isolation of mRNA

1. Proceed to mRNA isolation with 200–500 μg of total RNA for driver and tester (see Note 5). Save an aliquot of each total RNA sample to use in northern blots or RT-PCR analysis of expression of specific genes identified from the library. The experiment used for construction of the library should be considered as one of several replicate experiments required to reach a firm conclusion as to whether particular genes are differentially expressed.

2. Denature total RNA, mix with biotinylated oligo-dT, and affinity-purify using streptavidin-coated paramagnetic particles (Promega polyATtract system) (see Note 5). PolyA + enriched RNA is isolated from both the driver and tester total RNA preparations. This polyA + RNA is used for library preparation and, optionally, for the synthesis of radio-labeled cDNA (step 6 of Subheading 3).

3.5. SSH Library Construction

A SSH library is prepared according to the manufacturer's instructions (Clontech PCR-Select cDNA Subtraction Kit). It is important to follow closely the steps in the manual. In the typical application described here, the tester population is from leaves infected with the wild type population, and the driver, leaves infected with the mutant (see Note 6).

1. At the second stage of PCR amplification, a fairly uniform fragment size distribution is obtained (Fig. 1a) (see Note 7). A slight shift downward in average size occurs at the RsaI digestion step (see Clontech manual).

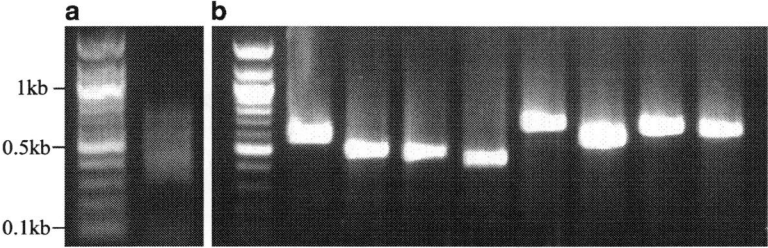

Fig. 1. Size distribution of SSH-derived clones. (**a**) Clone pool resulting from SSH. (**b**) Random sampling of cloned library: SSH-derived clone pool was cloned into pUC57 vector. Eight random clones were PCR-amplified using adaptor primers 1 and 2R (see Clontech PCRselect manual). Data from (16).

2. After the second amplification, excise the smear of PCR products from the gel and purify it using the Promega Wizard gel extraction kit for cloning.

3. Clone the PCR products, representing transcripts expressed differentially in wild type and mutant, in to vector pTZ57R/T (or equivalent) using InsT/Aclone PCR Product Cloning Kit (MBI Fermentas) and transform *E. coli* with the ligation.

4. Amplify cloned fragment directly from bacterial colonies using primers 1 and 2R from the SSH kit. An example of amplification of library clones is shown in Fig. 1b.

3.6. Analysis of Libraries

1. Pretreat NYTRAN SuPerCharge nylon membranes with printed 96-well grid for 5 min with water and then 5 min with 6× SSC, and allow to dry.

2. Denature the PCR products (from 3.5 to 4) by heating at 95°C for 5 min and spot onto the membranes.

Generate first strand cDNA:

3. Mix 0.5 µg of mRNA with 0.5 µg of oligo-dT primer, incubate at 70°C for 5 min, and immediately transfer to ice.

4. To each tube add 2 µL reaction buffer, 20 units of RNAse inhibitor, 1 µL 10 mM dNTPs without dCTP (dNTA, dNTG, dNTT), 47 µCi of α^{32}PdCTP, and 40 units MMLV reverse transcriptase, complete to 20 µL with ddH$_2$O (enzyme and reagents from Stratagene) and incubate 1 h at 37°C.

5. Stop the reaction by adding 5 µL of 0.5 M EDTA and 25 µL of 0.6 N NaOH.

6. Hydrolyze the RNA for 30 min at 70°C.

7. Prepare a Sephadex G-100 gravity column: mix Sephadex beads with the excess of TE buffer, pH 8, and allow to swell. Clog a Pasteur pipette with glass wool and fill it with the Sephadex beads slurry to form a column.

8. Mix the labeled probe with dextran blue (fast-moving, tracking molecules excluded from the column) and orange G (slow-moving, tracking small molecules including unincorporated radioactive nucleotides) and load onto the column.

9. Add TE buffer and collect the flow through. The flow of the TE buffer throw the column has to be constantly maintained to avoid drying. Keep the blue fraction containing most of the labeled cDNA for the subsequent hybridization.

10. Prewash the membranes with 5× SSC, 0.5% SDS, 1 mM EDTA, prehybridize at 65°C for 1 h in 6× SSC, 0.3% milk powder.

11. Add the probe and hybridize over night at 65°C.

12. Remove the excess of probe removed by two washes, 15 min each, in 2× SSC, 0.1% SDS at 65°C.

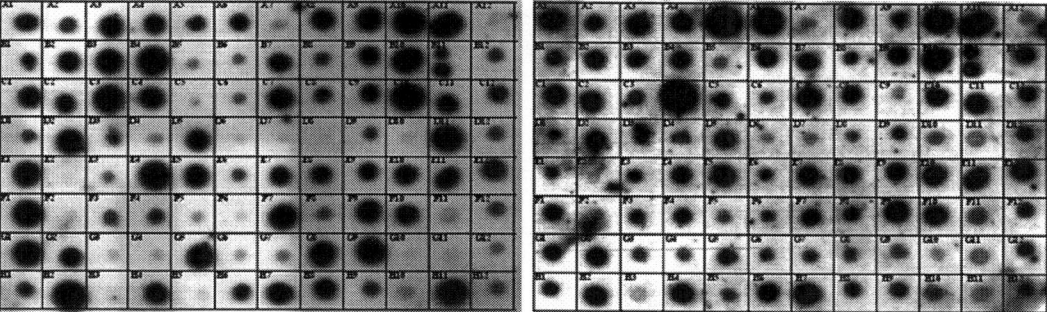

Fig. 2. Example of dot blot analysis. An array of 96 clones was spotted and sequentially hybridized with labeled cDNA probes synthesized from driver (*left*) or tester (*right*) mRNA samples. Data from (16). Clones giving clearly stronger signals upon hybridization with driver probe are likely to represent up-regulated transcripts.

Hybridizations are performed with cDNA pools representing tester (wild type-infected leaves with uninfected leaf mRNA added), driver (mutant-infected leaves), wild type-infected leaves, wild type and mutant grown in shake culture. Clones showing clearly stronger hybridization with the tester probe as compared to the driver probe are sent for direct sequencing using standard primers M13–20 forward and M13 reverse (see Note 8). An example of this analysis is shown in Fig. 2.

4. Notes

1. The procedure is described for the *C. heterostrophus*–maize interaction and can be adapted to any plant–pathogen or indeed host–pathogen interaction, by substituting the appropriate growth media and conditions.
2. DEPC is very toxic! Observe precautions as noted on the reagent package; treat solutions in chemical fume hood and avoid exposure to vapor until the reagent is destroyed by autoclaving of the solutions.
3. The procedure compares wild type and a nonsporulating mutant; for sporulating strains, inoculation with spores is more convenient.
4. Alternatively to the northern blot, estimation of fungal RNA portion in the mixed samples can be performed using semi-quantitative RT-PCR.
5. Although this system works well, when the yield is low, precipitation becomes a critical step, since the following steps require having about 2 μg RNA in a very small volume. It is essential to have a SpeedVac apparatus that concentrates the sample rapidly to avoid degradation. We find that one way to

ensure that the RNA remains intact is to dry the sample down to about 50 μL volume, and then complete the concentration by ethanol precipitation: add 1/10 volume 3 M NaAc (RNAse–free), mix well, store overnight at −20°C, spin 10 min at 4°C, carefully remove supernatant, wash with 80% EtOH (prepared using a freshly opened bottle of ethanol and RNAse-free water), air dry, and redissolve in 5 μl. The described method of total RNA isolation is suitable for experiments where relatively large amount of the starting material is available. It is highly advisable to start mRNA isolation with at least 200 μg total RNA since mRNA is eluted in 250 μl of water. Small mRNA amount makes its concentration by vacuum evaporation and subsequent precipitation challenging.

6. The choice of the RNA populations for comparison is a very important part of the SSH experimental design. The comparison should be made in such as way as to optimize the detection of differentially expressed genes. For example, when comparing wild type and mutant, infection stages that are as similar as possible phenotypically should be chosen. If comparing in planta to saprophytic growth, RNA from uninfected plants is added to the driver RNA population, at a ratio of 3:1 for example, to reduce the background of plant cDNAs isolated.

7. For PCR amplification of the library, it is highly advisable to use Clontech Advantage PCR system since it provides hot-start polymerase suitable for high fidelity amplification of complex templates. PCR products are compatible with T/A cloning methods. However, for optimal ligation efficiency, it is recommended to use PCR products immediately (<1 day) after amplification. The single 3′ A-overhangs on the PCR products will degrade over time, reducing the efficiency. Manual hot start is recommended if regular Taq polymerase is used for library amplification.

8. With the sequencing services now available, it has become much faster, and not very expensive, to sequence one or more sets of 96 clones even before testing to see which are differentially expressed. An advantage is that the sequence of clones that appear interesting by their homology to the databases can be used immediately to design primers for RT-PCR. These primers are then used to confirm differential expression on any number of RNA samples, including those from the original experiment from which the library was produced.

References

1. Madhani HD, Galitski T, Lander ES, Fink GR (1999) Effectors of a developmental mitogen-activated protein kinase cascade revealed by expression signatures of signaling mutants. Proc Natl Acad Sci USA 96:12530–12535
2. Roberts CJ, Nelson B, Marton MJ et al (2000) Signaling and circuitry of multiple MAPK pathways revealed by a matrix of global gene expression profiles. Science 287:873–880
3. Velculescu VE, Kinzler KW (2007) Gene expression analysis goes digital. Nat Biotechnol 25:878–880
4. Diatchenko L, Lau YF, Campbell AP et al (1996) Suppression subtractive hybridization: a method for generating differentially regulated or tissue-specific cDNA probes and libraries. Proc Natl Acad Sci USA 93:6025–6030
5. Diatchenko L, Lukyanov S, Lau YF, Siebert PD (1999) Suppression subtractive hybridization: a versatile method for identifying differentially expressed genes. Methods Enzymol 303:349–380
6. Xue C, Park G, Choi W, Zheng L, Dean RA, Xu JR (2002) Two novel fungal virulence genes specifically expressed in appressoria of the rice blast fungus. Plant Cell 14:2107–2119
7. Lev S, Horwitz BA A (2003) mitogen-activated protein kinase pathway modulates the expression of two cellulase genes in *Cochliobolus heterostrophus* during plant infection. Plant Cell 15:835–844
8. Schulze Gronover C, Schorn C, Tudzynski B (2004) Identification of *Botrytis cinerea* genes up-regulated during infection and controlled by the Ga subunit BCG1 using suppression subtractive hybridization (SSH). Mol Plant-Microbe Interact 17:537–546
9. Cramer RA, Lawrence CB (2004) Identification of *Alternaria brassicicola* genes expressed in planta during pathogenesis of *Arabidopsis thaliana*. Fungal Genet Biol 41:115–128
10. Yakovlev IA, Hietala AM, Steffenrem A, Solheim H, Fossdal CG (2008) Identification and analysis of differentially expressed *Heterobasidion parviporum* genes during natural colonization of Norway spruce stems. Fungal Genet Biol 45:498–513
11. Brechenmacher L, Weidmann S, van Tuinen D et al (2004) Expression profiling of up-regulated plant and fungal genes in early and late stages of *Medicago truncatula-Glomus mosseae* interactions. Mycorrhiza 14:253–262
12. Carpenter MA, Stewart A, Ridgway HJ (2005) Identification of novel *Trichoderma hamatum* genes expressed during mycoparasitism using subtractive hybridisation. FEMS Microbiol Lett 251:105–112
13. Morissette DC, Dauch A, Beech R, Masson L, Brousseau R, Jabaji-Hare S (2008) Isolation of mycoparasitic-related transcripts by SSH during interaction of the mycoparasite *Stachybotrys elegans* with its host *Rhizoctonia solani*. Curr Genet 53:67–80
14. Leach J, Lang B, Yoder O (1982) Methods for selection of mutants and in vitro culture of *Cochliobolus heterostrophus*. J Gen Microbiol 128:1719–1729
15. Sambrook J, Russel D (2001) Molecular cloning: a laboratory manual. CSHL Press, Cold Spring Harbor, NY
16. Lev S (2003) Signal transduction and gene expression in *Cochliobolus heterostrophus* during infection of maize. Technion – Israel Institute of Technology, Haifa, Israel

Chapter 9

Quantification of Fungal Infection of Leaves with Digital Images and Scion Image Software

Paul H. Goodwin and Tom Hsiang

Abstract

Digital image analysis has been used to distinguish and quantify leaf color changes arising from a variety of factors. Its use to assess the percentage of leaf area with color differences caused by plant disease symptoms, such as necrosis, chlorosis, or sporulation, can provide a rigorous and quantitative means of assessing disease severity. A method is described for measuring symptoms of different fungal foliar infections that involves capturing the image with a standard flatbed scanner or digital camera followed by quantifying the area, where the color has been affected because of fungal infection. The method uses the freely available program, Scion Image for Windows or MAC, which is derived from the public domain software, NIH Image. The method has thus far been used to quantify the percentage of tissue with necrosis, chlorosis, or sporulation on leaves of variety of plants with several different diseases (anthracnose, apple scab, powdery mildew or rust).

Key words: Disease quantification, Fungi, Plant disease, Scion image

1. Introduction

While visual estimations of disease severity can be an effective means of assessing disease resistance or pathogen virulence in molecular biology studies of plant-pathogen interactions, it is subjective thus potentially limiting its reproducibility. Furthermore, it may not always be sufficiently sensitive to distinguish subtle, yet still significant, changes in the amount of tissue being affected. An alternative would be to use digital images of plants to directly and objectively quantify the amount of tissue with and without symptoms. An example of quantifying plant health with digital images was the monitoring of plant nutrition using a flatbed scanner to take digital images of sugar maple leaves followed by analysis of the images for the percentage of green or red leaf area using the Scion Image for Windows software package (Scion Corporation, Frederick, MD) (1, 2). Scion Image for Windows is a PC version of NIH Image, which is a public domain program created by the National Institutes of Health (US Department of Health and Human Services) to analyze, enhance, annotate, and output digital images. Scion Image for Windows or MAC is freely available from Scion Corporation (www.scioncorp.com).

We showed that analysis of digital images of leaves with Scion Image can quantify fungal disease severity by measuring differences in leaf color produced during the development of symptoms of the disease or signs of the pathogen. The different types of colors produced in the infected leaves were a darker color due to necrosis, a lighter color due to chlorosis, or a completely different color due to fungal sporulation. Scion Image was effective in quantifying the foliar diseases of anthracnose (Fig. 1), apple scab, powdery mildew (Fig. 2), and rust (Fig. 3), thus showing that it is effective in measuring a wide range of signs and symptoms (3). As leaves must be detached from the plant to be imaged using a flatbed scanner, a variant of the procedure was created using a digital camera to take images of leaves that remained attached to the plant. This makes it possible to take repeated images and measurements of the same leaf as symptoms develops (Fig. 4).

2. Materials

2.1. Image Acquisition

1. Flatbed Scanner that is capable of capturing at least 1,000 pixels of resolution in the larger dimension (either width or height), such as a Perfection 1250U Epson flatbed scanner (Epson Canada Limited, Toronto, ON).
2. Digital Camera that is capable of capturing images of at least 1 megapixels, such as a Nikon CoolPix800 digital camera (Nikon Corporation, Tokyo, Japan).
3. Graphics program that can manipulate images (contrast, color balance, brightness, and size) and specific file formats (uncompressed TIFF), such as Paint Shop Pro 7 (Corel Corporation, Ottawa, ON).

Fig. 1. Images of leaves of *Convallaria majalis* (lily-of-the-valley) with anthracnose symptoms. Images (**a–d**) were the original images obtained with a flatbed scanner. Images (**e–h**) were obtained after modification with Paint Shop Pro. Images (**i–l**) were processed with Scion Image. The leaf in image (**a**) was healthy, and 0% of the leaf area with symptoms were detected with Scion Image (image **i**). Images **b–d** were infected and determined to have 6, 32, and 49% of the leaf area with symptoms using Scion Image (images **j–l**, respectively). The *dark color* on the Scion Image processed leaf image represents the leaf area with necrosis (Reproduced from (3)).

Quantification of Fungal Infection of Leaves with Digital Images 127

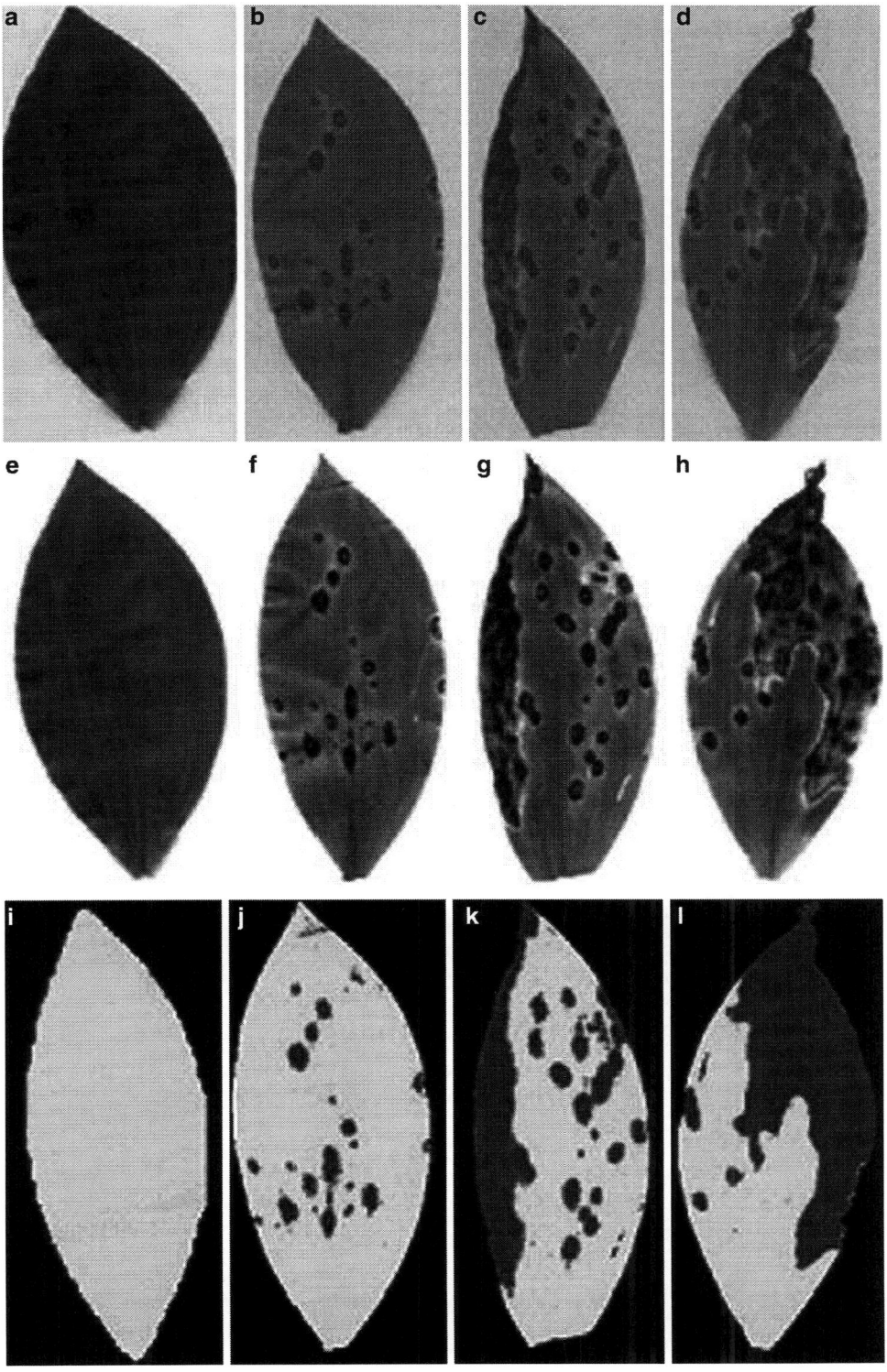

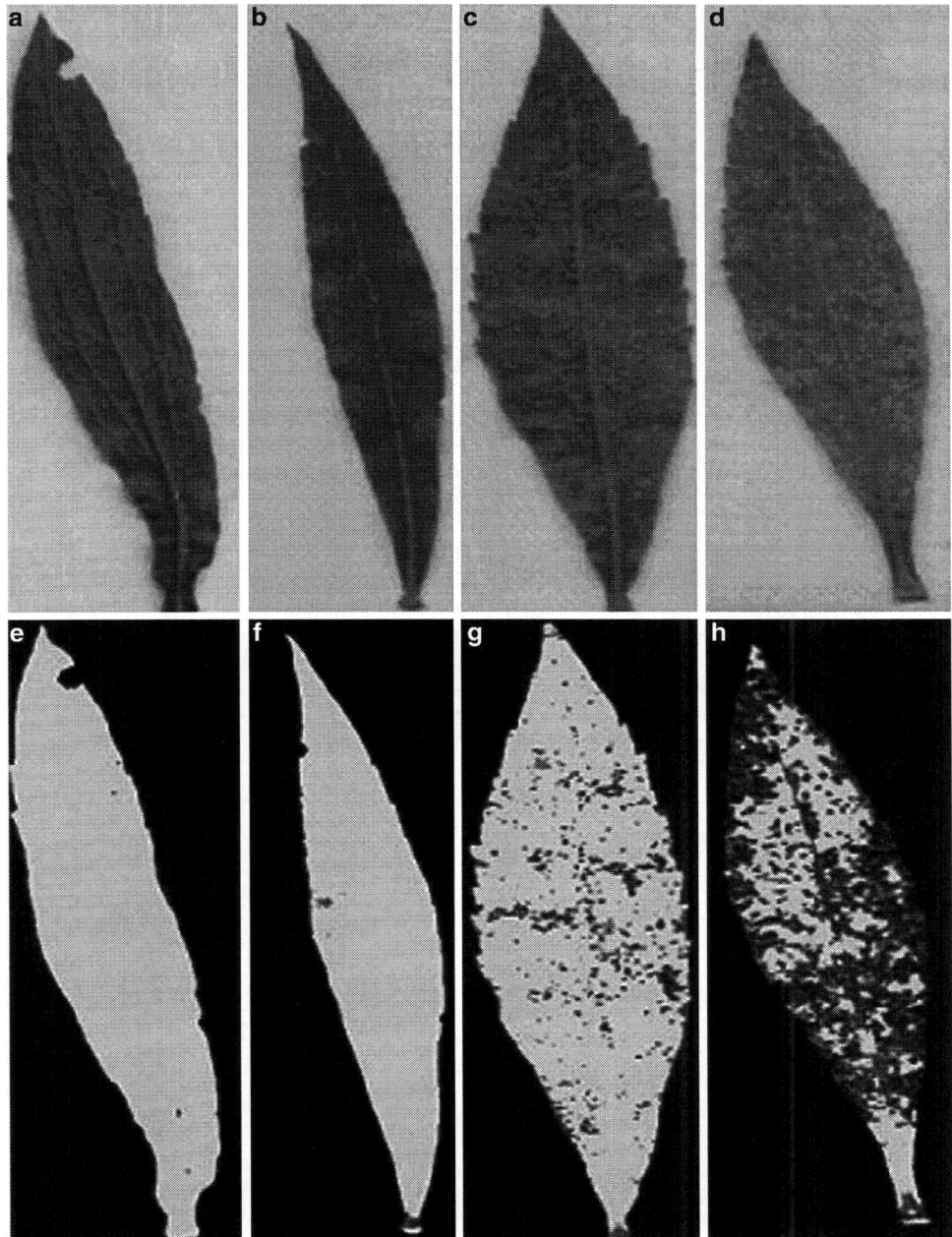

Fig. 3. Images of leaves of *Solidago canadensis* (goldenrod) with rust. Images (**a–d**) were the original images from a flatbed scanner. Images (**e–h**) were processed with Scion Image. Images (**a–d**) were determined to have 3, 4, 15, and 65% of the leaf area with uredia using Scion Image (images (**e–h**), respectively). The dark color on the Scion Image processed leaf image represents the leaf area with uredia (Reproduced from (3)).

◄───

Fig. 2. Images of leaves of *Panus peniculata* (phlox) with powdery mildew. Images (**a–d**) were the original images from a flatbed scanner. Images (**e–h**) were the modified versions obtained with Paint Shop Pro. Images (**i–l**) were processed with Scion Image. The leaf in image (**a**) was healthy, and 0% of the leaf area with powdery mildew was detected using Scion Image (image **i**). Images (**b–d**) were infected and determined to have 1, 17, and 63% of the leaf area with powdery mildew using Scion Image (images (**j–l**), respectively). The *dark color* on the Scion Image processed leaf image represents the leaf area where the fungus was visible (Reproduced from (3)).

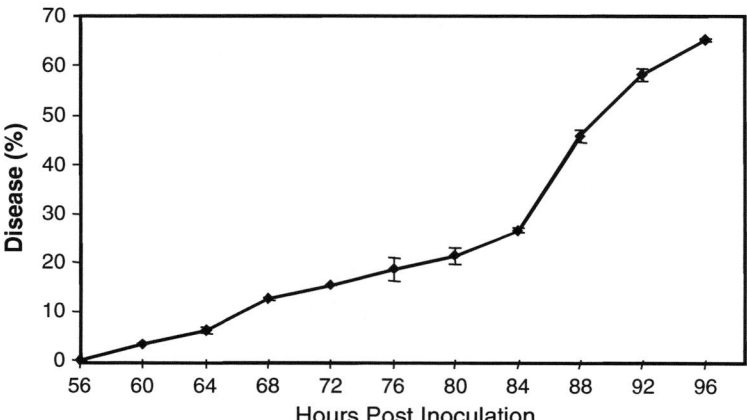

Fig. 4. Percentage of the leaf area with anthracnose symptoms on individual leaves of intact *N. benthamiana* plants inoculated with *Colletotrichum destructivum*. The percent symptomatic leaf area was determined by Scion Image analysis after repeatedly imaging the leaf every 4 h with a digital camera. The results show that no symptoms were visible at 56 h postinoculation, and most of the leaf was necrotic by 96 h post inoculation. The changes in necrotic leaf area over time showed that the spread of symptoms developed more slowly from 56 to 84 h post inoculation than from 84 to 96 h post inoculation. The means with standard error bars were calculated from 12 inoculated leaves on separate plants (Reproduced from (3)).

2.2. Image Analysis

1. Scion Image, a freely available software package (Scion Corporation, Frederick, MD) from www.scioncorp.com, version 4.03 for Windows 95 to XP, or version 1.63 for MacOS 7.5–9.2.2 (see Note 1). Documentation for Scion Image can be obtained from http://www.york.ac.uk/depts/biol/help/scion/sscion.htm.

3. Methods

3.1. Acquire and Adjust the Color Image to Input into Scion Image

1. Take image of leaf with a standard flatbed scanner or digital camera. Flat images without shadows are best, such as those obtained with a flatbed scanner, but images from a digital camera also can be used as long as the leaf can be kept flat, such as by resting it on a flat surface, like a supported piece of clear plastic, to avoid any shadows. Use backlighting or a white background as this helps give a strong contrast to the leaf and makes it easier to differentiate the healthy leaf color from the color of the affected tissue.

2. Open the image in software, such as Paint Shop Pro, to adjust the brightness and contrast so the symptoms stand out from healthy leaf. Standardize the amount and type of image adjustment for each disease under study, so that the changes

are consistent. For example, the images of rust of goldenrod were not adjusted, but the brightness and contrast were both increased by 25% for apple scab, and images were "Sharpened More" and then increased in brightness by 25% and contrast by 75% for powdery mildew of phlox (3).

3. Save the file as an uncompressed TIFF. Files saved as JPG and most other formats are not accepted in Scion Image (see Note 2). The size of the image should be no more than what will fit in a quarter of the screen in order to see the diseased leaf area measurements clearly (common screen resolutions is 1,024 by 768, and so make the image size 512 by 384).

3.2. Convert the Input TIFF Color Image into a Binary Image

1. Open the input image in Scion Image by clicking "File" then "Open." Two images are automatically created. The one in color is called the "Indexed color" image, and the other is a black and white version of that image which retains the original file name as in step 3 of Subheading 3.1, but with a suffix added (e.g., "image.tif*(Red)").

2. Select this black and white version by clicking on the window to bring it to the foreground. In the main menu at the top, click "Stacks" then "Stack to Windows." Three cascaded windows will appear on the screen labeled 001, 002, and 003. Close windows 001 and 002, leaving only window 003 visible.

3. With window 003 selected, click "Process" and then "Rank Filters" (Note: This step was not performed for the analysis of powdery mildew of phlox (3)). A dialog box will be displayed containing a list of options, including "Median (Reduce Noise)," which is the default selection, and "Iterations:" with default "1" in the box. Click "OK."

4. Click "Edit" on the main menu at top and then "Invert." This reverses the black and white parts of the leaf on the screen.

5. Click "Options" in the main menu, and then "Threshold." This separates objects of interest (e.g., the whole leaf) from the image background based on grayscale.

6. Click "Process" then "Binary," and then "Make Binary." At this point, processing of the black and white image to a binary image is complete. This binary image (003) will be used in step 1 of Subheading 3.4, in the "Image Math" procedure, where two images are combined.

3.3. Convert the Input TIFF Color Image into an Adjusted Color Image

1. Select the window for the "Indexed Color" image (from Subheading 3.2, step 1) by using "Windows," then "Indexed Color" or by clicking on the title bar of the image. Select "Stacks" from the main menu and then "8-bit Color to RGB." Select "Stacks" again and then "RGB to HSV."

2. A warning message will appear stating that "RGB to HSV color conversion is undoable." Click OK.
3. Click "Stacks" again and then "Stack to Windows."
4. Three cascaded windows will appear labeled 001, 002, and 003. Close windows 002 and 003. Select window 001, which is the adjusted color image. (For rust of goldenrod, window 003 is selected instead of window 001 (3)). This adjusted color image (001) will be used in step 1 of Subheading 3.4, in the "Image Math" procedure where two images are combined.

3.4. Combine the Binary Image from step 6 of Subheading 3.2, and the Adjusted Color Image from step 4 of Subheading 3.3, using the Image Math Procedure

The adjusted color image (001) and the binary image (003), that were created in step 4 of Subheadings 3.3, and step 6 of Subheading 3.2, respectively, will now be combined.

1. Select the adjusted color image (001).
2. Click "Process" and then "Image Math." A dialog box appears called the "Image calculator," which contains several boxes in it. Set the top box to 003, set the two boxes directly below to + on the left and 001 on the right, enter 1 in the box with X next to it, enter 0 in the box with + next to it, and leave the box with = next to it as "Result." Leave the "Real Result" box unchecked.
3. Click "OK." This creates a "Result*" image.
4. Reposition all the image windows by using "Windows" and then "Tile Images." This displays the windows adjacent to each other.
5. The same areas of the leaf should be visible in both the "Indexed color" image and the "Result*" window. If they are not similar, the two images should be moved within their window using the hand tool. To do this, use the "Hand" tool in the "Tools" toolbox to move the images within each window. It is important to see the same areas of the leaf in each image in order to successfully compare the color of the images.

3.5. Adjust the LUT Bar to Edit the "Result" Image from Subheading 3.4.6 to Mark the Symptomatic, Non-symptomatic or Total Area of the Leaf*

1. Edit the selected "Result*" image with the Look- up table (LUT) tool, which is the color spectrum bar that appears on the upper left next to the "Tools" toolbox. In the "Tools" toolbox, select the LUT tool, which is represented by a horizontal line, with an upward and downward arrow.
2. Double click this tool which will call up a red band in the middle of the LUT box. This red band is adjustable in height, and the color will be used to replace all spectrum of colors that it overlaps.

3. To change the color of the menu from red to another color (especially if red already appears in the "Result*" image), click on the "eyedropper" tool in the "Tools" toolbox, and then place the cursor over the red portion of the LUT box. Double-click and a "Color" menu window comes up. Select a color that is not normally found with the leaf under study. For example, bright yellow can be selected with Hue 40, Sat 240, Lum 120, Red 255, Green 255 and Blue 0.

4. Select the horizontal line in the LUT box with the two arrows (single click only). Move the bottom border of the red bar (or whichever color was selected) to near the bottom of the LUT box, usually near position 254. Move the top border of the red bar until either the symptomatic or nonsymptomatic areas of the leaf become entirely red (or the color that was selected in step 3 of Subheading 3.5). This creates the edited "Result*" image. Record the numeric values in the LUT box for the top and bottom boundaries. These numbers can be found by clicking on the word "white" beneath LUT bar. In the info box below that, the words lower and upper will appear followed by numbers. Testing a number of leaves should make it possible to standardize these values for each disease under study, so that the analysis is consistent between leaves. For example, the boundaries of the LUT bar were set at 27×254 for apple scab, 50×254 for anthracnose of *Nicotiana benthamiana*, and 115×254 for rust of goldenrod (3).

5. The leaf area of the symptomatic or nonsymptomatic area (depending on which was colored in step 4 of Subheading 3.5) can now be measured according to Subheading 3.7.

6. In addition to obtaining the symptomatic area, this process can also be used to obtain the entire leaf area. In the LUT box, set the boundaries of the LUT bar to their maximum (1×254), so that the entire leaf area is colored red (or whichever color was selected in step 3 of Subheading 3.5) in the same image. This edited "Result*" image can be used in Subheading 3.7 to measure total leaf area.

3.6. Use of Paintbrush or Paintbucket Tools

For samples such as powdery mildew, where the symptomatic tissue is much lighter in color than the healthy tissue, use the paintbrush tool or paintbucket tool to adjust the images

1. In images of powdery mildew, the leaf margins should be marked using the "Paintbrush" in the "Tools" window, lining the leaf margins with it. The "Paintbrush can be resized to cover a smaller or larger area by double-clicking on the paintbrush tool and changing the pixel size.

2. To obtain the total leaf area of leaves with powdery mildew, the "Paint Bucket" tool can then be used to fill the area within the margins.

3.7. Measure the Leaf Area

Measurement of the leaf area is performed using the metric scale according to O'Neal et al. (4)

1. Select the edited "Result*" image from step 4 of Subheading 3.5, by clicking on that window.
2. For quantification of the symptomatic or nonsymptomatic areas of the leaf, use the main menu item "Analyze." In the "Analyze" menu box, select "Options" to increase the number of significant digits by entering a number (e.g., 4) in the box next to "digits right of decimal point" to get the desired number of significant digits.
3. In the "Analyze" menu box, select "Set Scale." This opens a dialog box, where the units can be changed to centimeters.
4. Areas covered with red (or the color chosen in step 3 of Subheading 3.5) in step 4 of Subheading 3.5, can now be measured by the Scion Image program. To do this, click "Analyze" and then "Measure." Then click "Analyze" again, followed by clicking "Show Results." The leaf area is given in cm^2.
5. For quantification of the total leaf area, repeat Subheading 3.7 with the edited "Result*" image described in step 6 of Subheading 3.5, where the entire leaf area had been colored. First, click on the "Result*" image again, and then adjust the top border of the red bar following Subheading 3.5.
6. Calculate the ratio of the diseased to total leaf area to obtain the percent disease.
7. For leaves where the healthy leaf area is measured, the diseased leaf area is calculated by subtraction of healthy leaf area from the total leaf area.

4. Notes

1. Scion Image for Windows is derived from the original NIH Image originally written at the U.S. National Institutes of Health. Scion Corporation revised the program to support its frame grabber boards, but the program works without the frame grabber boards installed. When run for the first time, Scion Image will ask about the presence of a frame grabber board, and this message can be turned off permanently for future uses of the program.

2. Although Scion Image can open BMP (bitmap format) files, they cannot be processed with the methods described here, since only a single color or black and white image appears when a BMP file is first opened with Scion Image.

Acknowledgments

Funding for this study was provided by the Natural Science and Engineering Research Council of Canada and the Ontario Ministry of Agriculture, Food and Rural Affairs.

References

1. Schaberg PG, Van Den Berg AK, Murakami PF, Shane JB, Donnelly JR (2003) Factors influencing red expression in autumn foliage of sugar maple trees. Tree Physiol 23:325–333
2. Murakami PF, Turner MR, Van den Berg AK, Schaberg PG (2005) An instructional guide for leaf color analysis using digital imaging software. USDA Gen. Tech. Rep. NE-327. Newtown Square, PA: U.S. Department of Agriculture, Forest Service, Northeastern Research Station. 33 p. http://www.nrs.fs.fed.us/pubs/6995
3. Wijekoon CP, Goodwin PH, Hsiang T (2008) Quantifying fungal infection of plant leaves by digital image analysis using Scion Image software. J Microbiol Methods 74:94–101
4. O'Neal M, Landis DA, Isaacs R (2002) An inexpensive, accurate method for measuring leaf area and defoliation through digital image analysis. J Econ Entomol l95:1190–1194

Chapter 10

Expression Profiling of Fungal Genes During Arbuscular Mycorrhiza Symbiosis Establishment Using Direct Fluorescent *In Situ* RT-PCR

Pascale M. A. Seddas-Dozolme, Christine Arnould, Marie Tollot, Elena Kuznetsova, and Vivienne Gianinazzi-Pearson

Abstract

Expression profiling of fungal genes in the arbuscular mycorrhiza (AM) symbiosis has been based on studies of RNA extracted from fungal tissue or mycorrhizal roots, giving only a general picture of overall transcript levels in the targeted tissues. Information about the spatial distribution of transcripts within AM fungal structures during different developmental stages is essential to a better understanding of fungal activity in symbiotic interactions with host roots and to determine molecular events involved in establishment and functioning of the AM symbiosis. The obligate biotrophic nature of AM fungi is a challenge for developing new molecular methods to identify and localize their activity *in situ*. The direct fluorescent *in situ* (DIFIS) RT-PCR procedure described here represents a novel tool for spatial mapping of AM fungal gene expression simultaneously prior to root penetration, within fungal tissues in the host root and in the extraradical stage of fungal development.

In order to enhance detection sensitivity of the *in situ* RT-PCR technique and enable localization of low abundance mRNA, we have adopted direct fluorescent labeling of primers for the amplification step to overcome the problem of low detection associated with digoxigenin or biotin-labeled primers and to avoid the multiplicity of steps associated with immunological detection. Signal detection has also been greatly improved by eliminating autofluorescence of AM fungal and root tissues using confocal microscopy.

Key words: Arbuscular mycorrhizal fungi, Direct fluorescent *in situ* RT-PCR, Confocal microscopy, Fungal gene expression, Symbiosis, *Glomus intraradices, Medicago truncatula, Pisum sativum*

1. Introduction

Arbuscular mycorrhiza (AM) represent an ubiquitous symbiosis between roots of the large majority of land plants and Glomeromycota fungi (1). Several well-defined developmental stages characterize the life cycle of AM fungi: spore germination, presymbiotic hyphal growth, appressorium formation at the root surface before penetration, proliferation in internal cortical tissues with differentiation of intracellular haustoria (arbuscules) and vesicles, and formation of an extraradical mycelium with the generation of new spores (2). Determination of genes or proteins that are implicated in the symbiotic stages of AM fungal development is still difficult due to the obligate symbiotic character of AM fungi (which cannot be cultured *ex planta*) and by the fact that they cannot be genetically transformed in a stable way (3, 4).

Expression profiling based on RNA extracted from fungal or mycorrhizal tissues gives only limited information about genes associated with complex developmental processes such as those involved in AM development and function (5–11). However, such gene expression monitoring does not provide information about the spatial distribution of transcripts at the tissue level, especially in AM where stages of fungal-plant interactions are not uniform within roots.

Identification of fungal genes that are specifically related to AM symbiosis development and functioning requires spatial mapping of gene expression in the different fungal structures during interactions with host roots. Whereas on the plant side, *in situ* hybridization (12–19) or reporter gene technology (13, 20–26) have been used to locate gene expression in the AM symbiosis, targeting of AM fungal gene products in symbiotic interactions has relied on protein localization by immunocytochemistry (27–29), and more recently microdissection has identified transcripts in arbuscules (30, 31). Gene discovery by high throughput DNA sequencing is a developing area in the field of AM fungi (32, 33), creating a requirement for the development of sensitive techniques to examine gene expression at the cellular level.

The direct fluorescent *in situ* RT-PCR (DIFIS RT-PCR) methodology described here, associated with confocal microscopy to eliminate autofluorescence of plant tissues, allows localization of low abundance transcripts in individual structures of an AM fungus and detection of gene expression during different stages of interactions with host roots (34). This methodology was adapted from a protocol first applied to lentiviruses inside cells (35) and to plant tissues (see for example ref. (36)). It greatly improves on a recently described *in situ* RT-PCR approach based on the immunofluorescent detection of amplification products used to localize transcripts in an ectomycorrhizal fungus, *Hebeloma cylindrosporum*, but which is applicable only to pure culture mycelium and not to symbiotic tissues (37). This new DIFIS RT-PCR method applied to obligatory symbiosis fungi offers new perspectives for monitoring gene expression in different structures of fungal tissues and for associating organ anatomy with transcription patterns. It can thus make an important contribution to the developmental process knowledges and to gene expression pattern analysis in complex biological systems, such as plant-fungal associations. This DIFIS RT-PCR is an interesting useful tool in AM fungi research for coupling gene transcription and cellular localization. This methodology will provide gene expression localization, notably from the *G. intraradices* genome sequencing project (32, 38).

2. Materials

2.1. Presymbiotic and Symbiotic Fungal Structures

1. Percoll gradient.
2. Bottom: 9 mL Percoll diluted with 1 mL of 1.5 M NaCl. Top: 4.5 mL Percoll washed with sterile water and diluted with 0.5 mL of 1.5 M NaCl and 5 mL water.
3. Apigenin (Sigma): Prepare apigenin in water at 225 nM diluted in 0.02% (v/v) ethanol.
4. CO_2: Maintained at 2% by injection in a CO_2 incubator (IG150, Jouan).
5. Plants: Wild-type *M. truncatula* cv. Jemalong, line J5, and corresponding mycorrhiza defective mutants *dmi1* (Y6, INRA-Toulouse, France), *dmi2/Mtsym2* (TR25) and *dmi3/Mtsym13* (TRV25) (INRA-Dijon, France) (39), *P. sativum sym40* mutant (SGEFix.-1, ARRIAM, St. Petersburg, Russia), which is characterized by a rapid turn-over of arbuscules (40).
6. Sterile soil: mixture of Terragreen (OilDri-US special, Mettman, Germany) and a neutral γ- irradiated clay soil (2:1, vol/vol).
7. Inoculated soil: mixture (2:1, vol/vol) of Terragreen and a soil-based inoculum of *G. intraradices* BEG 141.
8. Growth substrate: 1:1 (v/v) sand:clay-loam soil (pH 7.8).
9. Sand (Special Aquarium, Quartz, Nr. 3, Zolux, France).
10. Bactoagar (Difco Laboratories, Detroit).
11. Nutrient solution: Long Ashton solution without phosphate and a double quantity of nitrate (41). For 1 L solution, dilute 4.04 g KNO_3, 9.44 g $Ca(NO_3)_2 \cdot 4H_2O$, 1.84 g $MgSO_4 \cdot 7H_2O$, 5 mL solution A (for 100 mL solution A, dilute 0.22 g $MnSO_4 \cdot 4H_2O$, 25 mg $CuSO_4 \cdot 5H_2O$, 29 mg $ZnSO_4 \cdot 7H_2O$, 0.31 g H_3BO_3, and 0.59 g NaCl), 0.5 mL solution B (for 100 mL solution B, dilute 88 mg $(NH_4)6Mo_7O_{24} \cdot 4H_2O$), and 0.11 g FeNa EDTA.

2.2. Sample Fixation

1. Fixation solution: 67% (v/v) ethanol, 23% (v/v) acetic acid, 10% (v/v) DMSO.

2. Wash solution: 67% (v/v) ethanol and 23% (v/v) acetic acid.
3. DEPC-treated water: Add 1 mL of DEPC to 1 L of water and incubate under stirring at room temperature for 16 h, then sterilize at 120°C for 20 min.

2.3. Probe Labeling and Fluorochrome Selection

2.3.1. Selected G. Intraradices Genes

1. The large ribosomal subunit gene (LSU rRNA) of *G. intraradices* was used to develop the DIFIS RT-PCR methodology because of ribosomal transcript abundance in cells; previously published primers were used as reported in (34).
2. Three protein-encoding genes were selected to cover different fungal activities (Table 1): stearoyl-CoA-desaturase (*DESAT*, lipid turn-over), peptidylprolyl isomerase (*PEPISOM*, protein turn-over) and Cu/Zn superoxide dismutase (*SOD*, anti-oxidative metabolism).
3. PCR primers (Table 1) were developed from fungal sequences identified amongst EST clusters from *M. truncatula* mycorrhizal roots (42), and their fungal specificity was confirmed by PCR on *G. intraradices* RNA extracted from spores (34).
4. Presence of fungal transcripts in RNA from inoculated *M. truncatula* and *P. sativum* roots was confirmed by RT-PCR (43).

2.3.2. Fluorochrome Selection

1. We use a LEICA TCS SP2 AOBS (Leica Microsystems, Germany) confocal laser microscope.
2. To determine the wavelengths between which autofluorescence of plant and fungal tissues was abolished in order to define the nature of the fluorochrome, the excitation laser wavelength and the window of emission signal collection, fixed samples without labeling were wavelength scanned using two lasers (see Note 1):
   – An argon laser (100 mW) with an excitation wavelength at 488 nm.
   – A He/Ne laser (1.5 mW) with an excitation wavelength at 594 nm.

Plant and fungal structures showed autofluorescence whatever the emission signal collection used with the argon laser, whereas no auto-fluorescence was detected between 606 nm and 640 nm with the He/Ne laser. Consequently, each gene-specific primer was 5′-labeled with Texas Red and the 594 nm laser was used for excitation.

2.3.3. Probe Labeling

5′-labeled probe with Texas Red was carried out at MWG Biotech.

2.4. Permeabilization of Cell Walls and Digestion of Genomic DNA

1. Chitinase: from *Streptomyces griseus* (FLUKA).
2. Pectinase: from *Aspergillus niger* (FLUKA).

Table 1
Putative function of selected genes of *Glomus intraradices*, cluster identity (ID) in MENS, primer sequences, annealing temperature and amplicon size

Protein category	ID MENS	PCR primers		Ann. temp (°C)	Amplicon (bp)
		Name	Sequence		
Peptidylprolyl isomerase (*PEPISOM*)	MtC00626_GC	PEPISOM for	GATGTTCATGCCGGTAAAAG	55	235
		PEPISOM rev	ACTGGATGAACCCAATGTCT		
Lipid metabolism					
Stearoyl-CoA desaturase (*DESAT*)	MtC91345_GC	DESAT for	TCGTGTTCCTGAAAATGAAG	55	269
		DESAT rev	GCTTTAGTGGAGTCTTTACC		
Anti-oxidative metabolism					
Cu/Zn Superoxide dismutase (*SOD*)	MtD23013_GC	SOD for	CTGGACCTCATTTTAACCCA	55	273
		SOD rev	CCGATAACACCACAAGCAA		

3. 25 mM and 50 mM Tris–HCl, pH 7.6.
4. Proteinase K (39 U/mg, Sigma): use at 1 μg/mL in 50 mM Tris–HCl, pH 7.6.
5. Restriction enzymes: *Hae* III, *Hpa* II (Promega).
6. RNAsin (Promega).
7. DNAse (Boehringer Mannheim).
8. RNAse A (Q-Biogene).

2.5. cDNA Synthesis and PCR

1. dNTPs: 2.5 mM each of dATP, dTTP, dCTP, and dGTP were mixed (Invitrogen).
2. First strand cDNA synthesis: cDNA was prepared by adding 0.25 μM of each unlabeled for-primers (see Table 1), dNTPs (0.25 mM each) and made up to a final volume of 11.5 μL with sterile distilled water. RNA was denatured for 5 min at 70°C, placed on ice, and 5 μL of MMLV 5 reaction buffer, 300 U of MMLV reverse transcriptase (Promega), and 80 U of RNase inhibitor (Promega) were added. First strand cDNA was synthesized at 42°C for 1 h.
3. PCR Primers: 5′-labeled for- and rev- primers (see Table 1) with Texas Red was used at 0.25 μM each.
4. PCR mix: 1 μL PCR mix (Invitrogen) supplemented with 1.5 mM $MgCl_2$, 0.25 μM of each primers, 0.125 mM of each dNTP, and 1 U Taq DNA polymerase (Invitrogen).
5. 100% Ethanol.

2.6. Confocal Microscopy

1. A LEICA TCS SP2 AOBS (Leica Microsystems, Germany) confocal microscope was used with a 40× oil (iris 1.25–0.75) horizontal resolution 0.156 μm objective.
2. Antifading medium (DAKO).
3. Camera and software are those from LEICA TCS SP2 AOBS (Leica Microsystems, Germany).

3. Methods

3.1. Presymbiotic and Symbiotic Fungal Structures

1. Spore purification: collect spores of *G. intraradices* in 18 mL water, deposit spores on a percoll gradient, then centrifuge at $1,000 \times g$ for 10 min and collect the spores at the interface of the two percoll solutions.
2. Incubate spores (2,000) in 225 nM apigenin for 7 days at 25°C in 2% CO_2. At this time, the spores will germinate and develop hyphae (43).

3. Seed sterilization: incubate seeds in 2 mL of pure H_2SO_4 for 6 min, then in pure ethanol for 5 min, then wash in tap water.

4. Seed germination: For *M. truncatula* incubate seeds on a 0.7% (w/v) Bactoagar (Difco Laboratories, Detroit) plate in dark for 72 h at 25°C. For *P. sativum*, incubate seeds for 3 days on humid filter papers in Petri dishes at 24°C in the dark.

5. Symbiotic fungal structures: Inoculate *M. truncatula* or *P. sativum* seedlings with *G. intraradices* isolate BEG 141, as described in (43).

6. For uninoculated plants, transfer individual seedlings of *M. truncatula* into a 75 mL sterilized (autoclaved once at 120°C during 7 h) mix of Terragreen and a neutral γ- irradiated clay soil (2:1, vol/vol). For inoculated plants, transfer seedlings into a 75 mL mix (2:1, vol/vol) of Terragreen and a soil-based inoculum of *G. intraradices* BEG 141. Grow plants under constant conditions (420 µE/m²/s, 24 and 19°C, 16 and 8 h, day and night, respectively, 70% humidity). Fertilize with 5 mL/pot of Long Ashton solution without phosphate and a double quantity of nitrate twice a week (43).

7. Transfer individual *P. sativum* seedlings into 200 g mix of a 1:1 (v/v) sand:clay-loam soil (pH 7.8). Sterilize soil by γ-irradiation (10 kGy). For the sand, wash for 4 h in water then heat sterilize for 4 h at 180°C.

8. Production of mycorrhizal plants: replace soil with a soil-based inoculum of *G. intraradices* from 10-week-old leek pot cultures. Grow plants in a constant environment chamber (24/19°C, 16 h light, 420 µE/m²/s, 70% relative humidity) and fertilize with 10 mL/pot of Long Ashton solution without phosphate three times a week.

9. Harvest plants 5 and 14 days after inoculation (DAI) (wild-type *M. truncatula*), 7 DAI (*M. truncatula* mutants) or 21 DAI (*P. sativum*). Wash the roots and immediately (1) fix for DIFIS RT-PCR (see Subheading 3.2), or (2) freeze in liquid nitrogen for RNA extraction (34).

3.2. Sample Fixation

1. Place about 200 *G. intraradices* spores, or 30 pieces (0.5 mm) of inoculated *M. truncatula* long or *P. sativum* roots in RNAse- and DNAse-free treated 500 µL-Eppendorf tubes.

2. Fix the tissue by incubating for 1 h at 4°C in 200 µL of fixative solution under vacuum.

3. Remove the fixative solution by pipetting, then add 200 µL of fresh fixative solution to the roots or spores, and incubate for 16 h at 4°C.

4. Remove the fixative solution, then wash twice by pipetting with 0.5 mL of 67% (v/v) ethanol and 23% (v/v) acetic acid and once with 0.5 mL of DEPC-treated water. If not used immediately, samples may be stored up to 3 days at 4°C.

3.3. Probe Labeling and Fluorochrome Selection

Each gene-specific primer was 5′-labeled with Texas Red (MWG Biotech), a 594 nm laser was used for excitation and the resulting signal was collected between 606 nm and 640 nm for germinated spores and inoculated root tissues (see Note 2). Qualifying signal of fungal gene expression could also be detected depending on inoculated plant (see Note 3).

3.4. Permeabilization of Cell Walls and Digestion of Genomic DNA

1. Enzymes: dilute enzymes in 25 mM Tris–HCl, pH 7.6 to a final concentration of 0.1 U chitinase and 6 U pectinase per 100 µL.
2. Incubate for 10 min at 37°C.
3. Remove the enzyme-containing solution by pipetting, wash samples three times by pipetting with 100 µL of 25 mM Tris–HCl, pH 7.6, then with 100 µL of DEPC-treated water and finally with 100 µL of 50 mM Tris–HCl, pH 7.6.
4. Incubate samples with proteinase K (0.1 µg/100 µL of 50 mM Tris–HCl, pH 7.6) for 30 min at 37°C, then wash three times with 100 µL DEPC-treated water by pipetting.
5. Resuspend samples in 100 µL of multicore buffer (Promega) containing 2 U each of *Hae* III and *Hpa* II enzymes and 1 µL RNAsin and incubate for 1 h at 37°C.
6. After removal of enzyme-containing solution by pipetting, suspend in 100 µL of DNAse buffer (Boehringer Mannheim) with 8 U DNAse and incubate for 5 h at 37°C to eliminate genomic DNA. For RNA controls, incubate samples for 1 h at 37°C with 100 U RNAse A diluted in 100 µL of 25 mM Tris–HCl, pH 7.6, then wash three times with 100 µL of DEPC-treated water before performing the DNAse treatment. If needed, samples can be stored at this stage overnight.

3.5. cDNA Synthesis and PCR

1. After removal of DEPC-treated water by pipetting, incubate permeabilized, DNAse-treated samples at 65°C for 5 min in 40 µL of RT-mix with 0.25 µM of the specific unlabeled for-primers (Table 10.1).
2. Placed on ice for 5 min and then add 1 µL RNAsin, 5 µL RT buffer, 1 µL MMLV reverse transcriptase H and 3 µL DEPC-treated water into each tube, and immediately incubate at 42°C for 1 h.
3. Remove the RT-mix by pipetting and add 50 µL PCR-mix into each tube [1 µL PCR-mix (Invitrogen), 1.5 mM $MgCl_2$, 0.125 mM of each dNTP, 0.25 µM of each Texas Red-labeled gene-specific primer, and 1 U Taq DNA polymerase].

4. Perform PCR amplification using the following cycles: 2 min at 95°C for denaturation, then 30 cycles of {1 min at 95°C, 1 min at annealing temperature (Table 10.1), 1 min 72°C}, 2 min at 72°C. Controls were performed by omitting primers from the reverse transcription reaction.

5. Remove the PCR-mix by pipetting and postfixe samples in 100 µL of 100% ethanol for 10 min at room temperature to precipitate the nucleic acid.

6. Place samples on a microscope slide in a droplet of antifading medium. If not used immediately, samples may be stored at 4°C in the dark up to 21 days.

3.6. Confocal Microscopy

Red fluorescent-labeled amplified cDNA was observed *in situ* using confocal microscopy and 40× oil immersion (LEICA TCS SP2 AOBS; Leica Microsystems, Germany). A Nomarski image was taken in parallel each time. As indicated above (2.3), excitation was carried out with a 594 nm laser (27% maximum power, photomultiplicator at 705 V), and the resulting signal was collected between 606 nm and 640 nm. Each fluorescent image corresponded to the maximum projection of optical sections (20–254) from a z series, each projection being between 50 nm and 300 nm in depth depending on the object size and magnification.

4. Notes

1. It is important to predetermine the fluorochrome, laser and emission detection wavelengths in order to avoid background. Fig. 1 shows an example.

2. The DIFIS RT-PCR methodology can be used to couple differential gene expression with fungal morphogenesis event by using different probes and including light and fluorescent images. An example is shown in Figs. 2 and 3.

3. Fungal gene expression may change on different host plants. An example is shown in Fig. 4.

Acknowledgments

This work was supported by the Regional Council of Burgundy (Faber Project No. 05 512 AA O6 S 2452). Authors thank C. Humbert, Centre de Microscopie Appliqué à la Biologie, Burgundy University, for advice in confocal microscope analyses,

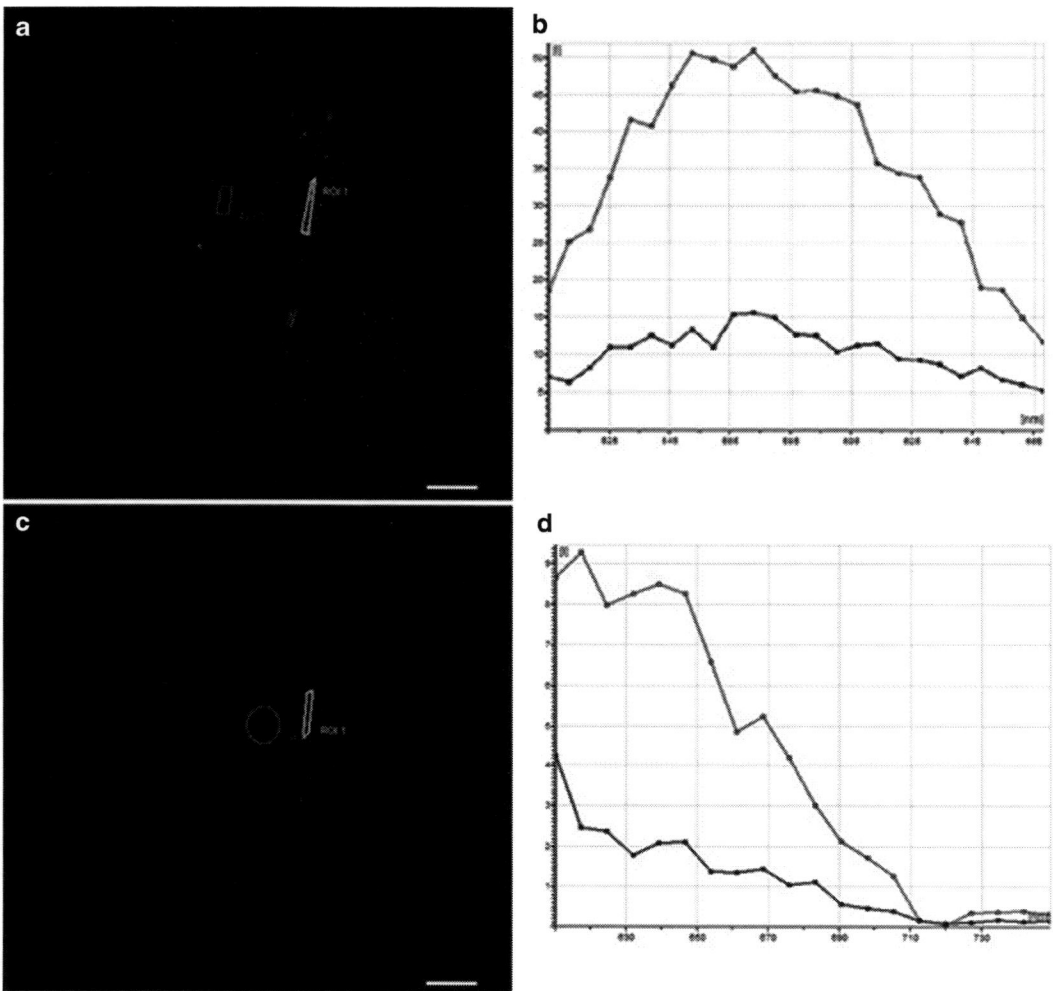

Fig. 1. Determination of fluorochrome, laser and emission detection wavelengths. Fixed *G. intraradices*-inoculated roots of *M. truncatula* 14 DAI were excited at 488 nm using an argon laser (**a**, **b**) or at 594 nm using a He/Ne laser (**c**, **d**). Scanning of the emission signal was carried out to determine the wavelengths of excitation and of emission. For the *G. intraradices/M. truncatula* symbiosis system, excitation was determined at 594 nm and emission signal was collected between 606 nm and 640 nm when autofluorescence of plant and fungal tissues was eliminated. Bar = 50 nm.

Fig. 2. (continued) gene expression was monitored. Transcripts accumulated in fungal structures, whereas no signal was detected in plant tissues. Nomarski pictures (**a**, **d**, **g**, **j**, and **m**) were taken in parallel to localize gene expression. Bar = 50 nm. *Hy* hypha, *Sp* spore, *Ap* appressorium, *Arb* arbuscule, and *Ve* vesicle. Transcripts of the two weakly expressed *SOD* and *PEPISOM* genes were localized in germinated spores. *SOD* transcripts were clearly detected within apigenin-elicited spores and subtending hyphae (**a–c**). *PEPISOM* transcripts, on the contrary were only weakly detected in spores, but the gene was highly expressed in the germinating hyphae (**d–f**). These observations illustrate the usefulness of the DIFIS RT-PCR methodology to couple differential gene expression with fungal morphogenesis event to detect molecular mechanisms at germination stage. *G. intraradices*-inoculated *Medicago truncatula* roots 5 DAI (**g–l**) or 14 DAI (**m–o**) were proceeded for DIFIS RT-PCR. *DESAT* transcripts were strongly detected in extraradical fungal structure such as hyphae (**g–i**) or appressoria (**j–l**) when fungus comes into contact with root, and also were in vesicle and more weakly in arbuscules, two fungal structures that are imbricated in root tissues (**m–o**) and which cannot be seen using Nomarski microscopy technology (**n**). This demonstrates that fungal gene expression can be specifically localized in extraradicular fungal structures and in intraradicular fungal structures inside root (see also ref. (34)).

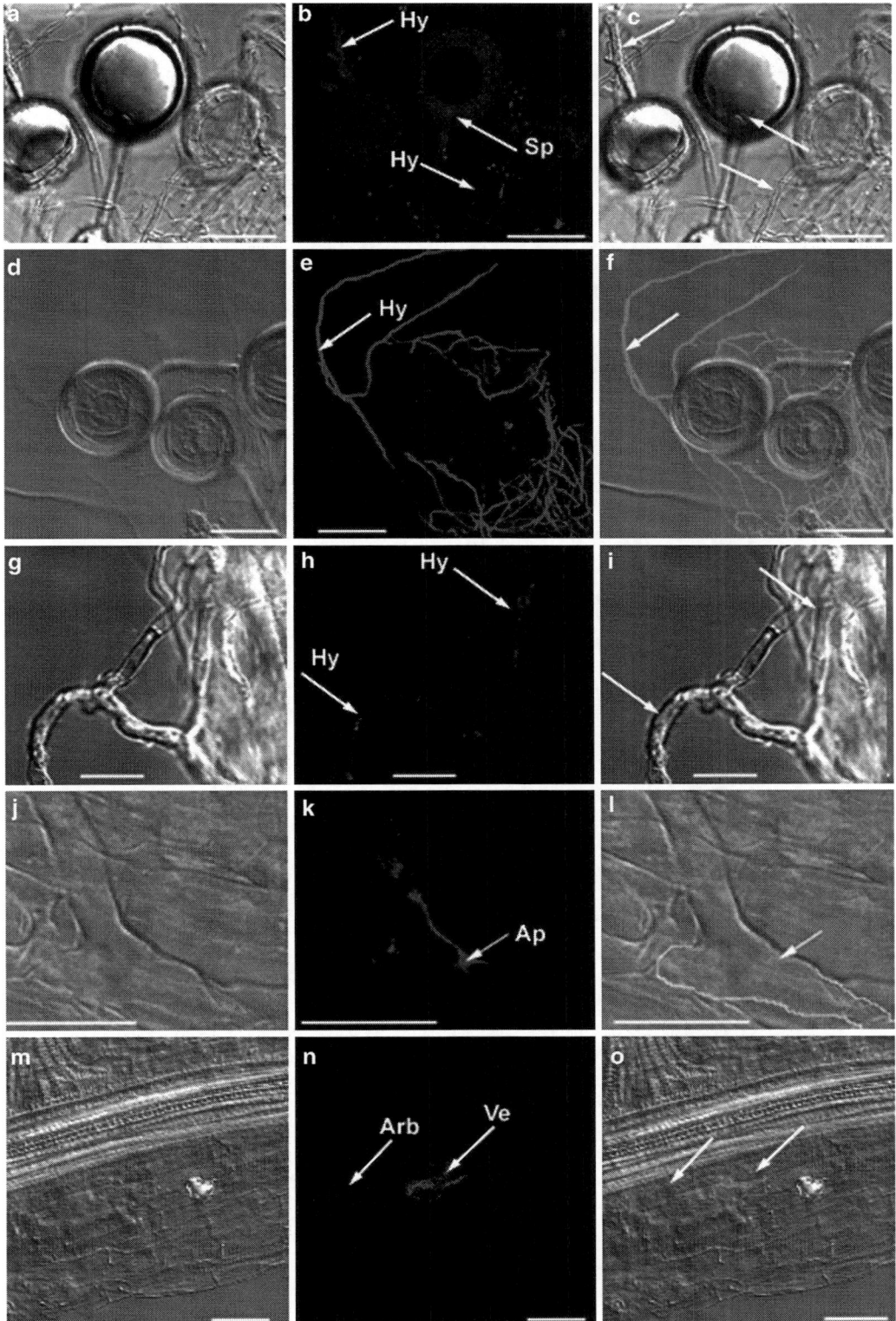

Fig. 2. Fungal gene expression in germinating spores and fungal structures associated with *M. truncatula* roots. Germinating spores of *G. intraradices* 7 DAI in 225 nM apigenin (**a–f**) or *G. intraradices*-inoculated *M. truncatula* roots 5 DAI (**g–l**) or 14 DAI (**m–o**) were processed for DIFIS RT-PCR. *SOD* (**b** and **c**), *PEPISOM* (**e** and **f**), and *DESAT* (**h, i, k, l, n,** and **o**)

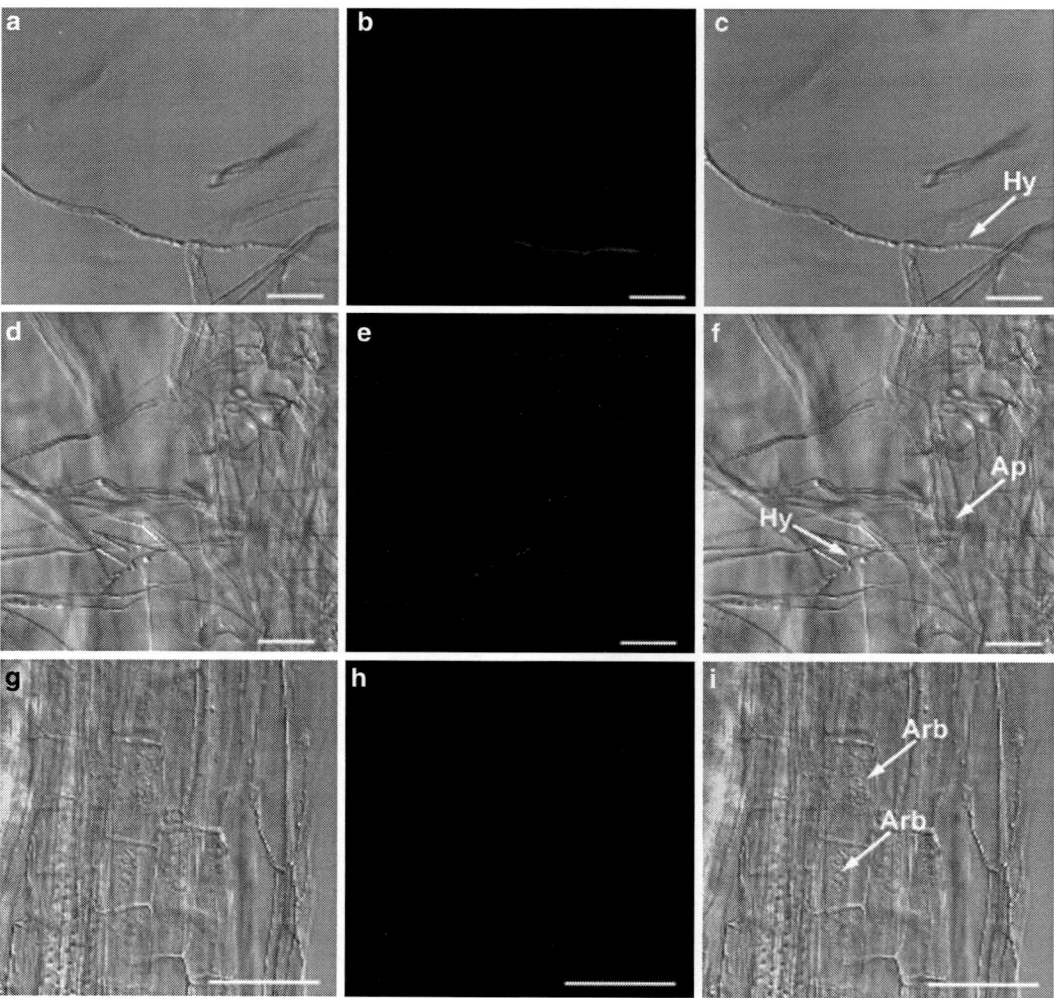

Fig. 3. Fungal gene expression in germinating spores and fungal structures associated with *M. truncatula* roots. To test the ubiquity of this methodology, we declined DIFIS RT-PCR using another host plant (*P. sativum*). *G. intraradices*-inoculated *sym40 P. sativum* roots were proceeded for DIFIS RT-PCR 21 DAI. *PEPISOM* (**b, c, e**, and **f**), and *SOD* (**h** and **i**) gene expression were monitored. Transcripts accumulated in fungal structures, whereas no signal was detected in plant tissues. Nomarski pictures (**a, d**, and **g**) were taken in parallel to localize gene expression. Bar = 50 nm. *Hy* hyphae, *Ap* appressoria, *Arb* arbucule. Fungal gene expression of *PEPISOM* has been detected in fungal structures: extraradical hyphae (**a–c**) and appressoria (**d–f**). *SOD* expression was also detected in extraradicular hyphae, but, As described in *M. truncatula* (34), this gene could not be detected in arbuscules (**g–i**), whereas arbuscules were active because activity of ribosomal Large Subunit could be detected (non illustrated).

Fig. 4. (continued) structures when *G. intraradices* was inoculated to wild-type *M. truncatula* J5 roots (compatible interaction, Fig. 4., ref. (34)). *DESAT* expression was compared in extraradicular hyphae of *G. intraradices* in compatible (wild-type, **a–c**) and incompatible (mutants *dmi1*, **d–f**; *dmi2/Mtsym2*, **g–i**; *dmi3/Mtsym13*, **j–l**) interactions with roots of *M. truncatula*. These DIFIS RT-PCR results corroborate with those obtained by *in vitro* RT-PCR (see chapter 10 in this book) (using RNA extracted from inoculated tissues (**m**) and underline the usefulness of the methodology to compare fungal gene expression under different biological conditions.

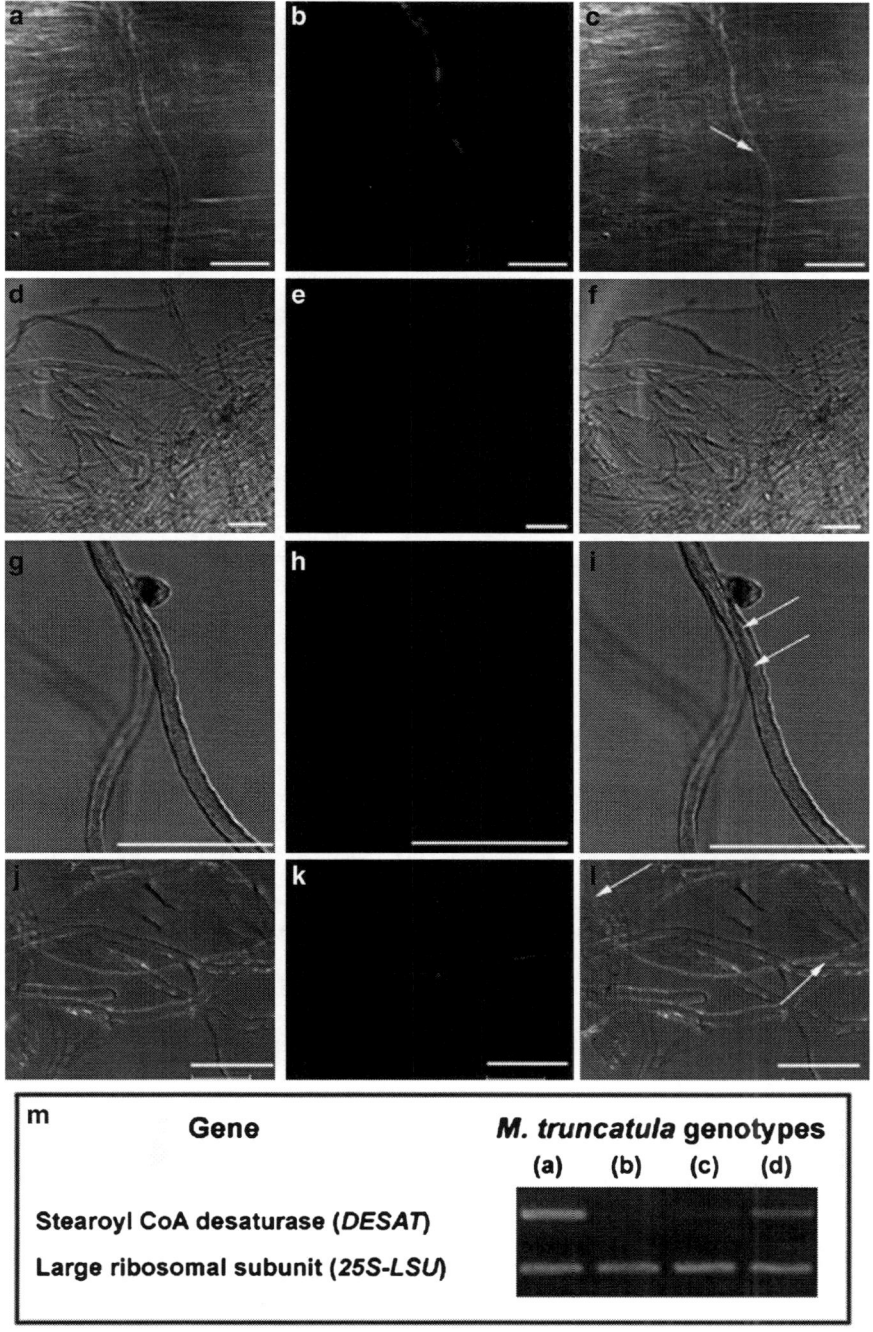

Fig. 4. Influence of plant genotype on fungal *DESAT* gene expression. *G. intraradices*-inoculated wild-type 5 DAI (**a**–**c**) or 7 DAI for *dmi1* (**d**–**f**), *dmi2/Mtsym2* (**g**–**i**) and *dmi3/Mtsym13* (**j**–**l**) *M. truncatula* roots were processed for DIFIS RT-PCR. Fungal *DESAT* gene expression was monitored. Level of transcript accumulation in fungal structures depended on the *M. truncatula* genotype, whereas no signal was detected in plant tissues. Nomarski pictures (**a**, **d**, **g**, and **j**) were taken in parallel to localize gene expression. (**m**) Gene expression monitored by *in vitro* RT-PCR using total RNA extracted from *G. intraradices*-inoculated roots of wild-type (**a**), *dmi1* (**b**), *dmi2/Mtsym2* (**c**), and *dmi3/Mtsym13* (**d**) genotypes of *M. truncatula*. Bar = 50 nm. Arrows indicate labeled hyphae. *DESAT* transcripts were clearly strongly detected in all fungal

Valérie Montfort and Annie Colombet and for producing inoculum of *G. intraradices* BEG 141, and Philippe Aubert for technical assistance in culture room monitoring.

References

1. Schüßler A, Schwarzott D, Walker C (2001) A new fungal phylum, the Glomeromycota: phylogeny and evolution. Mycol Res 105:1413–1421
2. Gianinazzi-Pearson V (1996) Plant cell responses to arbuscular mycorrhizal fungi: getting to the roots of the symbiosis. Plant Cell 8:1871–1883
3. Forbes PJ, Millam S, Hooker JE, Harrier LA (1998) Transformation of the arbuscular mycorrhizal fungus *Gigaspora rosea* by particle bombardment. Mycol Res 102:497–501
4. Helber N, Requena N (2008) Expression of the fluorescence markers DsRed and GFP fused to a nuclear localization signal in the arbuscular mycorrhizal fungus *Glomus intraradices*. New Phytol 177:537–548
5. Harrison MJ, van Buuren ML (1995) A phosphate transporter from the mycorrhizal fungus *Glomus versiforme*. Nature 378:626–629
6. Kaldorf M, Schmelzer E, Bothe H (1998) Expression of maize and fungal nitrate reductase genes in arbuscular mycorrhiza. Mol Plant Microbe Interact 11:439–448
7. Lanfranco L, Vallino M, Bonfante P (1999) Expression of chitin synthase genes in the arbuscular mycorrhizal fungus *Gigaspora margarita*. New Phytol 142:347–354
8. Ferrol N, Barea JM, Azcon-Aguilar C (2000) The plasma membrane H^+-ATPase gene family in the arbuscular mycorrhizal fungus *Glomus mosseae*. Curr Genet 37:112–118
9. Requena N, Breuninger M, Franken P, Ocon A (2003) Symbiotic status, phosphate, and sucrose regulate the expression of two plasmamembrane H^+-ATPase genes from the mycorrhizal fungus *Glomus mosseae*. Plant Physiol 132:1540–1549
10. Benedetto A, Magurno F, Bonfante P, Lanfranco L (2005) Expression profiles of a phosphate transporter gene (*GmosPT*) from the endomycorrhizal fungus *Glomus mosseae*. Mycorrhiza 15:620–627
11. Govindarajulu M, Pfeffer PE, Jin H, Abubaker J, Douds DD, Allen JW, Bücking H, Lammers PJ, Shachar-Hill Y (2005) Nitrogen transfer in the arbuscular mycorrhizal symbiosis. Nature 435:819–823
12. Bonfante P, Bergero R, Uribe X, Romera C, Rigau J, Puigdomenech P (1996) Transcriptional activation of a maize alpha-tubulin gene in mycorrhizal maize and transgenic tobacco plants. Plant J 9:737–743
13. Harrison MJ (1999) Molecular and cellular aspects of the arbuscular mycorrhizal symbiosis. Annu Rev Plant Physiol Plant Mol Biol 50:361–389
14. Gianinazzi-Pearson V, Arnould C, Oufattole M, Arango M, Gianinazzi S (2000) Differential activation of H^+-ATPase genes by an arbuscular mycorrhizal fungus in root cells of transgenic tobacco. Planta 5:609–613
15. Hildebrandt U, Schmelzer E, Bothe H (2002) Expression of nitrate transporter genes in tomato colonized by an arbuscular mycorrhizal fungus. Physiol Plant 115:125–136
16. Krajinski F, Hause B, Gianinazzi-Pearson V, Franken P (2002) *Mtha1*, a plasma membrane H^+-ATPase gene from *Medicago truncatula*, shows arbuscule-specific induced expression in mycorrhizal tissue. Plant Biol 4:754–761
17. Glassop D, Smith SE, Smith FW (2005) Cereal phosphate transporters associated with the mycorrhizal pathway of phosphate uptake into roots. Planta 222:688–698
18. Schaarschmidt S, Roitsch T, Hause B (2006) Arbuscular mycorrhiza induces gene expression of the apoplastic invertase LIN6 in tomato (*Lycopersicon esculentum*) roots. J Exp Bot 57:4015–4023
19. Massoumou M, van Tuinen D, Chatagnier O, Arnould C, Brechenmacher L, Sanchez L, Selim S, Gianinazzi S, Gianinazzi-Pearson V (2007) *Medicago truncatula* gene responses specific to arbuscular mycorrhiza interactions with different species and genera of Glomeromycota. Mycorrhiza 17:223–234
20. Strittmatter G, Gheysen G, Gianinazzi-Pearson V, Hahn K, Niebel A, Rohde W, Tacke E (1996) Infections with various types of organisms stimulate transcription from a short promoter fragment of the potato *gst1* gene. Mol Plant Microbe Interact 9:68–73
21. Hohnjec N, Perlick AM, Puhler A, Kuster H (2003) The *Medicago truncatula* sucrose synthase gene *MtSucS1* is activated both in the

infected region of root nodules and in the cortex of roots colonized by arbuscular mycorrhizal fungi. Mol Plant Microbe Interact 16:903–915

22. Elfstrand M, Feddermann N, Ineichen K, Nagaraj VJ, Wiemken A, Boller T, Salzer P (2005) Ectopic expression of the mycorrhiza-specific chitinase gene *Mtchit 3-3* in *Medicago truncatula* root–organ cultures stimulates spore germination of glomalean fungi. New Phytol 167:557–570

23. Fehlberg V, Vieweg MF, Dohmann EMN, Hohnjec N, Puhler A, Perlick AM, Kuster H (2005) The promoter of the leghaemoglobin gene *VfLb29*: functional analysis and identification of modules necessary for its activation in the infected cells of root nodules and in the arbuscule-containing cells of mycorrhizal roots. J Exp Bot 56:799–806

24. Nagy F, Karandashov V, Chague W, Kalinkevich K, Tamasloukht M, Xu GH, Jakobsen I, Levy AA, Amrhein N, Bucher M (2005) The characterization of novel mycorrhiza-specific phosphate transporters from *Lycopersicon esculentum* and *Solanum tuberosum* uncovers functional redundancy in symbiotic phosphate transport in solanaceous species. Plant J 42:236–250

25. Li HY, Yang GD, Shu HR, Yang YT, Ye BX, Nishida I, Zheng CC (2006) Colonization by the arbuscular mycorrhizal fungus *Glomus versiforme* induces a defense response against the root-knot nematode *Meloidogyne incognita* in the grapevine (*Vitis amurensis* Rupr.), which includes transcriptional activation of the class III chitinase gene *VCH3*. Plant Cell Physiol 47:154–163

26. Jentschel K, Thiel D, Rehn F, Ludwig-Mueller J (2007) Arbuscular mycorrhiza enhances auxin levels and alters auxin biosynthesis in *Tropaeolum majus* during early stages of colonization. Physiol Plant 129:320–333

27. Gianinazzi-Pearson V, Gollotte A, Lherminier J, Tisserant B, Franken P, Dumas-Gaudot E, Lemoine MC, van Tuinen D, Gianinazzi S (1995) Cellular and molecular approaches in the characterization of symbiotic events in functional arbuscular mycorrhizal associations. Can J Bot 73:S526–S532

28. Cordier C, Gianinazzi-Pearson V, Gianinazzi S (1996) An immunological approach for the study of spatial relationships between arbuscular mycorrhizal fungi in planta. In: Azcon-Aguilar C, Barea JM (eds) Integrated Systems: from Genes to Plant Development. European Commission, EUR 16728, Luxembourg, pp 25–30

29. Purin S, Rillig MC (2008) Immunocytolocalisation of glomalin in the mycelium of the arbuscular mycorrhizal fungus *Glomus intraradices*. Soil Biol Biochem 40:1000–1003

30. Balestrini R, Gómez-Ariza J, Lanfranco L, Bonfante P (2007) Laser microdissection reveals that transcripts for five plant and one fungal phosphate transporter genes are contemporaneously present in arbusculated cells. Mol Plant Microbe Interact 20:1055–1062

31. Gomez SK, Javot H, Deewatthanawong P, Torres-Jerez I, Tang Y, Blancaflor EB, Udvardi MK, Harrison MJ (2009) *Medicago truncatula* and *Glomus intraradices* gene expression in cortical cells harboring arbuscules in the arbuscular mycorrhizal symbiosis. BMC Plant Biol 9:10. doi:10.1186/1471-2229-9-10

32. Lammers PJ, Tuskan GA, DiFasio SP, Podila GK, Martin F (2004) Mycorrhizal symbionts of *Populus* to be sequenced by the United States Department of Energy's Joint Genome Institute. Mycorrhiza 12:67–74

33. Martin F, Gianinazzi-Pearson V, Hijri M, Lammers P, Requena N, Sanders IR, Shachar-Hill Y, Shapiro H, Tuskan GA, Young JPW (2008) The long hard road to a completed *Glomus intraradices* genome. New Phytol 180:747–750

34. Seddas PMA, Arnould C, Tollot M, Arias CM, Gianinazzi-Pearson V (2008) Spatial monitoring of gene activity in extraradical and intraradical developmental stages of arbuscular mycorrhizal fungi by direct fluorescent *in situ* RT-PCR. Fungal Gen Genet 45:1155–1165

35. Haase A, Retzel EF, Stakus KA (1990) Amplification and detection of lentiviral DNA inside cells. Proc Natl Acad Sci U S A 87:4971–4975

36. Pesquet E, Barbier O, Ranocha P, Jauneau A, Goffner D (2004) Multiple gene detection by *in situ* RT-PCR in isolated plant cells and tissues. Plant J 39:947–959

37. van Aarle IM, Viennois G, Amenc LK, Tatry MV, Luu DT, Plassard C (2007) Fluorescent *in situ* RT-PCR to localise gene expression in the ectomycorrhizal fungus *Hebeloma cylindrosporum*. Mycorrhiza 17:487–494

38. Kuster H, Vieweg MF, Manthey K, Baier MC, Hohnjec N, Perlick AM (2007) Identification and expression regulation of symbiotically activated legume genes. Phytochemistry 68:8–18

39. Morandi D, Prado E, Sagan M, Duc G (2005) Characterisation of new symbiotic *Medicago truncatula* (Gaertn.) mutants, and phenotypic or genotypic complementary informa-

tion on previously described mutants. Mycorrhiza 15:283–289

40. Jacobi LM, Petrova OS, Tsyganov VE, Borisov AY, Tikhonovich IA (2003) Effect of mutations in the pea genes *Sym33* and *Sym40* II. Dynamics of arbuscule development and turnover. Mycorrhiza 13:9–16

41. Hewitt EJ (1952) Sand water culture methods used in the study of plant nutrition. Commonwealth Agricultural Bureau Technical communication No. 22

42. Journet EP, van Tuinen D, Gouzy J, Crespeau H, Carreau V, Farmer MJ, Niebel A, Schiex T, Jaillon O, Chatagnier O, Godiard L, Micheli F, Kahn D, Gianinazzi-Pearson V, Gamas P (2002) Exploring root symbiotic programs in the model legume *Medicago truncatula* using EST analysis. Nucleic Acids Res 30:5579–5592

43. Seddas PMA, Arias C, Arnould C, van Tuinen D, Godfroy O, Aït Benhassou H, Gouzy J, Morandi D, Dessaint F, Gianinazzi-Pearson V (2009) Symbiosis-related plant genes modulate molecular responses in an arbuscular mycorrhizal fungus during early root interactions. Mol Plant Microbe Interact 22:341–351

Chapter 11

Application of Laser Microdissection to Study Plant–Fungal Pathogen Interactions

John Fosu-Nyarko, Michael G. K. Jones, and Zhaohui Wang

Abstract

Laser microdissection (LM) has become an important tool for isolating individual cells or cell types from suitably prepared tissue samples. The technique can be used to isolate both fungal and host plant cells after pathogen infection for molecular studies. Sample preparation is a crucial step in LM and involves fixing samples with appropriate fixatives to preserve the integrity of cell morphology and target metabolites (e.g., RNA), and embedding the fixed tissue in paraffin wax for sectioning onto microscope slides. The sample sections are then deparaffinised, rehydrated, and cells are dissected by a laser focused through a microscope. LM samples are collected into protective (e.g., RNAse-free) medium for isolation of RNA. The RNA can then be subjected to gene expression studies such as quantitative RT-PCR and microarray analysis after a linear RNA amplification process.

Key words: Laser microdissection, Fixation, Paraffin embedding, Fungal hyphae, RNA amplification

1. Introduction

Laser microdissection (LM) is currently the most accurate of several techniques developed to harvest single cell contents or groups of cells from plant or animal tissues without contamination from surrounding cells (1). Generally, LM systems are combined with microscopy to accurately delineate cells (or contents of cells) or tissues of interest which are dissected using the laser as a cutting device. The samples are collected by contact or non-contact transfer. Three main types of LM systems are available: they differ in the type of laser used for dissection and the mode of collection of dissected samples. The first is the PALM MicroBeam system (P.A.L.M. Microlaser Technologies AG, Bernried, Germany, www.palm-mikrolaser.com), which uses a pulsed UV-A laser of 337 nm to microdissect samples of interest. The pressure of a focused laser beam is then used to transfer dissected samples upwards into a cap placed above the section for non-contact sample collection (2, 3). The second system, the AS LMD laser microscope system (Leica Microsystems AG, Wetzlar, Germany, www.leica-microsystems.com), uses the same type of laser as the PALM MicroBeam system except that the microdissected samples fall into a cap placed beneath the mounting stage of a microscope (4). The third is the PixCell system (Arcturus Bioscience Inc., Mountain View, CA, USA, www.moleculardevices.com/pages/instruments/arcturusXT.html), which uses a near infrared laser to capture samples of interest and fuses them to a thermoplastic film (5).

LM has been applied successfully to study several aspects of plant biology including biochemistry, physiology, and responses to pathogen infection, such as in analysing gene expression in giant cells induced by the plant parasitic nematode *Meloidogyne javanica* on tomato roots (6, 7). LM has proved to be equally successful in studying host-fungal interactions: for example epidermal cells of leaves of Arabidopsis infected with powdery mildew (*Erysiphe cichoracearum*) have been isolated for gene expression studies (8). Fungal structures such as haustoria can also be separated accurately from infected host plant cells (9). This new application allows the biochemistry and structure of hyphal cells to be studied, and the interaction with host tissues.

This chapter describes how LM can be applied to obtain pure populations of host plant and fungal cells using the PALM system as an example. Isolation of RNA from LM samples and amplification of the RNA are also described for downstream molecular analysis such as gene expression studies in plant-fungal pathogen interactions.

2. Materials

2.1. Preparation of Plant Tissue

1. Farmer's fixative (3:1 (v/v) ethanol: acetic acid) prepare fresh at 4°C.
2. Graded series of ethanol ((v/v) 75, 85, 90, and 100%) for dehydration, keep at 4°C.
3. Ethanol: xylene series ((v/v) 3:1; 1:1; 1:3) for dehydration, keep at 4°C.
4. RNAseAway (Ambion Inc., Austin, TX, USA).
5. Diethylpyrocarbonate (DEPC)-treated water containing 0.1% DEPC, incubate at 37°C overnight and autoclave, store at room temperature.
6. Vacuum chamber (12×20 cm) and Vacuum pump, Dynavac Victoria (BetterVac Pty Ltd, Melbourne, Vic., Australia).

2.2. Sectioning of Fixed and Paraffin Wax-Embedded Plant Tissue

1. Glass microscope slides coated with polyethylene-naphthalate (PEN) (PALM, Bernried, Germany).
2. Liquid Paraplast embedding material (Tyco Healthcare UK Ltd, Gosport, United Kingdom) at 58°C.
3. Microtome (RM2235, Leica Microsystems, Nussioch, Germany).
4. Water bath filled with DEPC-treated water for tissue sections collection at 42°C.
5. Laminar Flow Work Station with UV of 254 nm wavelength (Gelman Sciences Inc, Michigan, USA).

2.3. Target Visualization and Laser Microdissection

1. 100% Xylene (Sigma-Aldrich).
2. Rehydration solutions (100, 90, and 70% ethanol), store at 4°C.
3. LM apparatus – PALM Robot-CombiSystem (PALM).
4. DEPC-treated water.
5. RNasin RNase inhibitor (Promega Corp, Sydney, Australia).
6. Collection cap of a 0.5 mL microcentrifuge tube.

2.4. RNA Isolation

1. Microcentrifuge.
2. Picopure RNA Isolation kit (Arcturus, Mountain View, CA, USA), store at room temperature.
3. RNAse-Free DNase Set (Qiagen Pty. Ltd, Doncaster, Australia) store at −20°C.

2.5. RNA Amplification

1. Superscript RNA Amplification System (Invitrogen Corporation, Carlsbad, CA, USA), references to reagents and solutions in the methods are as indicated in the kit. Store reagents at −20°C and solutions at room temperature.
2. Microcentrifige.
3. Savant Speed-vac (Selby Scientific and Medical, Perth, Australia).

3. Methods

LM of plant cells and fungal hyphae from infected tissues requires appropriate and effective sample preparation process to ensure preservation of cell morphologies for target cell visualization whilst enabling recovery of RNA with good quality for gene expression studies. The preparation process involves fixing infected tissues, embedding in paraffin wax and sectioning onto treated microscope slides. Before LM, sections from plant tissues are deparaffinised in xylene, rehydrated in a graded series of ethanol and air-dried. Targets are then accurately identified using the computer component of the LM facility, laser dissected and transferred into caps of collection tubes filled with protective medium. RNA can then be isolated from these samples and subjected to gene expression analysis.

3.1. Preparation of Plant Tissue

1. Trim infected plant tissues containing cells associated with hyphae from the source plant.
2. Transfer immediately into at least ten times the volume of freshly prepared Farmer's fixative for 4–6 h in a vial kept on a slow rotator at 4°C (see Note 1).
3. Remove the vial with samples from the rotator and place on ice in a small tub.
4. Place the tub containing the vial in a vacuum chamber and subject samples to 15 min of vacuum (400 mmHg) to assist infiltration (see Note 2).
5. Remove the tub and place the vial on the slow rotator at 4°C as before for 1 h.
6. Repeat the cycle of rotating and vacuum infiltration as in steps 3–5 for 4–6 h.
7. Dehydrate the fixed tissue at 4°C in a graded series of ethanol ((v/v) 75, 85, 90, 100, and 100%) for 15 min each time on the rotator.
8. Continue dehydration in a graded series of ethanol: xylene ((v/v) 3:1, 1:1, 1:3) followed by 100% xylene for 1.5 h (3× 30 min in each solution) (see Note 3).
9. Remove excess xylene and add liquid Paraplast embedding material to sample vials.
10. Replace liquid Paraplast at 10–15 min intervals for a 4–6 h embedding period.
11. Treat metal cases for embedding with RNAseAway and wash with DEPC-treated water.
12. To embed samples, align them in a metal case and pour liquid Paraplast until it covers all samples. Allow to harden on ice.
13. Allow embedded samples to set overnight at 4°C (see Note 4).

3.2. Sectioning of Fixed and Paraffin Wax-Embedded Plant Tissue

1. Treat all metals surfaces including blades for cutting with RNAseAway and DEPC-treated water.
2. Prepare PEN coated microscope slides: wash with RNaseAway, wash liberally with DEPC-treated water. Further sterilize PEN slides by placing the coated side up in a slide box and incubate in a laminar flow under 254 nm UV for 20 min.
3. Section embedded tissues with manual or rotary microtome into 10 μm ribbons (see Note 5).

4. Leave ribbons to float in a DEPC-treated water bath at 42°C to stretch while obtaining enough sections. Ribbons would normally fully stretch 1–2 min after contact with water. They could be removed when ready.

5. Mount sectioned ribbons on the PEN slides, leave on a warming plate at 42°C or air dry for about 10 min or until completely dry.

6. Store mounted slides in a slide box with a desiccant kept at 4°C. They should be laser dissected within 24 h.

3.3. Target Visualization and Laser Microdissection

1. Deparaffinise embedded sections in three changes of xylene for 3 min each time followed by rehydration with 100, 90, and 70% ethanol each for 2 min and then twice in DEPC-treated water and allow to air-dry.

2. Using the computer monitor, visualise and select cells of interest (see Note 6).

3. Use the laser (settings appropriate for the tissue type) to cut individual cells or tissues leaving a small bridging fragment, so that the dissected tissue is not blown away before catapulting.

4. To collect dissected samples, carefully place a flat cap of a 0.5 mL microcentrifuge tube containing 40 µL of DEPC-treated water with 40 units (U) of RNasin RNase inhibitor (see Note 7) upside down in the cap holder of the robotic hand of the LM facility making sure it is directly above, but not in contact with, the sample on the microscope slide.

5. Adjust the collection cap to come as close as possible to the glass slide. Increase the laser energy settings to catapult the dissected samples upwards into the cap.

6. The LM samples can be subjected to RNA isolation immediately. Otherwise, put the collection cap back onto the microcentrifuge tube, transfer into liquid nitrogen and store at −80°C until processed.

3.4. RNA Isolation

A combination of PicoPure RNA isolation kit and Qiagen RNAse-Free DNAse Set can be used to isolate RNA from LM samples (see Note 8).

1. Pipette 50 µL of Extraction Buffer (XB) from the PicoPure RNA isolation kit into a 0.5 mL microcentrifuge tube, and insert the sample collection cap onto the tube.

2. Invert the tube to ensure that the extraction buffer covers the samples in the cap. Incubate the samples in the extraction buffer for 30 min at 42°C.

3. Centrifuge the tube at $800 \times g$ for 2 min to collect all the extracts into the centrifuge tube.

4. Precondition the RNA purification column: pipette 250 µL of Conditioning Buffer from the PicoPure RNA isolation kit onto the purification column filter membrane and leave at room temperature for 5 min. Centrifuge the column inserted in a collection tube at $16,000 \times g$ for 1 min.

5. Add 50 µL of 70% ethanol into the LM sample extracts and mix well by pipetting up and down. Transfer the sample extract-ethanol mixture into the preconditioned purification column.

6. Bind the RNA to the column by performing a slow centrifugation step at $100 \times g$ for 2 min followed by a faster step at $16,000 \times g$ for 30 s to remove all flow-through.

7. Add 100 µL of Wash Buffer 1 from the PicoPure RNA isolation kit to the purification column and wash by centrifuging at $8,000 \times g$ for 1 min.

8. Add 5 µL of DNAse I Stock Solution to 35 µL of Buffer RDD from the Qiagen RNAse-Free DNAse Set and gently mix by inversion.

9. Pipette the DNAse I mixture onto the purification column and incubate at room temperature for 15 min.

10. Add 40 µL of Wash Buffer 1 from the PicoPure RNA isolation kit to the purification column and centrifuge at $8,000 \times g$ for 15 s.

11. Perform another wash with 100 µL of Wash Buffer 2 from the PicoPure RNA isolation kit to the purification column and centrifuge at $16,000 \times g$ for 2 min.

12. Remove residual wash buffer from the purification column by centrifuging at $16,000 \times g$ for 1 min.

13. To elute RNA, transfer the purification column to a 0.5 mL microcentrifuge tube and add 20–30 µL of Elution Buffer from the PicoPure RNA isolation kit directly to the membrane. Allow the column to stand at room temperature for 1 min.

14. Centrifuge the column at two speeds: first at $1,000 \times g$ for 1 min to distribute the elution buffer evenly in the column, and then at $16,000 \times g$ for 1 min to elute RNA. Use RNA immediately or store at −80°C.

3.5. RNA Amplification

Normally, the amount of RNA recovered from LM samples is very small. This section describes a T7 RNA polymerase-based linear RNA amplification method using the Superscript RNA Amplification System (Invitrogen), which involves first and second cDNA strand synthesis from the LM RNA sample, cDNA purification followed by *in vitro* transcription and purification of the amplified RNA (aRNA).

3.5.1. cDNA Synthesis

1. Mix 9 µL of LM RNA with 1 µL of T7-Oligo(dT) primer in a 0.2 mL RNase-free tube, incubate the tube at 70°C for 10 min, and then place on ice for at least 1 min (see Note 9).
2. Centrifuge the tube briefly to collect the contents, and add the following reagents at room temperature: 4 µL of 5× First-Strand buffer, 2 µL of 0.1 M DTT, 1 µL of 10 mM dNTP Mix, 40 U of RNaseOUT and 200 U of SuperScript III reverse transcriptase to make a total reaction volume of 20 µL.
3. Mix the contents gently, centrifuge the tube briefly and incubate at 46°C for 2 h and then at 70°C for 10 min to inactivate the reverse transcriptase.
4. Add the following components to the reaction tube on ice: 91 µL of DEPC-treated water, 30 µL of 5× Second-Strand Buffer, 3 µL of 10 mM dNTP Mix, 4 µL of DNA Polymerase I, 1 µL of DNA Ligase, 1 µL of RNaseH, to make a total volume of 150 µL.
5. Mix the contents gently by pipetting up and down and incubate at 16°C for 2 h.
6. After incubation, place the tube containing double-stranded cDNA on ice or store at −20°C until the cDNA is purified in the following steps.

3.5.2. cDNA Purification

1. Add 500 µL of cDNA Loading Buffer to the reaction tube to make a total volume of 650 µL and mix thoroughly by pipetting up and down.
2. Load the cDNA/buffer solution directly onto the Spin Cartridge pre-inserted into a collection tube. Centrifuge at 12,000 × g at room temperature in a microcentrifuge for 1 min. Remove the collection tube and discard the flow-through.
3. Place the Spin Cartridge in the same collection tube and add 700 µL of cDNA Wash Buffer to the column. Centrifuge at 12,000 × g at room temperature for 2 min. Remove the collection tube and discard the flow-through.
4. Place the Spin Cartridge in the same collection tube and centrifuge at 12,000 × g at room temperature for an additional 4 min. Remove the collection tube and discard the flow-through.
5. Place the Spin Cartridge into a recovery tube. Add 100 µL of DEPC-treated water to the center of the Spin Cartridge and incubate at room temperature for 2 min.
6. Centrifuge at 12,000 × g at room temperature for 1 min to collect the purified cDNA.

7. Place the eluted cDNA in a speed-vac, evaporate at low to medium heat until the sample volume is reduced to ≤20 μL and proceed to *in vitro* transcription.

3.5.3. In Vitro Transcription

1. Add DEPC-treated water to the purified cDNA to bring the total volume to 23 μL.
2. Add the following to the tube at room temperature: 1.5 μL 100 mM each of ATP, CTP, GTP, and UTP, 4 μL of 10× T7 Reaction Buffer and 7 μL of T7 Enzyme. Mix to make a total reaction volume of 40 μL.
3. Gently mix and centrifuge briefly to collect the contents of the tube. Incubate at 37°C for 6–16 h (see Note 10).
4. Add 2 μL of DNaseI to the tube. Gently mix and centrifuge briefly to collect the contents of the tube, then incubate at 37°C for 30 min (see Note 11).

3.5.4. aRNA Purification

1. Add 160 μL of aRNA Binding Buffer to the reaction tube to make a total volume of 200 μL. Mix thoroughly by pipetting up and down.
2. Add 100 μL of 100% ethanol and mix thoroughly by pipetting up and down.
3. Load the entire aRNA/buffer solution directly onto the Spin Cartridge pre-inserted into a collection tube. Centrifuge at 12,000 × g in a microcentrifuge for 15 s at room temperature. Remove the collection tube and discard the flow-through.
4. Place the Spin Cartridge in the same collection tube and add 500 μL of aRNA Wash Buffer to the column. Centrifuge at 12,000 × g for 15 s at room temperature. Remove the collection tube and discard the flow-through. Repeat the washing step once.
5. Place the Spin Cartridge in the same collection tube and centrifuge at full speed for an additional 2 min to dry the column. Remove the collection tube and discard the flow-through.
6. Place the Spin Cartridge into a recovery tube. Add 100 μL of DEPC-treated water to the center of the Spin Cartridge and incubate at room temperature for 1 min. Centrifuge at 12,000 × g for 2 min at room temperature to collect purified aRNA (see Note 12).

4. Notes

1. Fixatives other than Farmer's fixative could be used. Precipitative fixatives (e.g., Farmer's fixative, 100% acetone)

are preferred when higher yields of RNA is expected, whereas cross-linking fixatives (e.g., 4% paraformaldehyde or Methacarn (absolute methanol/chloroform/glacial acetic acid 6:3:1) will preserve the morphology of tissues for accurate identification of target cells but will result in lower recovery of RNA (10). Inorganic-based fixation methods can be used with delicate plant tissues such as leaves of Arabidopsis, where conventional fixation and paraffin sectioning methods do not preserve histological features. One of such is the microwave fixation method, which involves dissecting tissue/samples in 10 mM of Sørensen's phosphate buffer (pH 7.2) and then subjecting the samples in the solution to microwave at 450 W for 15 min three times. Comparatively this method is quicker than conventional fixation methods.

2. Mature plant tissue samples and/or those larger than 2 mm in diameter may need to be fixed longer to preserve the tissue well. Also vacuum infiltration could be done for 20 min instead of 15 min at a time at 400 mmHg on ice or at 4°C at hourly intervals during the fixation coupled with mixing on a slow rotator. Samples could also be left in the fixative for longer (e.g., overnight) to improve fixation.

3. Fixed tissue intended for paraffin embedding must be well dehydrated with a graded series of ethanol and a further dehydration step with organic solvents if necessary. In place of graded solutions of ethanol and xylene, ethanol-isopropanol (1:1) could be used followed by 100% isopropanol. However, if 100% acetone is used for fixation this may not be required because many plant tissues can then be embedded directly.

4. For mature plant sections, paraffin wax embedding could be done more efficiently in a vacuum chamber (305 mmHg) and at temperature specific for the molten Paraplast being used. Alternatively, plant tissues could be processed by cryosectioning of frozen samples. Although frozen tissues used in LM after crysectioning may yield more RNA with better quality, fixed and paraffin embedded tissues always provide much better cell morphologies in tissue sections for accurate target localisation and dissection.

5. Plant cells have often been laser microdissected from paraffin sections 6–10 µm thick. To reduce processing times and facilitate the number of cells that could be laser dissected and collected, the thickness of sections can be increased up to 50 µm. However, in cases where such thick samples could be contaminated with unwanted cells underneath sections as thin as 3 µm could be used.

6. A quick staining of the tissue section with 0.25% toluidine blue at room temperature can help visualization of histological structures.

The use of Green Fluorescent Protein (GFP)-labelled fungi as a cell maker also enhances target cell identification for accurate dissection when fluorescence microscopy and LM are combined (11).

7. It is imperative that microdissected samples should be collected into a protective/RNAase-free medium. Media that could be used include buffers later used for RNA isolation. However, during LM, these RNA extraction buffers in the collection cap evaporate and crystallise quickly because of illumination and heating by the bright field light. Up to 50 μL of DEPC-treated water with RNAse Inhibitors is recommended for long periods of dissection and collection of more samples in the same cap.

8. The resin spin column-based method used in this protocol (Picopure RNA isolation kit) is known to give high yields of RNA from LM samples. However, kits based on the same method can be obtained from other manufactures such as Nanoprep (Stratagene, Cedar Creek, TX, USA) and RNeasy micro kit (Qiagen). In addition, phenol-based TRIzol reagent (Invitrogen) methods could provide RNA of sufficient quantity and quality for molecular analysis.

9. Prior to cDNA synthesis, the quality of the LM RNA samples can be checked using the Agilent Bioanalyzer 2100 and the RNA 6000 LabChip kit (Agilent, Palo Alto, CA, USA) to ensure successful amplification with T7 RNA polymerase and down-stream molecular analysis.

10. Longer incubation at 37°C increases the final yield of the aRNA, and maximum yield is achieved at 16 h.

11. There are other RNA amplification approaches including the PCR-SMART technology (Clontech Laboratories, Inc., Mountain View, CA, USA) and the Ribo-Spia method (NuGEN Technologies, Inc., San Carlos, CA, USA). In PCR amplification, primer sites are integrated into the 3′ and the 5′ ends of the PCR template during the reverse transcription reaction. The Ribo-Spia method is similar to the T7 RNA polymerase-based *in vitro* transcription system used in this protocol except that the amplification step takes place during cDNA synthesis.

12. A single round of RNA amplification may not result in enough yield of aRNA for molecular applications such as microarray studies. The aRNA obtained can be used as a starting material for the next round of amplification. This results in exponential increase in yields.

References

1. Day RC, Grossniklaus U, Macknight RC (2005) Be more specific! laser-assisted microdissection of plant cells. Trends Plant Sci 10:397–405
2. Schütze K, Lahr G (1998) Identification of expressed genes by laser-mediated manipulation of single cells. Nature Biotechnol 16:737–742
3. Westphal G, Burgemeister R, Friedemann G, Wellmann A, Wernert N, Wollscheid V, Becker B, Vogt T, Knüchel R, Stolz W, Schütze K (2002) Noncontact laser catapulting: a basic procedure for functional genomics and proteomics. Methods Enzymol 356:80–99
4. De Souza AI, McGregor E, Dunn MJ, Rose ML (2004) Preparation of human heart for laser microdissection and proteomics. Proteomics 4:578–586
5. Emmert-Buck MR, Bonner RF, Smith PD, Chuaqui RF, Zhuang Z, Goldstein SR, Weiss RA, Liotta LA (1996) Laser capture microdissection. Science 274:998–1001
6. Nelson T, Tausta SL, Gandotra N, Liu T (2006) Laser Microdissection of plant tissue: what you see is what you get. Annu Rev Plant Biol 57:181–201
7. Ramsay K, Jones MGK, Wang Z (2006) Laser capture microdissection: a novel approach to microanalysis of plant and microbe interactions. Mol Plant Pathol 7:429–435
8. Inada N, Wildermuth MC (2005) Novel tissue preparation method and cell-specific marker for laser microdissection of *Arabidopsis* mature leaf. Planta 221:9–16
9. van Driel KGA, Boekhout T, Wösten HAB, Verkleij AJ, Müller WH (2007) Laser microdissection of fungal septa as visualised by scanning electron microscopy. Fungal Genet Biol 44:466–473
10. Jiang K, Zhang S, Lee S, Tsai G, Kim K, Huang H, Chilcott C, Zhu T, Feldman LJ (2006) Transcription profile analyses identify genes and pathways central to root cap functions in maize. Plant Mol Biol 60:343–363
11. Tang W, Coughlan S, Crane E, Beatty M, Durick J (2006) The application of laser microdissection to in planta gene expression profiling of the maize anthracnose stalk rot fungus *Colletotrichum graminicola*. MPMI 19:1240–1250

Chapter 12

Multiplex Gene Expression Analysis by TRAC in Fungal Cultures

Jari J. Rautio

Abstract

For an increasing number of microorganisms of scientific and industrial interest, the genome sequences have become available, which in turn has enabled genome-wide microarray studies. Global level transcriptomic analysis has flooded the research community with gene expression data from diverse biological states. One of the key aspects of this research is that in many cases the analysis of thousands of genes leads to the discovery of significantly smaller sets of genes, from a few to a few hundred, which provide the essential information about biological systems of interest. As a consequence, the requirement for technologies enabling rapid, cost-effective and quantitative detection of specific gene transcripts has increased. A method named TRAC (Transcript analysis with aid of affinity capture) is a novel solution hybridization and bead-based assay enabling multiplex mRNA target detection simultaneously from large sample numbers. Functionality of TRAC has been shown in a number of applications including microbial quantification and gene expression-based monitoring of biotechnical processes as well as cell-based cancer marker gene screening and siRNA validation.

Key words: Gene expression, RNA analysis, Functional genomics, Bioprocess monitoring, Filamentous fungi, Protein production, Gene marker

1. Introduction

Monitoring of microbial cultures and biotechnical processes are the application fields where TRAC has been most widely used to date. These applications range from use of TRAC in evaluation of gene expression stability in chemostat cultures (1, 2), monitoring of culture performance by analyzing marker genes relevant for a given process (3–5), and studying expression of protein folding related genes in strains producing foreign proteins (4, 6).

Microorganisms in biotechnical process encounter constantly changing environmental surroundings, to which they have to adapt by changing their internal physiology. The performance of the production strains has an essential effect on the performance of the process, which has made process monitoring and control strategies based on the physiological status of the culture more popular. Expression analysis of process-relevant marker genes provides a potential way for process physiology monitoring, since physiological events are rapidly detectable at the mRNA level and setting up assays that detect multiple targets in a single assay is simpler for nucleic acids than for metabolites or enzymes.

The TRAC method has been used for monitoring of different types of fungal cultures. In the studies performed, altogether 100–200 genes of filamentous fungus *Trichoderma reesei* and yeast *Saccharomyces cerevisiae* have been identified with presumed relevance to various biological pathways and were subsequently tested in culture conditions (3–5). Several of the evaluated genes have shown potential in prediction of physiological effects and process performance, such as growth and protein production rate, as well as availability of nutrients and oxygen. Furthermore, TRAC has been applied for evaluation of the stability of gene expression during the steady state phases of chemostat cultures used for producing biomass for genome or proteome wide studies (1). These data function as quality assurance prior selection of samples for costly systems levels studies.

For TRAC analysis, (1) the cellular material is first collected and lysed with a buffer that degrades the cells and destroys all enzyme activity; (2) hybridization is directly performed in solution from the resulting cell lysate. Each chosen gene of interest is recognized by a specific fluorophore-labeled probe. A pool of dozens of probes with different lengths or types of label is used. Biotin-oligo-dT captures polyA RNA from the sample; (3) The sample processing is conveniently performed by an automated magnetic bead processor in 96-well plate format. The probe-target hybrids are captured by magnetic beads and unbound material is washed off and finally probes are eluted off from the beads; (4) The probe pools are analyzed by capillary electrophoresis, which resolves the probes according to size and label. Probes are quantitated by fluorescence signal (Fig. 1).

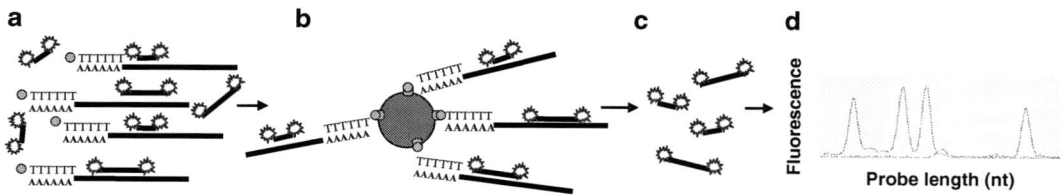

Fig. 1. In TRAC method, the target RNAs are hybridized with a probe pool consisting of differently sized oligonucleotides with double fluorophore labels and a biotinylated oligo(dT) probe (**a**) Hybridized targets are immobilized by affinity capture to streptavidin-coated magnetic beads (**b**) Unbound material is washed off and the probes are eluted from the beads (**c**) The probes are identified and quantified by capillary electrophoresis (**d**).

2. Materials

2.1. Cell Cultures

1. *Trichoderma reesei* Rut-C30 transformant pLLK13/295 producing *Melanocarbus albomyces* laccase (7).
2. Fermentor medium (g/L): lactose 40, peptone 4, yeast extract 1, KH_2PO_4 4, $(NH_4)_2SO_4$ 2.8, $MgSO_4 \cdot 7H_2O$ 0.6, $CaCl_2 \cdot 2H_2O$ 0.8, $CuSO_4 \cdot 5H_2O$ 0.025 and 2 mL/L of 2× trace element solution (8), pH 5.5–6. In fed-batch cultures, feeding of 24% (w/v) lactose solution was used.
3. Foaming was controlled by automatic addition of Struktol J633 polyoleate antifoam agent (Schill & Seilacher, Germany, Hamburg) or polypropylene glycol (mixed molecular weights; (9)).

2.2. Sample Preparation and Lysis

1. Sample filtration: Glass-fibre filter disks (Whatman GF/B 47 mm Ø, Kent, UK).
2. Lysis buffer: 5× SSC (750 mM sodium chloride, 75 mM sodium citrate), 2% (w/v) SDS and 66 U/mL RNA guard RNase inhibitor (Amersham Biosciences, Buckinghamshire, UK).
3. Lysis beads: Acid-washed glass beads, RNase free (Sigma).
4. FastPrep cell homogenizer (ThermoSavant, Dreiech, Germany)

2.3. Transcript Analysis with Aid of Affinity Capture (TRAC)

1. Hybridization buffer: 4 pmol biotinylated oligo(dT) capture probe (Promega), 1 pmol each detection probe (Promega, ThermoFisher Scientific) labeled with 6-carboxy fluorescein (6-FAM) or 2,7′,8′-benzo-5′-fluoro-2′,4,7-trichloro-5-carboxy-fluorescein (NED), 5× SSC, 0.2% SDS, 1× Denhardt solution (0.02% Ficoll, 0.02% polyvinyl pyrrolidone, 0.02% BSA), 3% dextran sulphate (see Note 1 for internal standard). The fluorescence labels are light sensitive, thus avoid unnecessary light exposure of the hybridization buffer.
2. Magnetic beads: 50 µg of streptavidin-coated MyOne DynaBeads (Dynal, Oslo, Norway).
3. Wash buffer: (a) 1× SSC, 0.1% (w/v) SDS. (b) 0.5× SSC, 0.1% (w/v) SDS. (c) 0.1× SSC, 0.1% (w/v) SDS.
4. Elution buffer: HiDi formamide (Applied Biosystems, Foster City, CA) with 1:1,000 dilution of size standard (see Subheading 2.4).

2.4. Capillary Electrophoresis Analysis

1. Runing buffer: Buffer 10× with EDTA (Applied Biosystems).
2. Polymer: 3100 POP-6 Performance optimized polymer (Applied Biosystems), Also POP-4 and POP-7 applicable with TRAC.

3. Capillaries: 3130×l/3100 Genetic Analyzer Capillary Array, 50 cm (Applied Biosystems).

4. Size standard: GeneScan 120LIZ Size Standard (Applied Biosystems).

3. Methods

To avoid any changes in the gene expression profile and RNA degradation the sample withdrawal from the fermentations and the lysis of the cells should be performed as quickly as possible. Quick filtration of the biomass is preferred over centrifugation as a harvesting method. Both untreated and lysed biomass can be stored at −80°C for more than 2 years. Biomass lysed with the lysis buffer is stable in room temperature at least for 4 days. However, this material is not suitable for gene expression assays requiring enzymatic steps in the protocol such as RT-qPCR or microarray.

In multiplex hybridization, shaking during the hybridization step is preferred for enhanced reaction kinetics. For the magnetic bead processing following the hybridization, automated instruments are recommended for optimal reproducibility and to facilitate the assay. The final analysis of the probe intensities is most conveniently performed with commonly available sequencer type capillary electrophoresis instruments.

3.1. Biomass Harvesting

1. Biomass is harvested from cultures for TRAC analysis by withdrawing medium containing 50–150 mg fresh biomass. This gives sufficient amount of biomass for more than 50 TRAC assays.

2. Biomass is most conveniently separated from medium by quick filtration with glass-fibre filter disks.

3. The biomass should be immediately washed with RNAse-free (dimethyl pyrocarbonite (DMPC)-treated) water, after which the biomass can be transferred, e.g., by using a pipet tip, in tarred screw-cap tubes to liquid nitrogen and stored at −80°C. This sampling procedure should not take more than 5 min to minimize any effects to the expression profiles.

3.2. Cell Lysis and Sample Preparation

1. Weight the sample tubes to give the fresh biomass. The amount of lysed biomass used in the hybridization reaction can also be used as basis for signal normalization (see Note 2).

2. Suspend frozen biomass samples (50–200 mg/mL fresh weight) in the lysis buffer and add 500 µL of acid-washed glass beads.

3. Disrupt biomass twice in a cell homogenizer using, e.g., settings 6 m/s for 45 s. Keep the sample tubes on ice between the two treatments in the homogenizer to avoid heating of the sample.

4. Collect the crude cell lysate by puncturing a hole to the bottom of the screw cap tube and placing it on top of 1.5 mL microtube.

5. Spin shortly (2–3 s) the screw cap tube on top of microtube in a centrifuge. To make the tubes fit into the centrifuge, do not use the lid of the rotor during the short spin. Caps of the tubes can also be removed before the spin. Be careful, however, not to spin for longer than a few seconds at a time.

6. Wash the glass beads and the remaining biomass with 200 µL of 5×SSC and repeat step 5 to combine the liquids into the same collection microtube. Throw away the screw cap tube and put the collection tube with the lysate on ice.

7. The lysed biomass can at this point be either frozen at −80°C for later use or used directly in hybridization. The lysate has to be mixed properly before using in TRAC hybridization. Although the resulting lysate may be dense, no centrifugation of the sample is needed prior use in hybridization (see Note 3).

3.3. Probe Design and Pooling

1. Oligonucleotide design is recommended to be performed by using the algorithms presented in Kivioja et al. (10, 11). This algorithm designs the probe sequences according to user defined criteria and arranges them into a minimum number of pools. The quality criteria used in probe selection have typically been the following: melting temperature (T_m) limits 60–70°C, GC% limits 38–62, maximum free energy change in hybridization ΔG_H −15 kcal/mol (12) and minimum target energy change A_c −10 kcal/mol (13). A maximum repeat size of 15 nt and a maximum similarity of 80% were used as probe specificity criteria. T_m values were calculated with the nearest-neighbour method (11) using 10 nM nucleic acid and 750 mM salt concentrations. The minimum length difference between adjacent probes should be at least 2 nt (see Note 4).

2. The custom made and purified oligonucleotide detection probes are mixed into minimal number of pools, according to their migration in capillary electrophoresis and the used label. Probes should have at least one unit size difference in their migration size (size related to ladder Liz120) (see Note 4).

3.4. Hybridization

1. Mix together the hybridization buffer and sample in 100 µL total volume. For one 96-well plate assay, prepare 10 mL of fresh hybridization buffer and add 90 µL into each well. Add

10 μL sample into 100 μL total volume and mix well. The sample can contain 50–300 μg of lysed biomass, 100–300 ng polyA RNA, or 2–4 μg of extracted RNA (see Note 5).

2. Perform the hybridizations in 96-well plates at 60°C for 30–40 min with shaking at 650 rpm. With extracted RNA as sample material, 2 h hybridization is preferred. Hybridizations can be performed also using conventional heat blocks or PCR instruments, but that decreases the hybridization performance.

3.5. Washing and Elution

The steps following hybridization, including affinity capture, washing, and elution are most conveniently performed with automated with a magnetic bead particle processors (such as KingFisher 96, ThermoFisher Scientific, Vantaa, Finland) using 96-well plates. The steps of the sample processing are the following:

1. Affinity capture of hybridized RNA targets to 50 μg of streptavidin-coated beads for 30 min at room temperature (RT). The beads should be washed prior to use by using wash buffer C.
2. Wash beads twice for 1.5 min each in 150 μL of washing buffer A.
3. Wash twice for 1.5 min in 150 μL of washing buffer B.
4. Wash once for 1.5 min in 150 μL of washing buffer C (see Note 6).
5. Elute probes with 10 μL of formamide (Applied Biosystems) for 20 min at 37°C.

3.6. Analysis

1. Analyze the eluates by capillary electrophoresis with an ABI PRISM 3100 Genetic Analyzer (Applied Biosystems) (see Note 7).
2. To calibrate the separation of the detection probes by size, add size standard to each sample (see Subheading 2.4). The identity of the probes is determined by the probe size, which is relative to the size standard, not absolute length of the probe.
3. The peak area is used for quantification of the respective mRNA expression level.
4. The expression levels of the target genes during the cultivations can be presented either as changes relative to total mRNA level (see Subheading 3.7 below), biomass used in the sample, or to house-keeping genes (see Note 2).

3.7. Quantification of PolyA mRNA

One alternative way to normalize expression of specific genes is to determine total polyA mRNA concentration in each sample after lysis step (see Subheading 3.2). The polyA mRNA amount added

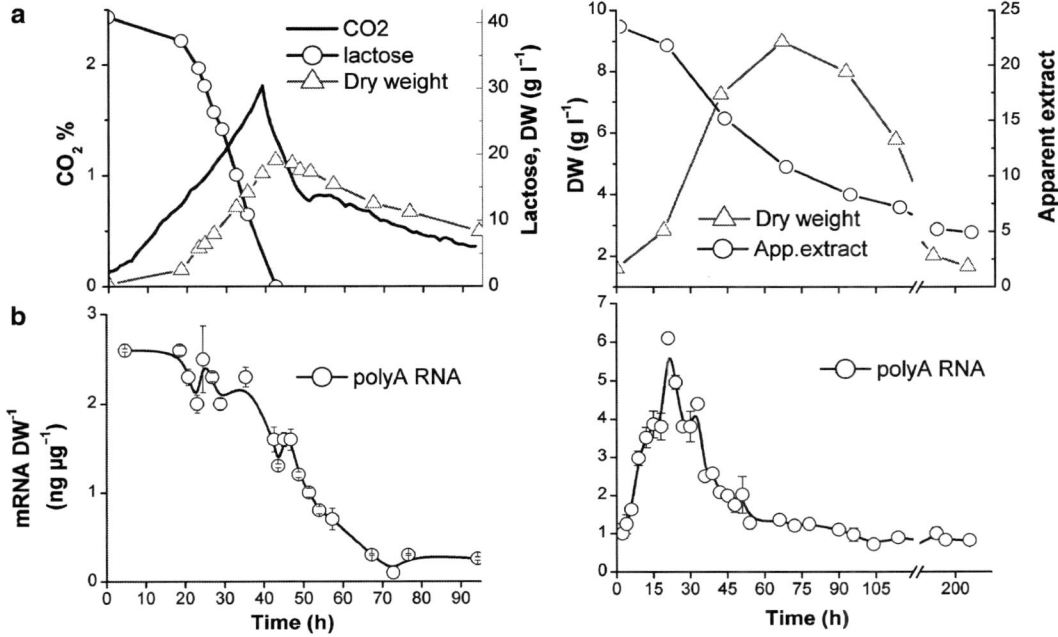

Fig. 2. (a) CO2, lactose and biomass dry weight during *Trichoderma reesei* batch fermentation (*left*) and biomass dry weight and apparent extract (specific gravity of wort) in brewers' yeast (*S. pastorianus*) (*right*) during beer fermentation. (b) PolyA RNA profiles during fermentations.

to each hybridization reaction should be equalized prior hybridization between 100 and 300 ng per reaction.

1. Quantification of total polyA RNA in lysates is performed with the above TRAC protocol without addition of detection probes.

2. The final elution of polyA RNA is performed with 50 µL of DMPC-treated water instead of formamide. RNA concentration in the eluate is quantified with a RiboGreen RNA quantitation kit (Promega). PolyA mRNA content in the biomass varies significantly between growth phases (see Fig. 2).

4. Notes

1. It is recommended to use an internal RNA or single stranded DNA standard in each hybridization reaction with corresponding probe sequence. By adding one or more of these standards to each reaction in equal amount, it is possible to use the corresponding signals to remove nonbiological variation in the assay. Approximately 1.5 fmol of in vitro transcribed

Escherichia coli TraT mRNA was used in many of the TRAC assays performed as the internal standard (see Subheading 3).

2. In most studies performed with fungi, the expression levels of specific genes have been related to the amount of biomass or the total polyA RNA content in the hybridization reactions. Relating marker gene expression to biomass can be more useful in predictions of culture performance such as specific growth and production rates, whereas normalization to overall mRNA expression (polyA RNA) predicts more accurately the physiological responses in metabolically active cells (see Fig. 2).

3. When working, e.g., with mammalian cell cultures, adding the lysis buffer results typically in viscous solution. The viscosity can be removed, e.g., by sonication or by applying shear force.

4. Test of probe pools is necessary prior to their use in the TRAC assay. Although 2 nt differences are used in the design of probes that are pooled together, the signals can anyway overlap in the capillary electrophoresis analysis.

5. Especially with industrial fermentation broths or with other media containing complex carbohydrates, extraction of RNA is recommended instead of direct use of cell lysates. Sample-dependent unspecific binding of probes potentially to steptavidin coated beads have been observed occasionally with such sample material. This unspecific binding can be observed also as high variation in the signal level of the internal standard.

6. To increase the stringency of the washing, wash buffer C can be used in all the washing steps. This decreases the unspecific signals levels but can also decrease the signals from specific targets by more than 30% (14).

7. Besides ABI 3100 (16-channels), ABI 310 (1-channel), also ABI 3130 (48-channels) have been tested and verified to be functional as analyzers for TRAC samples. In the electrophoresis settings, it is important to use maximum voltage (15 kV) to achieve the best possible sensitivity. The injection time should be adjusted for the used polymer and capillary length. Long injection time increases the sensitivity but also decreases the distance between two adjacent probe peaks in the electrophoregram.

8. The expression levels presented as molar amounts can be compared between conditions for a given gene but not between different genes, since the hybridization efficiency between probe-target pairs can vary.

9. Too high biomass concentration can lead to decreased signal levels. With less than $1\ \mu g/\mu L$ (wet weight of lysed mycelium), the hybridization inhibition is negligible.

Acknowledgments

The author would like to thank Professor Hans Söderlund, Kari Kataja, and Dr. Reetta Satokari for innovations and pioneer work related with the TRAC method. Dr. Teemu Kivioja and Dr. Mikko Arvas are acknowledged for the software development for probe design and data analysis. Michael Bailey, Dr. Marilyn Wiebe and Dr. Bart Smit are thanked for design and carrying out the bioreactor cultivations. Professor Merja Penttilä and Dos. Markku Saloheimo are thanked for design and coordination of the studies.

References

1. Rautio JJ, Smit BA, Wiebe M, Penttilä M, Saloheimo M (2006) Transcriptional monitoring of steady state and effects of anaerobic phases in chemostat cultures of the filamentous fungus *Trichoderma reesei*. BMC Genomics 7:247
2. Wiebe MG, Rintala E, Tamminen A, Simolin H, Salusjärvi L, Toivari M, Kokkonen JT, Kiuru J, Ketola RA, Jouhten P, Huuskonen A, Maaheimo H, Ruohonen L, Penttilä M (2008) Central carbon metabolism of *Saccharomyces cerevisiae* in anaerobic, oxygen-limited and fully aerobic steady-state conditions and following a shift to anaerobic conditions. FEMS Yeast Res 8:140–154
3. Rautio JJ, Huuskonen A, Vuokko H, Vidgren V, Londesborough J (2007) Monitoring yeast physiology during very high gravity wort fermentations by frequent analysis of gene expression. Yeast 24:741–760
4. Rautio JJ, Bailey M, Kivioja T, Söderlund H, Penttilä M, Saloheimo M (2007) Physiological evaluation of the filamentous fungus Trichoderma reesei in production processes by marker gene expression analysis. BMC Biotechnol 7:28
5. Rintala E, Wiebe MG, Tamminen A, Ruohonen L, Penttilä M (2008) Transcription of hexose transporters of Saccharomyces cerevisiae is affected by change in oxygen provision. BMC Microbiol 8:53
6. Gasser G, Maurer M, Rautio J, Sauer M, Bhattacharyya A, Saloheimo M, Penttilä M, Mattanovich D (2007) Monitoring of transcriptional regulation in *Pichia pastoris* under protein production conditions. BMC Genomics 8:179
7. Kiiskinen LL, Kruus K, Bailey M, Ylösmaki E, Siika-Aho M, Saloheimo M (2004) Expression of *Melanocarpus albomyces* laccase in *Trichoderma reesei* and characterization of the purified enzyme. Microbiology 150:3065–3074
8. Mandels M, Weber J (1969) The production of cellulases. Adv Chem Ser 95:391–413
9. Wiebe MG, Robson GD, Oliver SG, Trinci AP (1994) Evolution of *Fusarium graminearum* A3/5 grown in a glucose-limited chemostat culture at a slow dilution rate. Microbiology 140:3023–3029
10. Kivioja T (2004) Computational tools for a novel transcriptional profiling method, PhD thesis (http://ethesis.helsinki.fi/julkaisut/mat/tieto/vk/kivioja/). Department of Computer Science. University of Helsinki
11. Kivioja T, Arvas M, Kataja K, Penttilä M, Söderlund H, Ukkonen E (2002) Assigning probes into a small number of pools separable by electrophoresis. Bioinformatics 18:199–206
12. Le Novere N (2001) MELTING, computing the melting temperature of nucleic acid duplex. Bioinformatics 17:1226–1227
13. Luebke KJ, Balog RP, Garner HR (2003) Prioritized selection of oligodeoxynucleotide probes for efficient hybridization to RNA transcripts. Nucleic Acids Res 31:750–758
14. Rautio JJ, Kataja K, Satokari R, Penttilä M, Söderlund H, Saloheimo M (2006) Rapid and multiplexed transcript analysis of microbial cultures using capillary electophoresis-detectable oligonucleotide probe pools. J Microbiol Methods 65:404–416

Chapter 13

Amplification of Fungal Genomes Using Multiple Displacement Amplification

Simon J. Foster and Brendon J. Monahan

Abstract

The availability of genomic DNA of sufficient quality and quantity is fundamental to molecular genetic analysis. Many filamentous fungi are slow growing or even unculturable and current DNA isolation methods are often unsatisfactory. Multiple displacement amplification (MDA) is a technique that can be employed to reliably amplify whole genomes from such recalcitrant species. Template DNA obtained using traditional DNA extraction methods, glass bead-mediated disruption of fungal spores or alkaline lysis of mycelium can be used to produce DNA of sufficient quality to be used as a substrate in MDA. With the advent of next generation sequencing methods, the ability to utilize relatively small samples of DNA to achieve complete genome sequencing is now a possibility.

Key words: Multiple displacement amplification, Whole genome amplification, MDA, WGA, Phi29, Fungal genomes

1. Introduction

Complete genome sequences are now available for a number of filamentous fungi (1). Consequently, it is possible for researchers to perform high throughput functional genetic analyses with comparative ease. Fundamental to this capability is the availability of sufficient quantities of genomic DNA (gDNA). For certain filamentous fungi, this step can be a bottleneck. DNA extraction techniques for many fungi do not yield DNA in sufficient quantity and of sufficient quality for modern molecular analyses due to the co-purification of cellular material (e.g., polysaccharides) that interferes with downstream applications such as PCR, restriction digestion and sequencing. While efforts have been developed to optimize extraction methods (e.g., (2)), these are often time consuming and need to be modified for different fungal species. In addition, many species are slow growing in culture or may be unculturable obligate pathogens, parasites or mutualists of plants and animals, limiting the amount of material available from which DNA can be extracted. Whole genome amplification (WGA) can be used to overcome these limitations. Multiple displacement amplification (MDA) is a non-PCR-based WGA technique, which can be used to amplify whole genomes and has been shown to be particularly useful to obtain large quantities of DNA from fungal species for which traditional DNA extraction procedures are problematic (3). MDA can generate microgram quantities of product from as low as 1–10 copies of input DNA (4) or even from single fungal spores (5).

MDA exhibits less amplification bias and greater yield, product length and fidelity than alternative PCR-based WGA methods such as degenerate oligonucleotide-primed PCR (DOP-PCR) (6) or primer extension preamplification (PEP) (7, 8). MDA employs the properties of the DNA polymerase from bacteriophage Φ29 to amplify DNA fragments of up to 70 kb with an error rate of $<10^{-6}$ (9) in an isothermal reaction which is typically conducted at 30°C for 8–16 h. Unlike PCR-based WGA methods which require repeated heat denaturation of DNA into single strands, MDA reactions proceed by strand displacement, relying on the ability of Φ29 DNA polymerase to displace previously copied DNA strands (Fig. 1). Phosphorothioate-modified random hexamers act as primers in the reaction.

2. Materials

2.1. Preparation of Template DNA by Glass Bead Disruption of Fungal Spores

1. DNA-, RNA- and pyrogen-free sterile 1.5 mL screw-topped micro centrifuge tubes.

2. Spore suspension buffer: 10 mM Tris–HCl, pH 8.0, 0.01% (v/v) Triton X-100 (Sigma-Aldrich, Dorset, UK), 1 mM ethylenediaminetetraacetic acid (EDTA). Make up using previously unopened stocks of each component if possible and use molecular biology grade water. Filter sterilize through 0.2 μm syringe filters and store in 1 mL aliquots in 1.5 mL screw-topped micro centrifuge tubes at room temperature, ready for use in spore harvesting (see Note 1).

3. Acid-washed glass beads (425–600 μm; Sigma-Aldrich); wash glass beads in 50% concentrated HCl (~5.8 M) overnight in a glass beaker and rinse with ten complete changes of sterile distilled water. Leave beads in a glass bottle/beaker in an oven until completely dry.

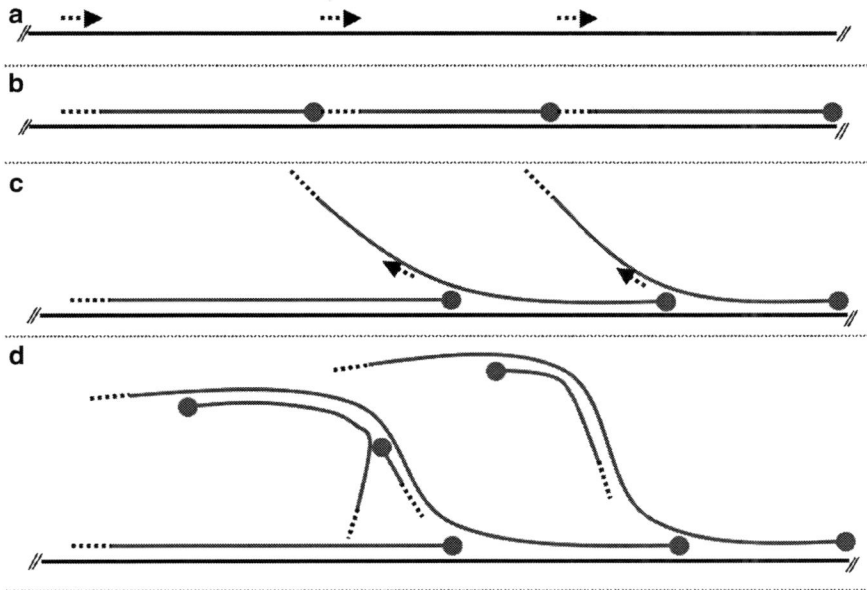

Fig. 1. Schematic representation of multiple displacement amplification. For simplicity, only a single DNA strand (*solid black line*) is shown. (**a**) Random hexamer primers (*dashed arrows*) bind to the denatured DNA. (**b**) The Φ29 DNA polymerase (represented by a *circle*) extends the primers until reaching newly synthesized double-stranded DNA. (**c**) The key feature of Φ29 DNA polymerase is that it now proceeds to displace the newly synthesized strand and continues replicating the DNA. Primers can now, in turn, bind to the single stranded DNA displaced by the action of the ploymerase. (**d**) DNA replication continues on the newly formed strands, forming a hyper-branched structure, and resulting in the amplification of the original DNA substrate.

2.2. Preparation of Template DNA by Alkali Treatment of Purified DNA

1. DNA-, RNA- and pyrogen-free sterile 1.5 mL screw-topped micro centrifuge tubes.
2. Denaturation buffer: 400 mM KOH, 10 mM EDTA. Filter sterilize through 0.2 μm syringe filters and store in 1 mL aliquots.
3. Neutralization buffer: 1 M HCl/1 M Tris–HCl, pH 7.5 (4:6 v/v). Filter sterilize through 0.2 μM syringe filters and store in 1 mL aliquots.

2.3. Preparation of Template DNA by Alkali Treatment of Fungal Mycelium

1. DNA-, RNA- and pyrogen-free sterile 1.5 mL screw-topped micro centrifuge tubes.
2. Denaturation buffer: (see Subheading 2.2, **item 2**).
3. Neutralization buffer: (see Subheading 2.2, **item 3**).
4. Disposable sterile, DNA-, RNA- and pyrogen-free plastic micro-pestles (for example item no. K-749520-0000 from Anachem, Luton, UK) (see Note 2).

2.4. Multiple Displacement Amplification Reaction

1. 5× TE: 50 mM Tris–HCl, pH 8.0, 0.5 mM EDTA.
2. 500 μM random hexamers with phosphorothioate bonds at the two 3′-most nucleotides (i.e., $5'\text{-}N_pN_pN_pN_{ps}N_{ps}\text{-}3'$)

(see Note 3). Resuspend dried down oligos using molecular biology grade water.

3. 10× MDA reaction buffer: 370 mM Tris–HCl, pH 7.5, 500 mM KCl, 100 mM $MgCl_2$, 50 mM $(NH_4)_2SO_4$. Filter sterilize through 0.2 μm syringe filters and store in 100 μl aliquots at −20°C.

4. Mixture of deoxynucleotidetriphosphates (dNTPs) such that each dNTP is present at a concentration of 25 mM. Make up by combining 100 mM stocks of each of the separate dNTPs (dCTP, dATP, dGTP, dTTP) (e.g., cat # 11969064001 from Roche Diagnostics Ltd, West Sussex, UK is a dNTP set that contains 100 mM tubes of each deoxynucleotide).

5. 100 mM dithiothreitol (DTT) (see Note 4).

6. Yeast pyrophosphatase (Roche Diagnostics Ltd) diluted to 0.5 units/mL with sterile molecular biology grade water.

7. Φ29 DNA polymerase, 10,000 units/mL (New England Biolabs, Ipswich, MA, USA).

3. Methods

Template DNA for MDA reactions can be obtained from a number of sources. Although previously purified DNA, for example by phenol/chloroform isolation or by using commercial kits, can be used as a template, in practice this would only be used as a template for rare, historical samples for which the viable biological source is no longer available. Much cruder DNA samples, prepared by the disruption of fungal spores or mycelium can also be used as reliable template; indeed, for many unculturable or slow growing fungi, this may be the only source and it is for these fungi that MDA is a particularly valuable technique.

The growth habit of filamentous fungi varies to such an extent between genera and species that it is outside the scope of this chapter to describe exactly how harvesting of fungal spores or mycelium for use in preparing MDA template DNA from a large range of fungi should be performed. It is presumed that most researchers will be experienced in methods used to obtain spores and mycelium from the particular fungi they work with. Here, we will describe the collection of spores from profusely sporulating filamentous fungi that grow well on agar plates. We also describe a method for preparation of template DNA from mycelia of filamentous fungi growing on agar plates. Both of these methods can be used to obtain crude yet reliable sources of DNA for amplification by MDA.

3.1. Preparation of Template DNA by Glass Bead Disruption of Fungal Spores

1. For the purposes of this method, we draw on our experience with *Penicillium paxilli* a profusely sporulating ascomycete that grows well on agar-based media.
2. From a culture plate, take approximately 0.5–1.0 cm^2 of sporulating mycelium and place into a 1.5 mL screw-topped micro-centrifuge tube containing 1 mL of spore suspension buffer.
3. Vortex tube for 30 s to facilitate release of spores into the buffer (see Note 5).
4. Transfer 200 μL of spore suspension to a fresh 1.5 mL screw-topped micro-centrifuge tube and add approximately 250 mg of acid washed glass beads (see Note 6).
5. Process samples using a FastPrep (Thermo Scientific, Waltham, MA, USA) at a speed setting of 4.0 for 20 s, or by using an equivalent instrument (see Notes 7 and 8). See Fig. 2 for an example of spore disruption and the DNA released as a consequence.
6. Remove samples from the FastPrep and immediately place on ice. If the processed spore samples are not to be used for MDA within 2 h, store them at –20°C to safeguard against DNA degradation.

3.2. Preparation of Template DNA by Alkali Treatment of Purified DNA

1. Into a clean 1.5 mL screw-topped micro-centrifuge tube, place 5 μL of previously purified DNA (purified for example using phenol/cholorofom extraction or commercial kits).
2. Add 5 μL of denaturation buffer, briefly vortex to mix and pulse spin in a centrifuge to collect contents to the bottom of the tube. Place tubes on ice for 10 min.
3. Add 5 μL of neutralization buffer, briefly vortex to mix and pulse spin in a centrifuge to collect contents to the bottom of the tube. Place samples back on ice; if the samples are not to be used for MDA immediately, store the denatured sample at –20°C to safeguard against DNA degradation.

3.3. Preparation of Template DNA by Alkali Treatment of Fungal Mycelium

1. Using a sterile scalpel blade, scrape approximately 0.5 cm^2 of fungal mycelium from a culture plate and place into a sterile screw topped micro-tube containing 100 μL of denaturation buffer.
2. Homogenize the mycelium using a sterile plastic micro-pestle. A few twists of the micro-pestle should be sufficient to disrupt the mycelium and expose the cells to the denaturation buffer.
3. Place the tubes containing homogenized mycelium on ice for 10 min.
4. Add 100 μL of neutralization buffer to the homogenized mycelium, vortex briefly to mix and keep samples on ice. If the samples are not to be used for MDA immediately, store

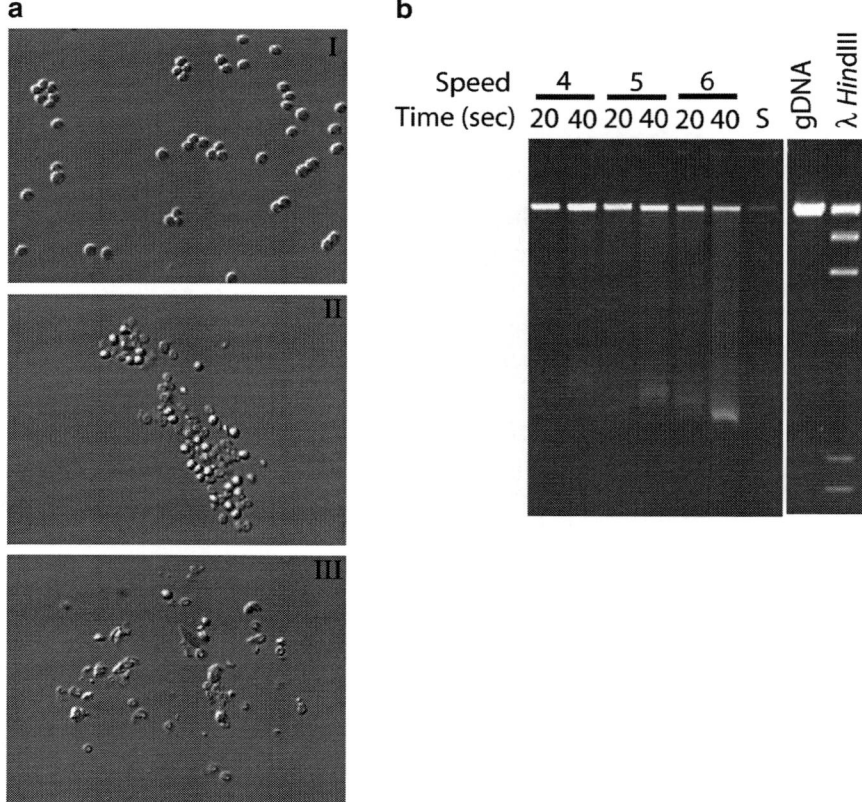

Fig. 2. Disruption of *P. paxilli* conidia to obtain crude DNA samples. (**a**) Microscopic analysis of *P. paxilli* spore suspensions (approximately 10^6 spores/mL) that were untreated (Panel I) or mechanically disrupted with glass beads using a FastPrep at speed setting 4.0 for 20 s (Panel II) and 6.0 for 40 s (Panel III). Pictures were taken at a magnification of 400×. (**b**) Twenty microlitre samples of disrupted spore suspensions were run on a 1% agarose gel and stained with ethidium bromide. The speed and duration of each treatment is indicated; (S) spores only treatment that contained no glass beads and underwent agitation at 6/ms for 40 s, (gDNA) *P. paxilli* genomic DNA (500 ng) purified using a conventional phenol/chloroform method, (λ *Hin*dIII) size marker with a top band of 23.1 kb. Portions reprinted from ref. (3) with permission from Elsevier Science.

the processed mycelial sample at −20°C to safeguard against DNA degradation.

3.4. Multiple Displacement Amplification Reactions

1. The MDA reactions are set up by preparing a separate sample mix (containing template DNA and random hexamers) and a reaction mix (containing the remaining components of the reaction). The two are then brought together to initiate the reaction. At all stages of the preparation, it is essential that tubes and pipette tips used are guaranteed to be DNA-, RNA- and pyrogen-free. Care should be taken to ensure that tubes are kept closed at all times between pipetting steps to minimize

the risk of cross-contamination between samples and also to prevent contamination from environmental DNA.

2. For each sample to be amplified, it is advisable to include a negative control (no DNA) reaction to monitor contamination (see Note 9).

3. Sample mix: If multiple samples are to be amplified, prepare a master mix to reduce the risk of contamination by reducing the number of pipetting steps. Sufficient sample mix should be prepared for $n+1$ samples in case of pipetting errors. For each sample to be amplified, the following reagents are combined in a 0.5 or 1.5 mL tube: 1 μL 5× TE, 2 μL 500 μM random hexamers and 1 μL molecular biology grade water. Vortex the sample mix and pulse-spin in a centrifuge briefly to collect contents at the bottom of the tube. For each reaction, pipette 4 μL of the sample mix into a 0.2 mL PCR tube.

4. Add 1 μL of template DNA (disrupted spores, alkali-treated mycelium or alkali-denatured purified DNA) to the tube containing 4 μL of sample mix. If using previously purified template DNA, which you have chosen not to denature using alkali treatment, the samples will need to be heat denatured prior to amplification; place the combined sample mix and DNA in a PCR block with a heated lid at 95°C for 5 min and then snap cool on ice for 10 min before proceeding. DNA samples prepared by glass-bead mediated disruption of spores or mycelium, or by alkali treatment of mycelium or DNA do not require prior heat denaturation (see Note 10).

5. Reaction mix: for each sample, combine the following reagents in a 0.5 or 1.5 mL tube. As above, for multiple samples make up a master mix for $n+1$ samples. For each sample, the reaction mix contains 1 μL 10 X MDA reaction buffer, 0.4 μL 25 mM dNTPs, 0.5 μL yeast pyrophosphatase, 0.4 μL 100 mM DTT, 2.2 μL molecular biology grade water and 0.5 μL Φ29 DNA polymerase. Vortex well to mix components.

6. For each sample to be amplified, add 5 μL of reaction mix to the previously prepared sample mix resulting in a final reaction volume of 10 μL.

7. Incubate samples for 12–16 h at 30°C (see Note 11).

8. Analyse amplified samples by running 0.5 μL on an agarose gel along with appropriate size markers such as λ DNA digested with *Hin*dIII (see Note 12). The amount of DNA synthesized can also be determined by spectrophotometric analysis or fluoromtery (see Note 13). See Fig. 3 for an example of DNA from *P. paxilli* amplified using MDA.

9. For use in PCR, no further treatment of the amplified DNA is necessary, as the denaturaton steps during the PCR cycles

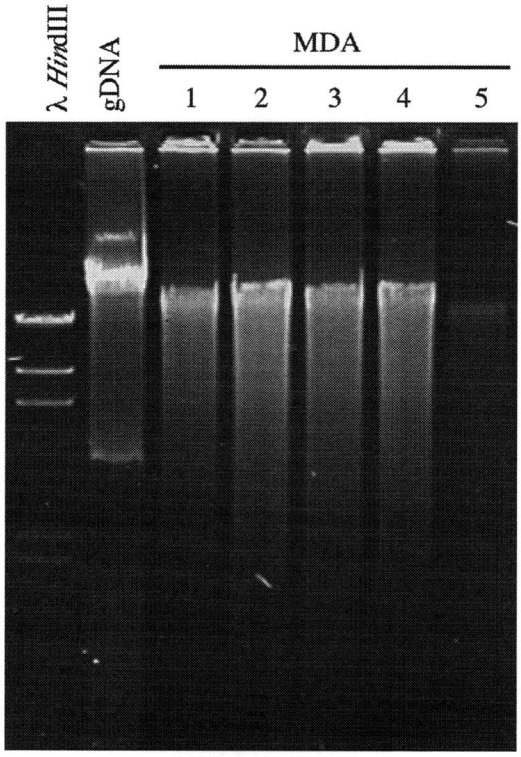

Fig. 3. Whole genome amplification of *P. paxilli* by multiple displacement amplification. Multiple displacement amplification (MDA) reactions were done using as starting template; 10 ng gDNA (lane 1), 1 μL of DNA from mechanically disrupted spores (lane 2) and 1:5 (lane 3) and 1:10 (lane 4) dilutions thereof. A negative control reaction (no template DNA) is also shown (lane 5). In each case, 0.5 μL of a 10 μL MDA reaction was loaded on the gel. A 500 ng *P. paxilli* genomic DNA sample, isolated using a conventional phenol/chloroform method, is shown (gDNA). Size marker λ HindIII has a top band of 23.1 kb. Reprinted from ref. (3) with permission from Elsevier Science.

will efficiently denature the Φ29 DNA polymerase. For applications such as Southern blotting, however, the MDA reaction should be heated to 65°C for 10 min to denature the enzyme.

4. Notes

1. MDA is a non-specific reaction and will amplify any DNA present in the reaction tube or sample. Hence, avoidance of contamination from previously processed samples and from environmental DNA is essential. All consumables and chemicals used to make up solutions should be new and unopened if possible to reduce the chance that they have become

contaminated with DNA from other sources. Unless otherwise stated all solutions should be made up using molecular biology grade water (e.g., catalogue # W4502 from Sigma-Aldrich). Upon making up solutions, we recommend to aliquot appropriate volumes into the tubes they will eventually be used in whenever possible so as to avoid additional manipulation and contact wherever possible. If possible, MDA reactions should be set up in an environment that is not normally used for DNA isolation or other molecular biology techniques, such as PCR, to reduce the risk of contamination with DNA from these procedures. If possible, set up reactions in a clean safety/flow cabinet that has previously been exposed to a UV sterilizing light.

2. Although the disposable micro-pestles can be autoclaved and reused, we strongly recommend against reuse for MDA reactions because of the risk of contaminating future reactions with DNA. The listed micro-pestles from Anachem (K-749520-0000) are supplied complete with a micro-centrifuge tube and hence are ideal as a single-use item.

3. Φ29 DNA polymerase has 3′–5′ exonuclease activity which will rapidly degrade standard oligonucleotides. Inclusion of phosphorothioate bonds at the last two nucleotide positions in the random hexamers protects the oligonuleotides from this activity.

4. The presence of active reducing reagent in the reaction buffer is critical for the activity of the Φ29 DNA polymerase enzyme. To avoid degradation of DTT by repeated freezing and thawing, we recommend storing small single use aliquots in the freezer and adding these to the 10× MDA reaction buffer as required.

5. Ideally, the spore concentration following harvesting should be approximately 10^6/mL. However, it is possible that lower spore concentrations could also be used; this will require empirical testing for the fungal strain being used. Alternatively, if less spores are released, they may be spun down and resuspended in a smaller volume of spore suspension buffer to achieve a concentration of 10^6/mL.

6. We use a home-made scoop to ensure that the same amount of glass beads are transferred into each tube. This can be easily fashioned from a cut-off bottom of a 1.5 mL micro-centrifuge tube (cut approximately 0.5 cm from the bottom of the tube) and either attached using super-glue to a piece of stiff wire (for example, an inoculating loop) or to a pipette tip by melting the end of the tip and allowing it to set in contact with the section of the micro-centrifuge tube.

7. Although we have used a FastPrep in our studies, alternative equipment for disrupting the spore samples with glass beads

is available. For example, the Retsch Mixer Mills MM 200 or MM 400 (Retsch GmbH, Haan, Germany) are suitable for the task and grinding jars which hold multiple 1.5 mL micro-centrifuge tubes are available for these devices (Retsch accessory #22.008.0005).

8. If processing spores from a particular fungal species for the first time, it is advisable to optimize the speed and duration of disruption step and check the degree of disruption either by microscopy or by running a 20 μL aliquot of the disrupted spores on a 0.8% agarose gel to confirm that DNA has been released and that it is not excessively damaged. See Fig. 2 for an example of spore disruption and the DNA released as a consequence. When assessing the optimal parameters for spore disruption, it is important to consider the balance between efficient spore disruption and resulting DNA damage by shearing. For example in Fig. 2, the level of disruption achieved at the lowest FastPrep settings used (speed 4.0 for 20 s) was considered optimal as this released sufficient DNA that did not suffer appreciably from shearing damage. Viewing the disrupted spores by microscopy showed that in most cases the spores had simply been broken open to release the DNA (panel II) rather than excessively disrupted (panel III).

9. We consistently observe amplification of DNA in MDA reactions that do not contain template DNA. This is expected as binding of the random hexamers to each other provides a template suitable for extension by the Φ29 DNA polymerase. Given the non-specific nature of the MDA technique, we advise that a negative control reaction (no template DNA) be included in all experiments employing MDA to ensure the genetic purity of the amplification product. PCR using primers specific to your target organism should be used to confirm that amplified DNA in negative controls does not arise from amplification of contaminating DNA.

10. For the MDA reaction to succeed, the random hexamers need to bind to single stranded regions of the template DNA. Although traditionally (for example in PCR), template denaturation is achieved by heating samples to 95°C or higher, we have found that alkali denaturation of template DNA results in MDA-amplified DNA that appears more similar (in restriction digest profiles) to that of purified genomic DNA (3). Therefore, we recommend alkali-denaturation as the method of choice, although heat denaturation may be used for previously purified DNA sample templates as described in Subheading 3.4, step 3.

11. Following incubation, the success of the MDA reaction can often be gauged by flicking the tube. If the reaction has worked successfully, then the contents will appear very viscous

compared with failed reactions, due to the concentration of DNA in the tube. A 10 µL reaction set up as described will often yield up to 10 µg of amplified DNA, regardless of the amount of initial input template DNA. If a higher yield of DNA is desirable, the reaction can be scaled up, and the amount of DNA amplified will increase linearly.

12. Note that the amplified DNA will appear as a smear on agarose gels (see Fig. 3) because of the presence of incomplete extension products which appear below the high molecular weight DNA down to approximately 1 kb.

13. Gel analysis of amplified samples is recommended prior to quantification by other means as substantial amounts of DNA can be amplified even in the absence of template DNA by non-specific annealing and extension of the random hexamers. This non-specific product can normally be distinguished from amplified template DNA by gel electrophoresis as it is generally smaller in size (see Lane 5, Fig. 3). Spectrophotometric or fluormetric methods, however, will not distinguish between non-specific DNA product and DNA arising from template amplification.

Acknowledgments

We wish to thank Prof. Barry Scott, Massey University, New Zealand who granted us the time and materials to conduct this work in his laboratory.

References

1. Foster SJ, Monahan BJ, Bradshaw RE (2006) Genomics of the filamentous fungi – moving from the shadow of the bakers yeast. Mycologist 20:10–14
2. Al-Samarrai TH, Schmid J (2000) A simple method for extraction of fungal genomic DNA. Lett Appl Microbiol 30:53–56
3. Foster SJ, Monahan BJ (2005) Whole genome amplification from filamentous fungi using *Phi29*-mediated multiple displacement amplification (MDA). Fungal Genet Biol 42:367–375
4. Dean FB, Hosono S, Fang LH, Wu XH, Faruqi AF, Bray-Ward P, Sun ZY, Zong QL, Du YF, Du J, Driscoll M, Song WM, Kingsmore SF, Egholm M, Lasken RS (2002) Comprehensive human genome amplification using multiple displacement amplification. Proc Natl Acad Sci U S A 99:5261–5266
5. Wang Y, Zhu M, Zhang R, Yang H, Wang Y, Sun G, Jin S, Hsiang T (2009) Whole genome amplification of the rust *Puccinia striiformis* f. sp. *tritici* from single spores. J Microbiol Methods 77:229–234. doi:10.1016/j.mimet.2009.02.007
6. Telenius H, Carter NP, Bebb CE, Nordenskjold M, Ponder BAJ, Tunnacliffe A (1992) Degenerate oligonucleotide-primed PCR – general amplification of target DNA by a single degenerate primer. Genomics 13:718–725
7. Zhang L, Cui XF, Schmitt K, Hubert R, Navidi W, Arnheim N (1992) Whole genome amplification from a single cell – implications for genetic analysis. Proc Natl Acad Sci U S A 89:5847–5851
8. Lasken RS, Egholm M (2003) Whole genome amplification: abundant supplies of DNA from precious samples or clinical specimens. Trends Biotechnol 21:531–535
9. Esteban JA, Salas M, Blanco L (1993) Fidelity of *Phi 29* DNA polymerase Comparison between protein-primed initiation and DNA polymerization. J Biol Chem 268:2719–2726

Part III

Microscopy and Protein Analysis

Chapter 14

Biochemical Methods Used to Study the Gene Expression and Protein Complexes in the Filamentous Fungus *Neurospora crassa*

Jinhu Guo, Guocun Huang, Joonseok Cha, and Yi Liu

Abstract

Biochemical approaches are powerful tools for investigating mechanisms of biological processes. Here, we describe several biochemical approaches that have been successfully in our laboratory to study the filamentous fungus *Neurospora crassa*. These approaches include protein extraction and western blot analysis, protein purification using epitope-tagged fusion protein, protein immunoprecipitation (IP) and Chromatin Immunoprecipitation (ChIP) assays. These methods can also be modified for use in other filamentous fungi.

Key words: *Neurospora crassa*, Protein extraction, Western blot, SDS-PAGE, c-Myc fusion protein purification, Immunoprecipitation (IP), Chromatin immunoprecipitation (ChIP) assay

1. Introduction

The filamentous fungus *Neurospora crassa* is an important model organism for research in genetics, biochemistry, and molecular biology (1, 2). *Neurospora* is easy to grow and has a haploid life cycle. Currently, *Neurospora* is used as model system to study circadian clock, gene silencing, fungal growth, and differentiation. To date, series of protocols have been developed for the molecular and biochemical studies of *Neurospora*. In our laboratory, we adopted the traditional methods for *Neurospora*. In addition, we also established and improved several protocols for the biochemcial analyses of *Neurospora*.

Here, we introduced some biochemical approaches used in the studies on *Neurospora* in our laboratory, which include protein extraction and western blot analysis (see Subheading 3.1), protein purification using epitope-tagged fusion protein (see Subheading 3.2), protein immunoprecipitation (IP, see Subheading 3.3), and Chromatin Immunoprecipitation (ChIP) assays (see Subheading 3.4). These methods, which are widely used for biochemical analyses in many other model organisms, were adapted for *Neurospora*.

Immunoblot analysis, also called western blot, is a simple and most commonly used method to detect specific proteins in tissue homogenates or protein extracts (3). Protein purification is also an approach routinely used to study the protein activity and to identify protein complexes. Similarly, immunoprecipitation (IP) is a method to specifically concentrate the protein of interest by using a specific antibody, which is widely used to study protein–protein interaction (PPI) and to identify novel protein(s) that interact with the protein of interest. In addition, we also developed the method of ChIP assay in *Neurospora* to investigate the association of a protein of interest with specific DNA regions (4). The methods described here can be modified and adapted for the experiments under different growth conditions and used for studies of other filamentous fungi.

2. Materials

2.1. Buffers for Protein Extraction and Western Analysis

1. Extraction Buffer: 50 mM 4-(2-hydroxyethyl)-1-piperazineethanesulfonic acid (HEPES, pH 7.4), 137 mM NaCl, 10% glycerol (v/v). Store at 4°C. Add protease inhibitors to the buffer before protein extraction.

2. Protease inhibitors. The concentrations of stocking solution of these protease inhibitors are: pepstatin A, 1 mg/mL in methanol; leupeptin, 1 mg/mL in H_2O; phenylmethylsulphonyl fluoride (PMSF), 100 mM in isopropanol, respectively. Store these solutions at –20°C.

3. Solutions for preparing gels: 1.5 M Tris–HCl (pH 8.8); 10% sodium dodecyl sulfate (SDS); 10% ammonium persulfate (APS); N,N,N′,N′-tetramethylethylenediamine (TEMED); 30% acrylamide/bis solution (37.5:1, 2.6% C, purchased from Bio-Rad). The recipe for four stacking gels: 5.0 mL 0.5 M Tris–HCl (pH 6.8), 2.7 mL 30% acrylamide/bis solution (w/v), 0.2 mL 10% SDS, 0.1 mL 10% ammonium persulfate (APS), 20 µL TEMED and 12.3 mL H_2O.

4. Gel loading buffer (2×). Mix 900 µL urea-containing protein loading buffer, 50 µL dithiothreitol (DTT) and 50 µL bromophenol blue solution in H_2O to 1 mL. Urea-containing protein loading buffer contains 100 mM Tris (pH 6.8), 4% SDS, 2 mM EDTA, 2% glycerol and 6 M Urea.

5. Gel running buffer (10×): 14.4% glycine (w/v), 3.03% Tris base (w/v), 1% SDS (w/v). Dilute it in H$_2$O to make 1× gel running buffer for electrophoresis.

6. Protein transfer buffer (5×): 115.2% glycine (w/v), 24.2% Tris base (w/v). Store at room temperature. 1× Transfer buffer for gel transfer is prepared before use by mixing 200 mL methanol, 200 mL 5× Transfer buffer and 600 mL H$_2$O to 1 L.

7. PBST (1×): PBS buffer (1×) with 0.05% Triton X-100 (v/v).

8. Incubation buffer: dissolve nonfat dry milk powder (Nestle or other vendors) in PBST buffer (1×) and the concentration of milk is 5% (w/v).

9. Membrane staining buffer: 40% methanol (v/v), 10% acetic acid (v/v), 0.1% Amido black 10B (w/v) (Sigma).

2.2. Buffers for Epitope-Tagged Protein Purification

1. 0.1% glucose media: 0.1% glucose (w/v), 0.17% arginine (w/v) in 1× Vogel's solution.

2. Extraction buffer and protease inhibitors: same as described above for nondenatured protein extraction (see Subheading 2.1).

3. Wash buffer: 50 mM HEPES, pH 7.4, 300 mM NaCl, and 20 mM imidazole. Store at 4°C.

4. Elution buffer: 50 mM HEPES, pH 7.4, 137 mM NaCl, 200 mM imidazole, and 20% glycerol (v/v). Store at 4°C.

5. High salt buffer: 20 mM Tris–HCl, pH 7.5, 500 mM NaCl.

6. Low salt buffer: 20 mM Tris–HCl, pH 7.5, and 50 mM NaCl.

7. Ni-NTA beads (Qiagen).

8. Anti c-Myc agarose beads (Santa Cruz Biotechnology).

9. 4-15% gradient SDS-PAGE gel (Bio-Rad).

2.3. Buffers for Immunoprecipitation (IP) Analysis

1. Extraction buffer and protease inhibitors: see Subheading 2.1.

2. GammaBind G Sepharose beads (Amersham Biosciences). The beads need to be preequilibrated with the extraction buffer before use.

2.4. Buffers for ChIP Assay

1. Lysis buffer: 50 mM HEPES, 137 mM NaCl, 1 mM EDTA, 1% Triton X-100 (v/v), 0.1% deoxycholate (w/v), 0.1% SDS (w/v), pH 7.5. Store at 4°C for no more than 6 months.

2. Washing buffer: 50 mM HEPES, 137 mM NaCl, pH 8.0.

3. Glycine stock solution: 1.25 M glycine, pH 7.5.

4. LNDET buffer: 0.25 M LiCl, 1% NP40 (v/v), 1 mM EDTA, 10 mM Tris–HCl, pH 8.0.

5. Low salt washing buffer: 0.1% SDS (w/v), 1% Triton X-100 (v/v), 2 mM EDTA, 150 mM NaCl, 20 mM Tris–HCl, pH 8.0.
6. High salt washing buffer: 0.1% SDS (w/v), 1% Triton X-100 (v/v), 20 mM Tris–HCl, 500 mM NaCl, pH 8.0.
7. Elution solution (freshly made): 1% SDS (w/v), 0.1 M $NaHCO_3$.
8. TE buffer: 1 mM EDTA, 10 mM Tris–HCl, pH 8.0.
9. Proteinase K 10 mg/mL: Proteinase K powder (Sigma) dissolved in the proteinase K buffer and stored at −20°C. Proteinase K buffer: 10 mM Tris, 5 mM EDTA, 0.5% SDS (w/v), pH 7.8.

3. Methods

3.1. Protein Extraction and Western Blot

Western blot analysis can detect specific protein(s) in tissue homogenate or protein extracts (3, 5–7). Western blot analysis utilizes gel electrophoresis to separate native or denatured proteins by either the molecular weight (denaturing conditions) or the 3-D structure and charges (native/non-denaturing conditions) of polypeptides. After protein extraction, proteins are transferred to a membrane (typically nitrocellulose or PVDF), then probed by using antibodies specific to the proteins of interest. The protocol described here utilizes SDS-PAGE denaturing gel to separate the proteins (8).

1. Harvest *Neurospora* samples by compressing tissues wrapped between paper towels or by vacuum to eliminate remaining medium in samples. Quickly drop the tubes containing the dehydrated tissues in liquid nitrogen. The samples can then be kept in −80°C freezer before extraction (see Note 1).
2. Grind the tissue into fine powders in liquid nitrogen with a mortar and pestle. Collect the powders into Eppendorf tubes with a spoon cooled in liquid nitrogen. Keep the tubes in liquid nitrogen or at −80°C prior to protein extraction.
3. Add ice-cold extraction buffer with freshly added protease inhibitors to the tissue powder. Before protein extraction, freshly add the three proteinase inhibitors to Extraction Buffer at the final concentrations of: Pepstatin A, 1 μg/mL; Leupeptin, 1 μg/mL; PMSF, 1 mM.
4. Mix gently by inverting the tubes and incubate on ice for 10 min.
5. Centrifuge cell homogenate at $15,000 \times g$ for 15 min at 4°C. Transfer the supernatant to a new tube, avoid the cell debris on the top and pellet at the bottom.

6. Measure protein concentration of the supernatant: mix 2 µL protein extract with 800 µL H$_2$O and 200 µL Bio-Rad protein assay dye, measure OD$_{595}$ to calculate protein concentration.

7. For casting gels. For the western analysis of FRQ protein (989 aa), we use 7.5% SDS-PAGE gel. The recipe for four separating gels (7.5%): 10.8 mL 1.5 M Tris–HCl, pH 8.8, 10.8 mL 30% Acrylamide/Bis solution, 0.44 mL 10% SDS, 0.22 mL 10% ammonium persulfate (APS), 22°µL N,N,N′,N′-tetramethylethylenediamine (TEMED), and 21.3 mL H$_2$O. The recipe for four stacking gels: 5.0 mL 0.5 M Tris–HCl pH 6.8, 2.7 mL 30% Acrylamide/Bis solution (w/v), 0.2 mL 10% SDS, 0.1 mL 10% APS, 20 µL TEMED, and 12.3 mL H$_2$O. The 30% Acrylamide/Bis solution (37.5:1, 2.6% C) can be purchased from Bio-Rad or other vendors.

8. Prepare protein samples for SDS-PAGE. For a small gel (Bio-Rad Protean III gel), mix 50 µg total protein, 20 µL 2× gel loading buffer and H$_2$O to 40 µL for each well. Heat protein-containing tubes on heat block (95–105°C) for 5 min to denature the proteins, and chill the tubes on ice before loading onto SDS-acrylamide gel. Electrophoresis is performed using 1× gel running buffer. Use prestained protein marker as size standards.

9. Gel Transfer. Label on the right top of PVDF membrane (Millipore) and soak it in 100% methanol for a few seconds, then soak it in Transfer buffer. Transfer the proteins in the gel to PVDF membrane (Immoblon-P, Millipore) at 80 mA for 3 h (2 h is enough for most proteins) with Genie Electrophoretic Blotter (Idea Scientific). After the transfer, remove the membrane and soak it in 100% methanol for 1 min and air dry at room temperature.

10. Immunoblotting analysis. Put the membrane in the Incubation buffer for 1 h with gentle mixing at room temperature before adding the primary antibody at a predetermined concentration. Incubate for additional 1–3 h at room temperature or overnight at 4°C. The concentration needs to be determined for each antibody and time of the incubation can be adjusted depending on the sensitivity of the antibody (see Note 2).

11. Rinse the membrane with 1× PBST, then wash it in 1× PBST, 3×5 min.

12. Incubate the membrane in the Incubation buffer with secondary antibody for 20 min to 1 h at room temperature. Dilute antibody according to experience. For commonly used secondary antibodies use the follwoing: Goat-anti-Rabbit (Bio-Rad) 1:3000, Goat-anti-Mouse (Bio-Rad) 1:3000, Sheep-anti-Mouse (Amersham) 1:2000.

13. Rinse the membrane with 1× PBST first, then wash it in 1× PBST for 1 × 15 min and 5 × 5 min.

14. Incubate membrane in a 1:1 mix of the ECL (Amersham) reagents. Drain the membrane and expose it to X-ray film. Adjust the exposure time to obtain optimal exposure (see Note 3).

15. To visualize the protein loading on the membrane, incubate the membrane in the membrane staining buffer for 2–5 min, then destain it with water. Western blot analysis using the tubulin antibody can also be used to determine equal loading.

3.2. Purification of Epitope-Tagged Proteins

Epitope-tagged proteins can greatly facilitate the purification of the protein of interest and lead to the identification of novel protein complexes. A lot of epitope tags have been developed, including FLAG, c-Myc, 6-His, GST (Glutathione S-transferase), TAP, etc. (9). We have previously generated a tandem 6-His and 5 c-Myc tag that can be used in *Neurospora* for efficient protein purification (10). By expressing the His-Myc-CSN2 protein in *Neurospora*, we were able to purify the entire *Neurospora* COP9 signalosome in a two-step purification protocol. For this purpose, a *Neurospora* expression construct containing the open reading frame of the protein of interest fused with the His–Myc tag was transformed into a wild-type *Neurospora* strain (10). Afterwards, western blot analysis using the c-Myc anitboby was used to identify transformants expressing the fusion protein. The protocol below describes the procedures for c-Myc-tagged protein purification. The protein purification products derived according to this protocol can be used for further analysis, e.g., Mass Spectrometry to identify associated proteins and protein activity assay.

1. Wash the empty column with MilliQ water and add 3 mL of MilliQ water into the column (see Notes 1, 5).

2. Add desired amount of Ni-NTA beads (0.5–2 mL) into the column, and let the column stand for 20–25 min, so all the beads can sediment to the bottom.

3. Wash the column with 10 packed column volume (CV) of MilliQ water. This removes the ethanol in the beads.

4. Equilibrate the column with 10 CV of extraction buffer. Now the column is ready for use and can be stored at 4°C.

5. Harvest *Neurospora* culture and grind into fine powders in liquid nitrogen.

6. Add extraction buffer with protease inhibitors to the tissue powders and mix well by inverting the tube.

7. Centrifuge at $12,000 \times g$, 15 min, 4°C and transfer the supernatant to a fresh tube.

8. Centrifuge again at 45,000×g, at 4°C for 30 min. Transfer the supernatant to a fresh tube.

9. Measure the protein concentration, dilute to 2–4 mg/mL, and add imidazole to 0–20 mM final concentration to reduce nonspecific binding. If the addition of imidazole significantly affects the binding of the protein, it can be omitted.

10. Load the protein extract to the Ni-NTA column and collect the flow – through fraction in a tube. Adjust the flow rate of the column to around 1 mL/min.

11. Add 10 CV of wash buffer and collect the wash fraction in a tube.

12. Add 1 CV of elution buffer, (optional: allow the column to stand for 5 min) and collect the elution fraction in Eppendorf tubes. Repeat the process three more times. Each elution fraction should be analyzed by western blot analysis using a c-Myc antibody. Combine the most enriched fractions for the following steps.

13. Mix 1 mL elution fraction with 40 μL anti-c-Myc agarose beads (10 μL bead volume) (5 μL of 30% NP-40; optional), and incubate on a rotator at 4°C for 4–5 h.

14. Centrifuge at 4,000×g at 4°C for 1 min, and remove the supernatant in a separate tube as the unbound fraction.

15. Wash the beads with 1 mL of high salt buffer, centrifuge at 4,000×g, at 4°C for 1 min, and gently remove the supernatant.

16. Wash the beads with 1 mL of low salt buffer as step 15.

17. Repeat steps 15 and 16 twice.

18. Resuspend the beads in 1 mL MiniQ water and transfer to a fresh tube. Centrifuge at 4,000×g, at 4°C for 1 min, and gently remove the supernatant.

19. The beads can be resuspended in 1× SDS-PAGE loading buffer to be analyzed in a SDS-PAGE gel. Alternatively, the beads can be resuspended in a different buffer for other purposes.

3.3. Immunoprecipitation (IP) Assay

The application of immunoprecipitation (IP) started in 1960s (11). By using an antibody that specifically binds to a protein of interest, immunoprecipitation can specifically precipitate a protein from protein extracts which containing many thousands of different proteins. Therefore, immunoprecipitation can be used to purify and concentrate a particular protein and study PPI. Immunoprecipitation requires the antibody to be coupled to solid beads at some point in the procedure.

In immunoprecipitation, the antibody-coated-beads are added to the homogenized tissue or protein extract, and the antibody can bind to the target proteins. Then, the antibody bound beads and the associated proteins can be separated from the protein extract after washing. Similar to c-Myc fusion protein purification, immunoprecipitated proteins can be used for protein activity assay and mass spectrometry. In addition, western blot analysis after immunoprecipitation is a commonly used method to examine the potential PPIs. Compared to protein purification described above, immunoprecipitation is often used to concentrate the target protein on a smaller scale.

1. Prepare a nondenatured extract as described in protein extraction (see Subheading 2.1).

2. After measuring protein concentration, dilute 1–2 mg of total extract in the extraction buffer with an appropriate amount of FRQ antibody. Incubate the mixture on a rotator at 4°C for 2–4 h. The concentration needs to be optimized for each antibody; very often, antibodies that work well for western blot analysis may not work efficiently in immunoprecipitation.

3. Add 10–20 µL (packed volume) preequilibrated GammaBind G Sepharose beads, and incubate for 1 h or more at 4°C on a rotator.

4. Centrifugate at $4,000 \times g$ for 1 min at 4°C, and gently remove the supernatant.

5. Wash the beads in 1 mL of the extraction buffer, centrifuge at $4,000 \times g$ for 1 min at 4°C, and gently vacuum out the supernatant. Repeat the washing steps 2–4 times.

6. Resuspend the beads in 1× SDS-PAGE loading buffer to be analyzed on a SDS-PAGE gel and western blot analysis (see Subheading 2.1).

3.4. Chromatin Immunoprecipitation (ChIP) Assay

The association of transcription factors and other proteins to target DNA sites plays important roles in the regulation of global gene expression and maintenance. A number of methods have been developed for investigating protein-DNA interactions in vivo (4). ChIP assay is one of these methods to determine whether a protein, e.g., certain transcription factors, is associated with a specific genomic region in living cells or tissues.

ChIP assay is based on the principle that formaldehyde reacts with primary amines on amino acids and DNA molecules, forming covalent cross-links between proteins and DNA on which they are situated. Following the cross-linking, the crude cell extracts are sonicated to shear the DNA into small, uniform sized chromatin fragments. Then, the chromatin is precleared with protein G beads to reduce nonspecific background, and an antibody of interest is then incubated with the precleared chromatin extracts. Protein G beads are then added to the antibody/chromatin mixture.

After extraction of the DNA from the protein G beads, the cross-linking is reversed by heating, subsequently the proteins are removed by proteinase K treatment. To detect the DNA fragments bound to the protein of interest, quantitative PCR analysis using the purified DNA as templates is performed by using primers spanning the potent DNA binding site. The precipitated DNA fragments can also be used for other analyses, for instance, ChIP-on-chip analysis (12).

For ChIP assay, either formaldehyde or UV light can be used for crosslinking of the protein and DNA (13). Here, we only describe the method of formaldehyde-induced crosslinking in this protocol. In our laboratory, we often conduct ChIP assay to test the binding of the White collar complex (WCC), which is composed of White collar 1 (WC-1) and WC-2, to the promoter of *frequency* (*frq*) gene that encodes the central component of the circadian negative feedback loop in *N. crassa* circadian system. In *N. crassa* circadian clock, WCC acts as positive element to activate the transcription of *frq* gene, and the protein product FRQ functions to repress the binding of WCC to *frq* promoter (8, 14–19). Therefore, ChIP assay allows us to determine the function of the circadian negative feedback loop and the in vivo DNA binding activity of the WCC.

Here, we describe the protocol using our WC-2 antibody to conduct ChIP assay.

1. Inoculate the wild-type *N. crassa* (87-3) conidia from a 5–7 day-old slant into Perti dishes. After growth at room temperature for 2 days, a thin layer of mycelia mat forms on the surface of the medium. Mycelia plugs are inoculated in flasks with 50 mL medium and grown for 2 days under desired experimental conditions.

2. *N. crassa* tissues under the experimental conditions are fixed by adding 1% formaldehyde for 15 min with shaking (directly add 37% formaldehyde to the medium). Stop the fixation reaction by adding 1.25 M glycine (to 125 mM final concentration) directly into the medium with shaking for another 5 min. Harvest cultures and grind samples into fine powders in liquid nitrogen.

3. Add 2 mL lysis buffer with proteinase inhibitors (10 mM PMSF, 1 mM leupeptin and 1 mM pepstatin A) to the tissue powder in a 15 mL centrifuge tube.

4. Sonicate the chromatin (see Note 6). Chromatin is sheared by sonication to approximately 500–1,000 bp fragments.

5. Centrifuge for 15 min at $10,000 \times g$ at 4°C.

6. Determine the protein concentration and dilute the protein to 2 mg/mL, apply 1 mL for ChIP assay, and keep 100 μL as the input DNA.

7. Block the protein G beads (Gammabind G Sepharose code No: 17-0885-01, GE Healthcare) in lysis buffer containing 100 μg BSA/mL and 100 μg/mL salmon sperm DNA overnight at 4°C by constant rotation (see Note 7).

8. Preclear the chromatin mixture with 50 μL blocked protein G beads for 4 h on a rotator at 4°C.

9. Spin down beads by centrifugation at $4,000 \times g$ for 2 min. The supernatant is transferred to a fresh tube and 2 μL WC-2 antibody is added (see Note 8), then incubate overnight at 4°C on a rotator. A control antibody or the chromatin mixture from a *wc-2* null strain can be used as a negative control.

10. Add 50 μL blocked protein G beads and incubate for another 2 h with constant gentle rotation at 4°C.

11. Spin down the beads and carefully remove the supernatant.

12. Wash the beads with 1 mL of the following buffers sequentially: low salt buffer, high salt buffer, LNDET buffer, and TE buffer, each for 5 min.

13. Wash beads with TE buffer for another 5 min.

14. Spin down the beads by centrifugation at $4,000 \times g$ for 2 min. Prepare elution buffer: 1% SDS, 0.1% $NaHCO_3$. The elution buffer needs to be prepared just prior to use.

15. Add 250 μL elution buffer to the pellet beads. Vortex briefly to mix and incubate at room temperature for 15 min with rotation.

16. Spin down the beads, and transfer the supernatant fraction to another fresh tube. Repeat step 3. Combine elutions together (final volume 500 μL).

17. Add 20 μL 5 M NaCl to the 500 μL ChIP elution fraction, and 4 μL 5 M NaCl to the 100 μL input DNA sample.

18. Place tubes in a 65°C water bath for 4 h.

19. Centrifuge briefly to remove condensation in the sample tubes.

20. For input samples, add TE buffer to final volume of 500 μL.

21. Add 2 μL of proteinase K (10 mg/mL) to the elution samples and input samples and incubate at 45°C for 1 h.

22. Centrifuge briefly to spin down all liquid to the bottom, and add the equal volume of phenol/chloroform/isoamyl alcohol (25:24:1) to extract the DNA. The supernatant is then mixed with two volumes of pure ethanol to precipitate the DNA. Wash the DNA pellet with 70% ethanol, and dissolve the DNA pellet in 50 μL TE buffer (see Note 9).

23. Take out 2 µL DNA sample for PCR analysis. Perform PCR reaction in a total volume of 20 µL. Follow the regular PCR program, but the appropriate number of PCR cycles would be determined empirically, so the reaction is within the quantitative range. We normally start with 28 cycles for studying the binding of WCC to *frq* promoter. The PCR primers for ChIP analysis should span no more than 250 nt, which covers the binding site of the protein of interest (see Note 10).

24. PCR products are resolved by eletrophoresis on a 2% agarose gel. Our ChIP assay results consistently show that WC-2 binds to the *frq* promoter rhythmically, peaking at DD14 (18, 19).

4. Notes

1. For some further analyses, e.g., immunoprecipitation (IP), fresh protein extracts are preferred over frozen protein sample.
2. For certain antibodies, a longer incubation time (e.g., overnight at 4°C) with the primary antibody is preferred.
3. ECL plus western blotting detection system kit (Amersham) may be used to improve western blot signal strength.
4. Each affinity tag is purified under its specific buffer conditions, which could affect the protein of interest. Therefore, to achieve the highest purification yield and purity, optimizing the binding and washing condition is highly recommended.
5. To avoid protein degradation during the purification process, all solutions should be kept at 4°C, and the entire purification procedure should be performed in a cold room (4°C).
6. For sonication, always keep the tubes on ice. Sonicate each sample at 25 pulses for three times with 30 s to 1 min interval between each 25-min pulses to avoid rising of the sample temperature. The parameters for the sonication pulses are: duty cycle 50, output 4.
7. The salmon sperm DNA solution needs to be heated by boiling for 10 min and immediately chilled on ice water before use.
8. The antibody concentration for ChIP assay is generally significantly higher than that used for western blot analysis.
9. After DNA precipitation, it is normal that the DNA pellet is very small and difficult to see visually.
10. Real-time PCR, when available, can also be used for analyzing the ChIP assay results, and is more quantitative. For quantitative real-time PCR, use a house-keeping gene (in the input) as control (20).

Acknowledgments

This work was supported by grants from National Institutes of Health and Welch Foundation to Yi Liu.

References

1. Horowitz NH (1991) Fifty years ago: the Neurospora revolution. Genetics 127: 631–635
2. Kushnirov VV (2000) Rapid and reliable protein extraction from yeast. Yeast 16:857–860
3. Burnette WN (1981) "Western blotting": electrophoretic transfer of proteins from sodium dodecyl sulfate–polyacrylamide gels to unmodified nitrocellulose and radiographic detection with antibody and radioiodinated protein A. Anal Biochem 112:195–203
4. Aparicio O, Geisberg JV, Struhl K (2004) Chromatin immunoprecipitation for determining the association of proteins with specific genomic sequences in vivo. Curr Protoc Cell Biol. Chapter 17:Unit 17.7. Wiley
5. Renart J, Reiser J, Stark GR (1979) Transfer of proteins from gels to diazobenzyloxymethyl-paper and detection with antisera: a method for studying antibody specificity and antigen structure. PNAS 76:3116–3120
6. Towbin H, Staehelin T, Gordon J (1979) Electrophoretic transfer of proteins from polyacrylamide gels to nitrocellulose sheets: procedure and some applications. PNAS 76:4350–4354
7. Sambrook J, Fritsch M (1989) Molecular cloning, 2nd edn. Cold Spring Harbor Laboratory Press, USA
8. Garceau N, Liu Y, Loros JJ, Dunlap JC (1997) Alternative initiation of translation and time-specific phosphorylation yield multiple forms of the essential clock protein FREQUENCY. Cell 89:469–476
9. Terpe K (2003) Overview of tag protein fusions: from molecular and biochemical fundamentals to commercial systems. Appl Microbiol Biotechnol 60:523–533
10. He Q, Cheng P, He Q, Liu Y (2005) The COP9 signalosome regulates the Neurospora circadian clock by controlling the stability of the SCFFWD-1 complex. Genes Dev 19: 1518–1531
11. Howard GC, Kaser MR (2006) Making and using antibodies: a practical handbook. CRC Press, Boca Raton
12. Ren B, Robert F, Wyrick JJ, Aparicio O, Jennings EG, Simon I, Zeitlinger J, Schreiber J, Hannett N, Kanin E, Volkert TL, Wilson CJ, Bell SP, Young RA (2000) Genome-wide location and function of DNA binding proteins. Science 290:2306–2309
13. Tollefsbol TO (2004) Epigenetics protocols. In: Methods in molecular biology, vol 287. Humana Press, USA
14. Cha J, Huang G, Guo J, Liu Y (2007) Posttranslational control of the Neurospora circadian clock. Cold Spring Harb Symp Quant Biol 72:185–191
15. Heintzen C, Liu Y (2007) The *Neurospora crassa* circadian clock. Adv Genet 58:25–66
16. Dunlap JC, Loros JJ (2004) The neurospora circadian system. J Biol Rhythms 19: 414–424
17. Cheng P, Yang Y, Heintzen C, Liu Y (2001) Coiled-coil domain-mediated FRQ-FRQ interaction is essential for its circadian clock function in Neurospora. EMBO J 20: 101–108
18. He Q, Liu Y (2005) Molecular mechanism of light responses in Neurospora: from light-induced transcription to photoadaptation. Genes Dev 19:2888–2899
19. Huang G, Chen S, Li S et al (2007) Protein kinase A and casein kinases mediate sequential phosphorylation events in the circadian negative feedback loop. Genes Dev 21: 3283–3295
20. Xing H, Vanderford NL, Sarge KD (2008) The TBP-PP2A mitotic complex bookmarks genes by preventing condensin action. Nat Cell Biol 10:1318–1323

Chapter 15

Measuring Protein Kinase and Sugar Kinase Activity in Plant Pathogenic *Fusarium* Species

Burton H. Bluhm and Xinhua Zhao

Abstract

As ubiquitous metabolic and signaling intermediaries, kinases regulate innumerable aspects of fungal growth and development. At its simplest, the enzymatic function of a kinase is to transfer a phosphate from a donor molecule (such as adenosine triphosphate) to an acceptor molecule, such as a protein, carbohydrate, or lipid. Kinase activity is intricately interwoven into signal transduction, and ultimately modulates gene expression, downstream phosphorylation events, and other mechanisms of posttranslational modification. Therefore, sensitive and reproducible techniques to measure kinase activity are crucial to elucidate cellular signaling and for fungal functional genomics.

Protein and sugar kinases regulate multiple aspects of pathogenesis in the mycotoxigenic, plant pathogenic fungi *Fusarium graminearum*, and *Fusarium verticillioides*. Here, we present protocols to (1) quantify phosphorylation of mitogen-activated protein kinases in *F. graminearum*, and (2) determine glucokinase activity in *F. verticillioides*. The mitogen-activated protein kinase phosphorylation assay utilizes immunological methods to quantify substrate phosphorylation, whereas the glucokinase assay is a coupled enzyme assay, in which phosphorylation of glucose by glucokinase is measured indirectly through the subsequent reduction of NADP+ to NADPH, a substrate more amenable for spectrophotometric detection.

Key words: Mitogen-activated protein kinase, Glucokinase, *Fusarium*, Phosphorylation, Signal transduction, Virulence, Kernel-rotting diseases, Coupled enzyme assay

1. Introduction

Kinases, also called phosphotransferases, transfer phosphates from high-energy donors such as adenosine triphosphate to substrates such as proteins or small molecules. Kinase activity is vital for primary metabolism; for example, phosphorylation of sugars is a critical component of glycolysis. Beyond primary metabolism, kinases serve important roles in signal transduction. Phosphorylation is a widespread posttranslational modification of proteins; as much as 30% of the fungal proteome is predicted to be regulated through phosphorylation (1). In fungi, kinases regulate a wide range of biological processes in response to changing environmental conditions, ranging from metabolism, growth, morphological development, and reproduction (2, 3).

Fungal kinases are broadly categorized by substrate. *Protein kinases* phosphorylate other proteins, typically by transferring a phosphate to a free hydroxyl group of an amino acid. Serine/threonine specific kinases, which phosphorylate serine and/or threonine residues in a compatible substrate, are broadly conserved across taxonomic kingdoms, and are prevalent in fungi. Among serine/threonine kinase superfamilies, mitogen-activated protein kinases (MAPKs) comprise an important group of signal transduction intermediaries. MAPKs frequently function in signal transduction cascades, in which the MAP kinases (MAPKs) are usually activated by MAPK kinases (MEKs) that are in turn activated by MEK kinases (MEKKs). The catalytic subunits of MAPKs are highly conserved across the fungal kingdom and other kingdoms of life. In addition to serine/threonine kinases, conserved types of protein kinases include tyrosine kinases and histidine kinases. *Small metabolite kinases* act upon a wide range of compounds, including amino acids, lipids, and carbohydrates, and are indispensable for primary metabolism. For example, by phosphorylating glucose to create glucose-6-phosphate, glucokinases prime the first step of glycolysis and serve as glucose sensors.

As fungal biology enters the postgenomics era, increasing effort is being placed on understanding signaling pathways from systems-based perspectives. Because protein and small molecule phosphorylation plays multifaceted roles in cellular signaling, understanding kinase function is crucial to unmask the complexities of fungal signal transduction and the roles of individual genes. To this end, sensitive and specific kinase assays are an indispensable tool for fungal molecular genetics. Here, we provide protocols to assay (1) the phosphorylation of MAPKs in *Fusarium graminearum*, and (2) glucokinase activity in *Fusarium verticillioides*. These protocols are easily adapted for a wide range of fungal species and growth conditions. Lists of required equipment are compiled on the assumption that equipment required for routine procedures in molecular biology is already available for potential users of the protocols. Product names and vendors are provided in some cases for informational purposes and do not represent specific endorsements.

2. Materials

2.1. MAPK Phosphorylation Assay

2.1.1. Equipment

1. Bead beater (Fulltech, UF-60D11).
2. Protein electrophoresis equipment (Protean II by BioRad® and power supply).
3. Ultracentrifuge.
4. Spectrophotometer and cuvettes.
5. Phosphor imager, imaging screen, and cassettes.

2.1.2. Supplies and Reagents

1. Antibodies: anti-p42/44 MAPK (Cell Signaling #9102), anti-phospho-p42/44 MAPK (Cell Signaling #9101S), and anti-actin (Sigma Aldrich #A5060).
2. Carboxymethylcellulose (CMC) medium (1 L): 15 g CMC (low viscosity, e.g., Sigma #419273), 1.0 g NH_4NO_3, 1.0 g KH_2PO_4, 0.5 g $MgSO_4 \cdot 7H_2O$, 1.0 g yeast extract. Dissolve CMC in 500 mL warm H_2O before adding remainder of ingredients.
3. Protein Extraction Buffer: 50 mM Tris–HCl pH 7.5, 100 mM NaCl, 50 mM NaF, 5 mM EDTA, 1 mM EGTA, 1% Triton X-100, 10% glycerol, 1 mM phenylmethylsulfonyl fluoride (PMSF) (add immediately before use), and Protease inhibitor cocktail for fungal extracts (Sigma P8215) (add 20 μL into 2 mL extraction buffer immediately before use).
4. SDS-PAGE gel:
   - Resolving gel buffer: 1.5 M Tris–HCl pH 8.0, 0.4% w/v SDS, store at room temperature.
   - Stacking gel buffer: 0.5 M Tris–HCl pH6.8, 0.4% w/v SDS, store at room temperature.
   - Acrylamide: use premixed stock (BioRad), 37.5:1 ratio acrylamide:bis-acrylamide
   - 10% resolving gel (10 mL): 4.2 mL water, 2.5 mL resolving gel buffer, 3.4 mL acrylamide; to polymerize add 45 μL of 10% ammonium persulfate (APS) and 10 μL of N, N, N,N′-Tetramethyl-ethylenediamine (TEMED).
   - Stacking gel (5 mL): water 3.75 mL, stacking gel buffer 0.75 mL, and acrylamide 0.5 mL. To polymerize, add 40 μL of 10% APS and 15 μL of TEMED, mix well. Fill gels to top and insert the comb. This step must be done promptly because the gel will polymerize quickly.
   - Running buffer (5×): 72 g/L glycine, 15 g/L Tris base, 5 g/L SDS.
5. Nitrocellulose membranes (0.45 μm pore size)

6. Blotting apparatus: A semidry blotting apparatus (e.g., Hoefer Semi-Dry Transfer unit) works well for this application.
7. 10× stock tris-buffered saline (TBS) buffer: 1.54 M NaCl, 0.1 M Tris–HCl pH 7.5
8. 5% dry milk in TBS + 0.1% Tween 20 (w/v)
9. Chemiluminescence (ECL): Use a commercial kit such as Amersham ECL Plus™ Western Blotting Detection Reagents (RPN2132).
10. Phosphor imager: Visualize chemiluminescence with a phosphor imager such as the Storm 840 (GE Healthcare Life Sciences).
11. Stripping Buffer: Use the Restore™ Western Blot Stripping Buffer (Pierce) or a traditional stripping buffer such as 0.1 M glycine–HCl pH 3.0.

2.2. Glucokinase Activity Assay

2.2.1 Equipment

1. Spectrophotometer (for Bradford assay and kinase assay) and cuvettes.

2.2.2. Supplies and Reagents

1. NADP.
2. 0.2 U glucose-6-phosphate dehydrogenase.
3. Adenosine triphosphate.
4. Bradford reagent.
5. Buffer AT: 50 mM Tris–HCl pH 7.8, 5 mM $MgCl_2$, 1 mM EDTA, and 1% Trition X-100.
6. Assay buffer: 100 mM HEPES-KOH, pH 7.8, 10 mM $MgCl_2$, 2 mM ATP, 0.8 mM NADP, 0.2 U glucose-6-phosphate dehydrogenase, and 10 mM glucose.

3. Methods

3.1. Quantifying Phosphorylation of MAPKs in F. graminearum

The activation of MAPKs through phosphorylation of specific amino acid residues in a conserved region known as the activation loop is well conserved from fungi to mammals. To detect MAPK phosphorylation, commercial antibodies are available that are prepared against synthetic peptides corresponding to phosphorylated regions of MAPK activation loops. These antibodies are highly specific for phosphorylated MAPK proteins, and because activation loops are well-conserved across taxa, commercial anti-MAPK antibodies can be used effectively to assess MAPK phosphorylation in *F. graminearum*.

In the wheat and maize pathogen *F. graminearum*, signaling pathways underlying growth, development, and virulence converge on MAPK signaling cascades mediated by Mgv1 and Gpmk1 (4, 5). Measuring the phosphorylation of specific MAPK proteins is a powerful tool for functional genomics and the elucidation of complex signal transduction pathways. For example, *RAS2* of *F. graminearum* was recently determined to be crucial for pathogenesis, and functions at least in part by modulating the activation of the Gpmk1 MAPK pathway (6), as determined using the protocol described below. This protocol is derived largely from protocols developed to study MAPK phosphorylation in *Magnaporthe grisea* (7, 8).

3.1.1. Fungal Cultures

1. To induce production of macroconidia, inoculate CMC liquid cultures (50 mL in a sterile 250 mL flask) with an agar plug. Grow for 3–5 days at room temperature with shaking at 150 RPM.
2. Collect macroconida by centrifugation. Inoculate 5× YEG cultures; shake at 150 rpm at room temperature for 3–4 days.
3. Harvest tissue by centrifugation or filtration with a filter funnel under vacuum. Rinse tissue twice with ice-cold sterile water and store on ice.

3.1.2. Protein Extraction and Quantification

1. Place approximately 1 g of glass beads (0.5 mm) into the bead-beater screw cap tube. Add 1 mL of Protein Extraction Buffer and place on ice.
2. Suspend mycelia (about 150–200 mg; collected by filtration or centrifugation) in a tube prepared in step 1. Add enough Protein Extraction Buffer to completely fill the tube. Screw the cap on, incorporating as little air as possible.
3. Homogenize with a bead beater five times, each for 40 s. With the Fulltech Mini Bead beater, use a time setting of 3 and a speed setting of 48. Place the tube on ice between each pulse.
4. Centrifuge at 20800 RCF in a microcentrifuge for 20 min at 4°C.
5. Transfer the supernatant to a centrifuge tube suitable for ultracentrifugation. Centrifuge supernatants at 165000 RCF for 1 h at 4°C in a Beckman Ti50 rotor.
6. Quantify total protein concentration with Bradford reagent. First, make a range of dilutions (1, 1:10, 1:100, 1:1,000) of protein samples with distilled water in a total volume of 100 μL. For the calibration curve, pipette duplicate volumes of 10, 20, 40, 60, 80, and 100 μL of 1 mg/mL standard solution into test tubes, and make each up to 100 μL with distilled water. Also include a reagent blank by adding 100 μL of distilled water only. Then, add 5 mL of protein reagent to each tube and mix well by inversion. Measure the A_{595} of

the samples and standards against the reagent blank 20 min after mixing.

7. Adjust volumes as needed to normalize concentrations among samples/replicates.

3.1.3. Protein Separation and Blotting

1. Separate total proteins (approximately 30 μg) on SDS-PAGE at 200 V for about 1 h in 1× running buffer.
2. Transfer proteins to nitrocellulose membranes (use a blotting apparatus) following standard Western blotting protocols (9).

3.1.4. Detection of Phosphorylation

1. Block the membrane with 5% blotto or dry milk in TBS/0.1% Tween 20.
2. Successively probe western blots with anti-p42/44 MAPK (to detect total MAPK proteins), antiphospho-p42/44 MAPK (to detect phosphorylated MAPK proteins) and antiactin (to verify equal loading among samples). Incubate for 2 h at room temperature in a sealed Tupperware container with gentle rocking. Rinse with TBS/0.1% Tween 20 three times (10 min each) with gentle shaking at room temperature. Probe with the secondary antibody and wash membranes with TBS/0.1% Tween 20 (see Note 3).
3. Visualize antibody binding with enhanced chemiluminescence (ECL) kit. Mix ECL Plus Solution A and ECL Plus Solution B at 40:1 (1 mL solution/10 cm² membrane); incubate the membrane in the mixed detection solution for 5–10 min. After addition of chemiluminescent substrate, wrap blot in plastic wrap, place in a phosphor imaging cassette with a freshly blanked imaging screen, and incubate for 5 min to 1 h, depending on the anticipated signal strength. Visualize chemiluminescence with a phosphor imager (see Note 2).
4. Determine the intensity of each band with software such as ImageJ (http://rsbweb.nih.gov/ij/). First, confirm equal loading among samples (as determined by antiactin hybridizations). Then, calculate the relative activation of individual MAPKs by dividing the intensity of the band(s) representing the phosphorylated MAPK by the intensity of the corresponding nonphosphorylated form, with the level in the controls arbitrarily set as 1.0.
5. Between hybridizations, strip blots with Restore™ Western Blot Stripping Buffer (Pierce) or a traditional stripping buffer for 5–30 min with gentle shaking at room temperature (see Note 1).

3.1.5. Example of Results

In the previous work, we disrupted *RAS2* of *F. graminearum* and determined the effect of the mutation on phosphorylation of MAPKs Mgv1 and Gpmk1 (6). Under the conditions tested,

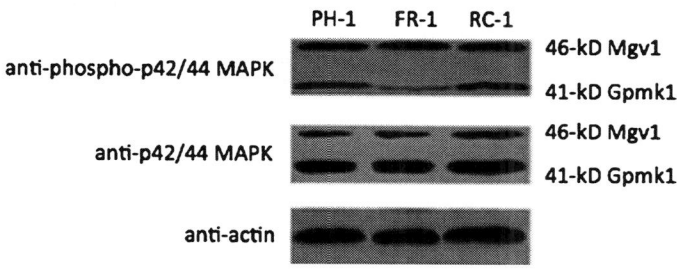

Fig. 1. Phosphorylation of MAPKs in *F. graminearum*. Total proteins were extracted from the wild type (strain PH-1), *RAS2*-disruption mutant (strain FR-1), and the mutant complemented with a wild-type copy of the *RAS2* gene (strain RC-1) and separated on SDS-PAGE gels. Phosphorylation of Gmpk1 was substantially reduced in the FR-1 strain, as indicated by the faint band detected with the anti-phospho-p42/44 MAPK antibody (*top panel*). In contrast, neither Gpmk1 expression, Mgv1 expression, nor Mgv1 phosphorylation was affected in the FR-1 strain. Modified from (6) with authors' permission.

disruption of *RAS2* did not reduce overall levels Mgv1 or Gpmk1, nor was phosphorylation of Mgv1 significantly altered (Fig. 1). However, disruption of *RAS2* substantially reduced the phosphorylation of Gpmk1, as indicated by a less intense band after hybridization with the anti-phospho-p42/44 MAPK antibody (Fig. 1). Normal phosphorylation of Gpmk1 was restored when the mutant strain was complemented with a wild-type copy of the *RAS2* gene, thus confirming that *RAS2* interacts with the Gpmk1 signaling cascade.

3.2. Measurement of Glucokinase Activity in *F. verticillioides*

Glucokinases catalyze the phosphorylation of glucose to glucose-6-phosphate. Directly monitoring the production of glucose-6-phosphate is technically challenging, however, because the product lacks distinguishing spectral properties. To circumvent this problem, glucokinase activity can be determined indirectly with a coupled enzyme assay. Glucose-6-phosphate produced by phosphorylation of glucose with glucokinase is used as the substrate for glucose-6-phosphate dehydrogenase, which in turn catalyzes the formation of 6-phosphgluconate through the reduction of NADP+ to NADPH (Fig. 2). Unlike NADP+, NADPH strongly absorbs UV light, thus facilitating spectrophotometric quantification of its production by monitoring absorbance at 340 nm.

The maize ear rot pathogen *F. verticillioides* produces fumonisin mycotoxins during kernel colonization. Although the regulation of fumonisin biosynthesis in the kernel environment is not fully understood, carbohydrate metabolism may play a key role (10). In the previous work, we determined total glucokinase activity in the wild-type strain and a defined mutant of *F. verticilloides* during kernel colonization to study the interplay between glucose

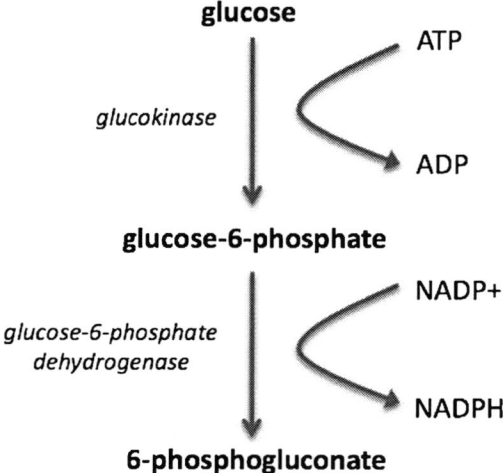

Fig. 2. Conceptual overview of the coupled enzyme assay for glucokinase activity.

metabolism, fumonisin biosynthesis, and the disruption of a transcription-factor encoding gene required for fumonisin biosynthesis during kernel colonization (11). A detailed protocol follows below, adapted in part from techniques described by Olsson et al. (12).

3.2.1. Culture Conditions

1. Inoculate PDA plates with a culture of *F. verticillioides*. Incubate at 28°C for 3–5 days. Under these conditions, the wild-type strain produces large amounts of microconidia (see Note 4).
2. Inoculate liquid media such as potato dextrose broth with 10^6 microconidia. Incubate at 28°C with shaking at 150 rpm for 48 h (see Notes 5 and 6).

3.2.2. Protein Extraction

1. Harvest fungal tissue from liquid cultures through centrifugation or filtration. Rinse tissue twice with sterile water to remove residual growth medium.
2. Grind tissue (1–5 g) under liquid nitrogen in a mortal and pestle. Add approximately 200 mg sterile glass beads (100–250 µm) to increase cell disruption. Grind vigorously, adding more liquid nitrogen as required, until samples become a fine powder.
3. Extract ground samples for 10 min in buffer AT at room temperature with gentle shaking.
4. Determine the total protein concentration for each extract with Bradford reagent. Adjust volumes of each sample as needed to normalize total protein concentrations.

3.2.3. Measuring Glucokinase Activity

1. Add protein (10 µg) to freshly prepared assay buffer in a final volume of 1 mL (see Note 7).

2. Incubate at 30°C for 20 min.

3. Measure A_{340} for each sample with a spectrophotometer. Blank with assay buffer.

4. Calculate units of glucokinase activity/mg total protein. A unit of glucokinase activity catalyzes the formation of 1 µmol NADPH/min. The millimolar extinction coefficient of β-NADPH at 340 nm = 6.22.

4. Notes

1. Suitable conditions to strip blots must be determined empirically. Start with a short incubation in stripping solution; expose blot to phosphor imaging screen to verify antibody removal. If residual signal is detected, repeat stripping procedure.

2. To get an optimal signal, the time of exposure to the imaging screen may need to be adjusted to reflect protein loading and/or phosphorylation levels.

3. If the signal of anti-phospho-p42/44 MAPK is not strong enough, increasing the concentration of anti-phospho-p42/44 MAPK will often yield a stronger signal. Increasing the secondary antibody, however, often increases the background. Another way to increase signal is to add Phosphatase Inhibitor Cocktail (Sigma P5726) in the protein extraction buffer.

4. Growth parameters (time, temperature) will vary widely among fungal species.

5. Liquid media can be varied as needed (carbon source, pH, etc.) to determine glucokinase activity in response to a wide range of environmental conditions.

6. When extracting fungal tissue grown on complex matrices such as maize kernels that may exhibit endogenous glucokinase activity, analyze noninoculated samples as negative controls.

7. The amount of protein added to the glucokinase assay buffer must be determined empirically and may vary considerably among developmental stages, growth substrates, and mutant strains.

References

1. Manning G, Plowman GD, Hunter T, Sudarsanam S (2002) Evolution of protein kinase signaling from yeast to man. Trends Biochem Sci 27:514–520
2. Chen RE, Thorner J (2007) Function and regulation in MAPK signaling pathways: lessons learned from the yeast *Saccharomyces cerevisiae*. Biochim Biophys Acta 1773: 1311–1340
3. MacCorkle RA, Tan TH (2005) Mitogen-activated protein kinases in cell-cycle control. Cell Biochem Biophys 43:451–461

4. Hou Z, Xue C, Peng Y, Katan T, Kistler HC, Xu JR (2002) A mitogen-activated protein kinase gene (MGV1) in *Fusarium graminearum* is required for female fertility, heterokaryon formation, and plant infection. Mol Plant Microbe Interact 15:1119–1127
5. Jenczmionka NJ, Maier FJ, Lösch AP, Schäfer W (2003) Mating, conidiation and pathogenicity of *Fusarium graminearum*, the main causal agent of the head-blight disease of wheat, are regulated by the MAP kinase gpmk1. Curr Genet 43:87–95
6. Bluhm BH, Zhao X, Flaherty JE, Xu JR, Dunkle L (2007) RAS2 regulates growth and pathogenesis in *Fusarium graminearum*. Mol Plant Microbe Interact 20:627–636
7. Bruno KS, Tenjo F, Li L, Hamer JE, Xu JR (2004) Cellular localization and role of kinase activity of PMK1 in Magnaporthe grisea. Eukaryot Cell 3:1525–1532
8. Xu JR, Hamer JE (1996) MAP kinase and cAMP signaling regulate infection structure formation and pathogenic growth in the rice blast fungus Magnaporthe grisea. Genes Dev 10:2696–2706
9. Sambrook J, Russell DW (2001) Molecular cloning: a laboratory manual, 3rd edn. Cold Spring Harbor Laboratory Press, New York
10. Bluhm BH, Woloshuk CP (2005) Amylopectin induces fumonisin B-1 production by *Fusarium verticillioides* during colonization of maize kernels. Mol Plant Microbe Interact 18:1333–1339
11. Bluhm BH, Kim H, Butchko RA, Woloshuk CP (2008) Involvement of *ZFR1* of *Fusarium verticillioides* in kernel colonization and the regulation of *FST1*, a putative sugar transporter gene required for fumonisin biosynthesis on maize kernels. Mol Plant Pathol 9:203–211
12. Olsson T, Thelander M, Ronne H (2003) A novel type of chloroplast stromal hexokinase is the major glucose-phosphorylating enzyme in the moss *Physcomitrella patens*. J Biol Chem 278:44439–44447

Chapter 16

A Detailed Protocol for Chromatin Immunoprecipitation in the Yeast *Saccharomyces cerevisiae*

Melanie Grably and David Engelberg

Abstract

Critical cellular processes such as DNA replication, DNA damage repair, and transcription are mediated and regulated by DNA-binding proteins. Many efforts have been invested therefore in developing methods that monitor the dynamics of protein–DNA association. As older techniques such as DNA footprinting, and electrophoretic mobility shift assays (EMSA) could be applied mostly in vitro, the development of the chromatin immunoprecipitation (ChIP) method, which allows quantitative measurement of protein-bound DNA most accurately in vivo, revolutionized our capabilities of understanding the mechanisms underlying the aforementioned processes. Furthermore, this powerful tool could be applied at the genomic-scale providing a global picture of the protein–DNA complexes at the entire genome.

The procedure is conceptually simple; involves rapid crosslinking of proteins to DNA by the addition of formaldehyde to the culture, shearing the DNA and immunoprecipitating the protein of interest while covalently bound to its DNA targets. Following decrosslinking, DNA that was coimmunoprecipitated could be amplified by PCR or could serve as a probe of a genomic microarray to identify all DNA fragments that were bound to the protein.

Although simple in principle, the method is not trivial to implement and the results might be misleading if proper controls are not included in the experiment. In this chapter, we provide therefore a highly detailed protocol of ChIP assay as is applied successfully in our laboratory. We pay special attention to describe every small detail, in order that any investigator could readily and successfully apply this important and powerful technology.

Key words: *Saccharomyces cerevisiae*, Chromatin immunoprecipitation, DNA binding proteins, Transcription, Heat shock

1. Introduction

1.1. Critical Cellular Processes Are Mediated by DNA-Binding Proteins

Some crucial cellular processes, such as DNA replication, transcription and DNA repair are carried out and regulated by DNA-binding proteins. These processes are dynamic, imposing dramatic, but rapidly changing, effects on chromatin organization.

Thus, the spectra of proteins associated with a specific region of the DNA are different at any given time. Trapping the protein–DNA complexes at any desired time-point in vivo, identification of the proteins associated with the DNA, and revealing the posttranslational modifications they possess, is crucial for understanding the mechanism of action and the regulation of transcription, replication, repair, and chromatin organization.

1.2. ChIP Technology Allows Identification of Proteins Associated with DNA at a Given Time In Vivo

The method described here, Chromatin immunoprecipitation (ChIP) is a powerful, accurate, and quantitative technique that allows the identification of proteins associated with DNA in vivo. The power and accuracy of the method stem from two elements. One, the first step of the ChIP procedure is treating the cells with formaldehyde. The purpose of this treatment is to covalently crosslink molecules that are in close proximity (including proteins associated with DNA), but at the same time formaldehyde instantly inactivates most or all cellular activities, thereby freezing the situation at the particular moment desired by the investigator. Two, the method is not restricted to the identification of one protein and one DNA fragment. Because of the initial crosslinking step, proteins that are not directly associated with DNA could be identified as well, protein complexes could be isolated and finally, the method could be readily applied in full genomic and proteomic scales (1–11).

1.3. The Principles of the Technique Are Rather Simple

Step 1: Crosslinking. At the desired time point, or following the experimental operation of interest, a yeast culture is treated with formaldehyde (Fig. 1a). All enzymatic activities immediately cease and protein–protein and protein–DNA that reside in close proximity are crosslinked.

Step 2: DNA shearing. Prepare whole cell lysate and mechanically cleave DNA (usually by sonication). This treatment provides relatively small DNA fragments with proteins still crosslinked to them (Fig. 1b).

Step 3: Immunoprecipitation. Using specific antibodies, immunoprecipitate the protein of interest (Fig. 1c). If the protein is crosslinked to DNA fragments, they will be coprecipitated along with the protein.

Step 4: Reverse crosslinking of protein–DNA in the immunoprecipitated complex and remove proteins to purify DNA. DNA that was coprecipitated is now separated from proteins purified and could be readily analyzed (Fig. 1d).

Step 5: Analyze DNA by PCR with primers that recognize DNA fragments of interest (e.g., enhancers, promoters, origin of replication, DNA-damaged sites) as is shown in Fig. 1e. Alternatively, the purified DNA could also be analyzed on a microarray chip.

If the immunoprecipitated protein is expected to bind DNA in higher quantities after the biological experimental operation

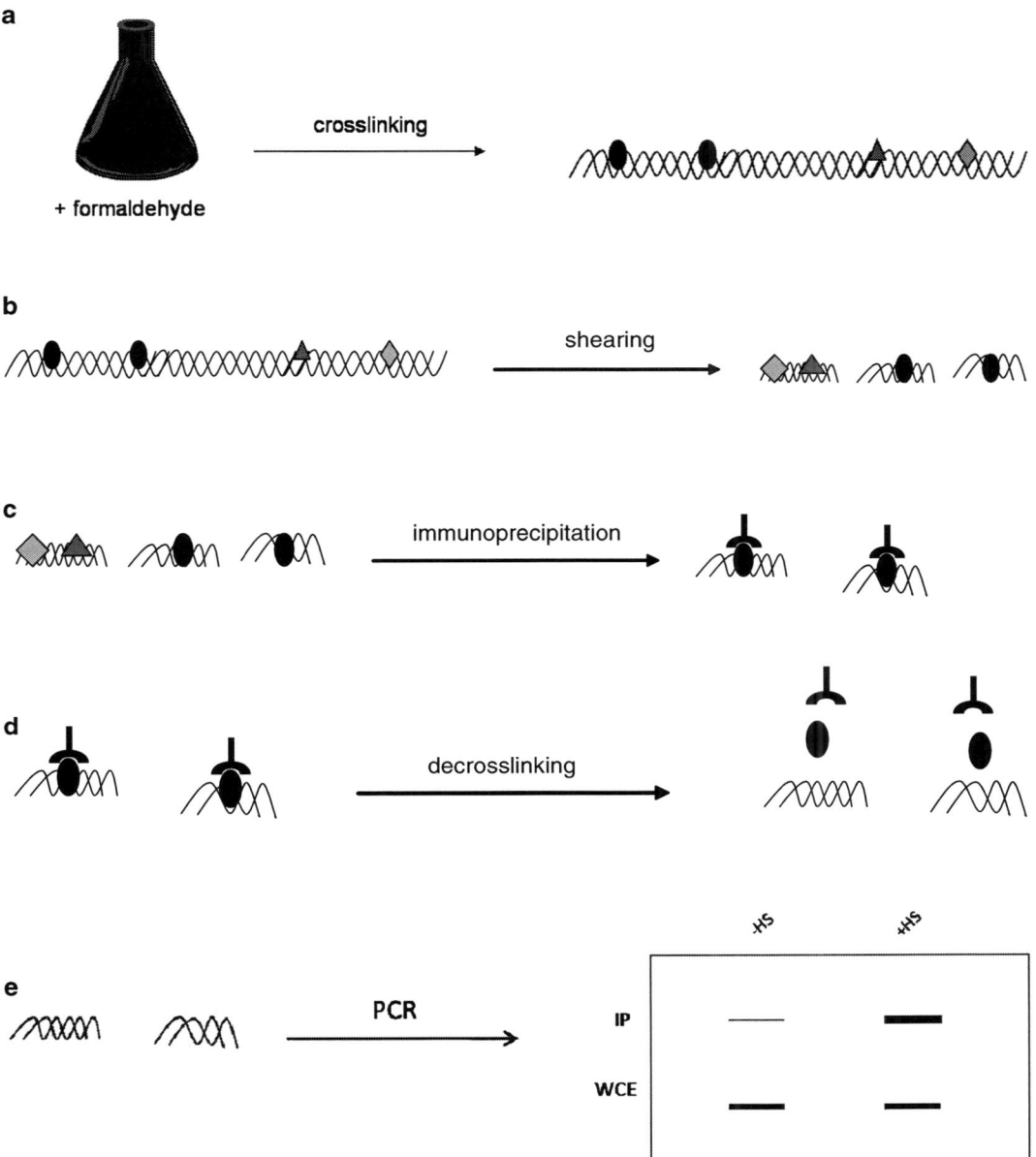

Fig. 1. Stages of ChIP: (**a**) Crosslinking, (**b**) DNA shearing, (**c**) Immunoprecipitation, (**d**) Reverse crosslinking, (**e**) Analysis of DNA products by PCR.

[such as heat shock (HS) as is the case in the figure above], then more DNA is coprecipitated and larger amounts of amplified DNA fragments are obtained (upper bands in the figure above). When the same PCR reaction is performed on nonimmunoprecipitated protein–DNA complexes (WCE) no change in DNA quantities and PCR products should be measured (lower bands).

1.4. Although Simple in Principle, Applying the ChIP Protocol Could Be Problematic

The power of the method on one hand and the relative simplicity of its implementation on the other made ChIP a widely used technology (12–15). Also, in many studies, the DNA and the proteins recovered by ChIP were further used in a more global manner to identify all DNA sequences occupied by a single protein (ChIP-chip) and all proteins occupying a specific DNA region (proteome ChIP). Consequently, the ChIP technology lead to the acquirement of unequivocal important and surprising data, including mapping the changes in histone modifications during the course of transcription (8, 10, 16–19), global views on changes in nucleosome organization (18, 20), revealing proteins recruited to DNA damage sites and (totally unexpected) identification of components that function in cytoplasmic signal transduction pathways (mainly kinases), as DNA binding proteins (21, 22).

Yet, in spite of the relative simplicity of the method in principle, it is not trivial to implement the ChIP procedure properly for the first time. Furthermore, the accuracy and quality of several steps in the procedure may vary significantly between laboratories, sometimes because of small differences between one laboratory set-up and the other. Experimental quality could differ even between one experiment and another because they are dependent on specific instruments (e.g., efficiency of sonicator, or properties of the PCR machine) or quality of reagents (mainly the specificity and avidity of the particular antibodies used). Applying ChIP for the first time could be assisted by the many protocols already published (23–26), or available in the internet, but in many cases fine-details are required to have the method work. Therefore, the purpose of the protocol provided below is to describe the ChIP method, as it is applied in our laboratory, in the highest resolution possible, so that any small detail is mentioned and could be repeated faithfully. In addition, we provide a large number of references covering in details various aspects of the ChIP method, such as setting the proper controls and different ways for quantitating the data.

2. Materials

2.1. Solutions Required for Chromatin Immunoprecipitation

1. Formaldehyde solution (should be prepared fresh): 11% formaldehyde (from 37% stock solution), 0.1 M NaCl, 1 mM EDTA, 50 mM Hepes-KOH, and pH7.5.
2. 3 M glycine (can be stored, preferably at 30°C, for a couple of months): Appropriate amount of glycine in DDW sterilized using 0.2 micron filter. Some heating may be required to dissolve glycine. We also found it best to store the 3 M glycine in a 30°C incubator.

3. TBS (can be prepare as 5 L stock and store at 4°C): 20 mM Tris–HCl pH7.5, 150 mM NaCl.

4. Lysis buffer: 50 mM Hepes-KOH pH7.5, 150 mM NaCl, 1 mM EDTA, 1% Triton X-100, and 0.1% sodium deoxycholate. Can be prepared in advance without PMSF and stored at 4°C. Before use add 1 mM PMSF and 0.1% or 0.5% SDS or 500 mM NaCl as required at time of use.

5. TE buffer (store at room temperature): 10 mM Tris–HCl pH8.0, 1 mM EDTA.

6. 2× pronase buffer (store at room temperature): 50 mM Tris–HCl pH7.5, 10 mM EDTA, and 1% SDS.

7. Pronase (20 mg/ml in DDW; Boehringer Mannheim).

8. LETS Buffer (prepare fresh): 0.1 M LiCl, 0.01 M EDTA pH8, 10 mM Tris–HCl pH7.4, and SDS 0.2% (see Note 1).

9. Acrylamide gel (for 1 gel): 6 mL TBE 5×, 7.5 mL 40% Bis-acrylamide (37:5:1), 16.5 mL DDW, 0.3 mL APS 10%, and 30 μL Temed. Run gels in 1× TBE buffer.

2.2. Equipment

1. Vortex.
2. Centrifuges: Sorvall RC-5 centrifuge with a GSA rotor (for 250 mL centrifuging bottles).
3. Centrifuge for 50 mL falcon tubes and 15 mL snap-cap tubes and micro eppendorf tubes SIGMA 4 K15, or Eppendorf 5810R cooled to 4°C.
4. Microcentrifuge cooled to 4°C and room temperature.
5. Ultra centrifuge with a Ti-50 rotor; we use Beckman L8-70 M.
6. Sonicator (Sonics & Materials, Inc. Ct, USA. Model VX760 Vibra cell™ with a probe model CV33).
7. PCR machine; any Biometra thermocycler with a silver-plated heating block (see Note 2).

3. Methods

The following protocol yields material for five immunoprecipitation reactions. The protocol is general, but follows a particular example of an experiment, routinely performed in our laboratory. The experiment monitors changes in the spectra of proteins occupying a specific DNA fragment (promoter of the *HSP104* gene) at different time points following exposure of various yeast strains to heat shock.

3.1. Preparation of Yeast Cells

1. Grow 100 mL of the desired yeast strains in YPD overnight at 30°C with shaking.

2. The following morning, dilute yeast cells to O.D.$_{600}$ = 0.25 and continue growth. Use 200 mL per time point or treatment. The total volume of the culture depends on the number of time points to be taken and the number of treatments one wishes to apply (see Note 3).

3. Grow culture to O.D.$_{600}$ ≅ 0.8, divide the culture to separate flasks (200 mL/flask) according to the number of treatments (see Note 4).

3.2. Crosslinking

1. At each desired time point (we routinely used 5, 10, and 15 min after applying heat shock), remove one flask of each strain and supplement with 20 mL 11% formaldehyde solution.

2. Further, shake the flask gently at room temperature (in a hood, as formaldehyde is quite toxic) for 20 min (see Note 5).

3. Quench crosslinking with 30 mL of 3 M glycine and further incubate at room temperature for 5 min.

4. Cool 250 mL centrifuge bottles by placing on ice (the ice should cover at least half of the bottle). The centrifuge bottles do not need to be sterilized.

5. Pellet cells by centrifugation at 4,000 rpm (2600 × g) in a GSA rotor of Sorvall RC-5 centrifuge in the cold bottles for 5 min at 4°C (see Note 6).

6. Wash the pellet twice with 100–150 mL cold TBS buffer (see Note 7).

7. At this step, it is possible to store the pellets in –70°C. If this is the case, perform a final wash with 10 mL cold TBS; transfer the cells to a 50 mL falcon tube, spin for 5 min in a table-top centrifuge, (such as SIGMA 4 K15, or eppendorf 5810R) at 4°C, 4,000 rpm (3200 × g) and decant the supernatant. The tubes with the pellet can be stored up to 1 week at –70°C.

3.3. Cell Breakage

1. Add 10 mL cold lysis buffer/0.1% SDS to the cell pellet in 50 mL falcon tubes, spin in a cooled table-top centrifuge at 4,000 rpm (3200 × g) for 5 min and decant the supernatant.

2. Add to the pellet 1 mL cold lysis buffer/0.5% SDS.

3. Split each sample to two glass tubes (~15 mL round-bottom, such as Kimble 16 × 100 mm) and add ~1.125 mL glass beads. Each tube thus contains about 0.750 mL cells and 1.125 mL glass beads.

4. Vortex the tubes for 15 min. These 15 min should be performed at rounds of 30 s at maximal speed with a minimum of 30 s intervals on ice (see Notes 8 and 9).

5. After vortexing leave the tubes on ice.

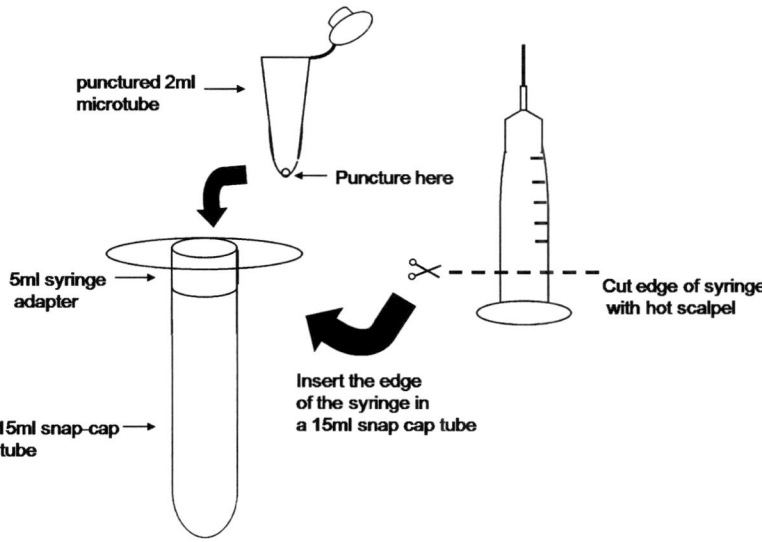

Fig. 2. The "lysate collecting apparatus" (see text for details on construction).

3.4. Collection of Cell Lysate

1. To the top of a 15 mL snap-cap tube insert (as an adapter) the top edge of a 5 mL syringe, which was cut out of the syringe with a hot scalpel blade (see Fig. 2).
2. Puncture the bottom of a 2 mL micro tube with a hot needle (this will allow for the passage of the cell lysates while trapping the beads in the micro tube; see Fig. 2).
3. Insert the micro tube into the adapter (placed on top of the 15 mL snap-cap tube) and place the whole ensemble on ice.
4. Add 1 mL of cold lysis buffer/0.1% SDS to the mix of broken cells and glass beads, mix in an "up and down" motion and transfer the cells to the 2 mL punctured micro tube on top of the "lysate collecting apparatus."
5. Repeat this step 4–6 times until washings come out clear from the beads.
6. Spin down the whole "ensemble" at 1,000 rpm (200 × g) for 2 min at 4°C.
7. Remove as much of the sup as possible to an appropriate ultracentrifuge tube and centrifuge in a Ti50 rotor at 45,000 rpm (150,000 × g) for 30 min at 4°C.

3.5. Chromatin Shearing

1. Decant the supernatant from the ultracentrifuge tube and with a very thin spatula scrape off the pellet and transfer it to a micro centrifuge Eppendorf tube.
2. Wash the ultracentrifuge tube with 1.2 mL cold lysis buffer/0.1% SDS and transfer to the microcentrifuge tube containing the pellet.

3. Sonicate each tube three cycles of 20 s with 1 min on ice interval (see Note 10). The goal of this step is to get DNA fragments ranging from 100 to 1,000 bp, while the majority should be of ~500 bp (see Note 11).

4. Centrifuge at 15,000×g for 15 min in a cooled (4°C) Eppendorf centrifuge and transfer the sup to a 15 mL tube.

5. To each tube add 3.25 mL cold lysis buffer/0.1% SDS. Aliquot 0.7 mL (set 0.25 mL aside), freeze in liquid nitrogen and store at −70°C. Material can be kept at this point up to 1 month.

3.6. Chromatin Verification

It is important to perform this step until consistency in chromatin preparation is acquired.

1. In order to verify whether the sonication and chromatin preparations were successful, take the 0.25 mL sample set aside (from step 5 above), add 0.25 mL 2× pronase buffer and 20 µL of 20 mg/mL pronase.

2. Incubate for 2 h at 42°C and then transfer to 65°C for 6 h or overnight.

3. Add 50 µL 4 M LiCl, extract once with phenol-chloroform (in LETS buffer) and then with chloroform.

4. To precipitate DNA, add 20 µg glycogen and 1 mL ethanol and incubate for 1 h. Centrifuge for 10 min at maximal speed in an eppendorf centrifuge, resuspend in 30 µL TE, add 1 µL RNase (10 mg/mL), and incubate for 30 min at 37°C.

5. Analyze samples by electrophoresis on 1% agarose gel. A smear should appear ranging from 100 to 1,000 bp, while the majority is concentrated at 500 bp.

3.7. Immunoprecipitation and Elution

For each immunoprecipitation reaction, use one of the 0.7 mL aliquots stored at −70°C.

1. Mix 0.5 mL of the 0.7 mL aliquot with antibodies, already prebound to protein G sepharose beads (Amersham), and incubate at 4°C on a rotator for 2 h (save aside the remaining 0.2 mL) (see Note 12).

2. Following the 2 h incubation, centrifuge at low speed (500 rpm (50 × g) in an Eppendorf 5810R table top centrifuge for 2 min, and remove the sup while taking care not to remove beads.

3. Wash the content of each tube as follows. In each step, incubate samples for 5 min at room temperature on a rotator. After each washing step, centrifuge at low speed for 2 min, remove the supernatant, taking care not to remove beads. Washing includes the following steps:

4. Twice in 1.4 mL lysis buffer/0.1% SDS

5. Twice in 1.4 mL lysis buffer/0.1% SDS, 0.5 M NaCl

6. Once in 1.4 mL 10 mM Tris–HCl pH 8.0, 0.25 mM LiCl, 1 mM EDTA, 0.5% NP-40, 0.5% Na-deoxycholate

7. Once in 1.4 mL TE.

8. Elute the immunoprecipitated material from the beads by heating for 10 min at 65°C in 0.25 mL 2× pronase buffer.

9. Pellet beads, transfer the eluted material to a new micro tube and add 0.25 mL TE buffer.

3.8. Decrosslinking

1. Transfer 50 µL of the remaining 0.2 mL of the crosslinked chromatin (that were not ran through the immunoprecipitation process) to a fresh micro tube, add 0.2 mL TE buffer and 0.25 mL 2× pronase buffer. This is the so called Whole Cell Extract (WCE) fraction of the experiment that will serve as a critical comparison control for the experiment.

2. To all tubes, (including those from Subheading 3.7) add 20 µL of 20 mg/mL pronase and incubate for 2 h at 42°C, then 6 h (or overnight) at 65°C.

3. Add 50 µL 4 M LiCl and extract with phenol-chloroform (in LETS buffer) and then with chloroform.

4. To precipitate DNA, add 20 µg glycogen [1 µL of pellet paint (Novagen) can also be used instead of glycogen] and 1 mL ethanol and incubate overnight at –20°C. Wash the pellet with 100% ethanol and resuspend in 0.2 mL TE.

3.9. PCR Analysis

We use 10 µL of IP'ed material for the PCR reactions, or 10 µL of 1:50 dilution (in TE) of the WCE.

1. PCR mix. Reactions are carried out in 50 µL in the presence of: 5 µL of 10 µM primers (1 µM final), 2 µL of 10 mM dNTPs (2.5 mM final), 5 µL 10× Taq Polymerase buffer with 25 mM $MgCl_2$, 3 µL of 25 mM $MgCl_2$ (0.3 mM final), 0.1 mCi/mL of [α-P32]dCTP (specific activity 3,000 Ci/mmol, we use 0.5 µL/reaction), and 0.5 µL of Taq DNA polymerase (SuperTherm) (see Note 13).

2. PCR steps. The following program has worked for us with many different primers, however, specific conditions must be carefully determined in each case (see Notes 14 and 15): 94°C 8 min, 29 cycles × [94°C 1 min, 47.9°C 1 min, 72°C 1 min, 72°C for 10 min] then a pause at 10°C.

3. Add 4 µL of 6× sample buffer to the PCR reaction tubes and load 30 µL onto a 8% acrylamide::bis-acrylamide TBE gel (no prerunning of the gel is necessary).

4. Run the gel at 100 V until the loading dye has run 2/3 of the gel.

5. Remove the gel from the gel running apparatus and place it on a 3 mm Whatmann paper, cut slightly larger than the size of the gel.

6. Cover the gel with saran wrap (only top side) and dry the gel for 1 h at 80°C under vacuum using a gel dryer (see Note 16).

7. Place film, expose for 1 h in –80°C, and develop the film. Expose for longer or shorter periods of time according to the results (see Note 17).

3.10. Controls and Results Analysis

Observing the PCR results, how could one assess whether the whole process was actually successful?

1. Check that the PCR reactions formed on the WCE fractions give similar signals to the samples. This is critical since the ChIP results are normalized to the signal obtained with the WCE.

2. Protein bound DNA should be specifically enriched following immunoprecipitation. This could be verified by using a control of a yeast strain lacking the relevant protein. ChIp applied on a lysate prepared from such a knockout strain should give no signal, verifying the specificity of the antibodies and the other reagents, as well as the accuracy of the procedure.

3. If using a tagged protein in the immunoprecipiation another elegant control would be an identical strain, harboring the same protein with no tag. A PCR product should not be obtained from the latter strain with ChIP performed with anti tag antibodies.

4. One should verify that the amplified DNA corresponding to the ChIP experiment is increased or decreased in response to the biological treatment, in particular where expected (see below).

5. An important control is to run a PCR reaction with the same DNA (obtained in the ChIP), but using primers recognizing nonrelevant genes, or nonrelevant regions. For example, when studying the binding of a transcription factor, one should also use primers recognizing a promoter region of a gene to which the transcription factor studied does not bind. No product should be obtained (see Note 18).

As an example for some of the controls required, we present an experiment in which we monitored the changes in association with DNA and covalent modification occurring in the histone H4 when bound to a promoter affected by heat shock (the *HSP104* promoter) (Fig. 3).

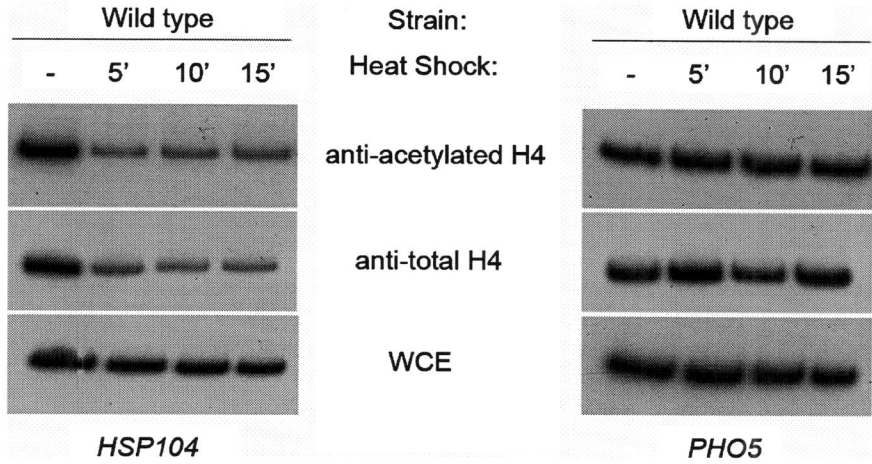

Fig. 3. Acetylated histone H4 undergoes dissociation from the *HSP104* promoter as part of chromatin remodeling in response to heat shock. ChIP from wild type yeast cells grown under heat shock conditions (5, 10, and 15 min at 39°C "-" indicates control at 30°C). Antibodies recognizing acetylated histone H4 and total histone H4 were used for these ChIP assays. On the left panels, primers of the *HSP104* promoter region were used and in the right panels primers the *PHO5* gene were used to amplify DNA immunoprecipitated with the antibodies mentioned above.

1. Lysates were prepared for ChIP assays from a culture collected before heat shock (-) and at the indicated time points after heat shock.

2. ChIP was performed using either anti-total histone H4 antibody or anti-acetylated H4 antibody.

3. Following ChIP, PCR reactions were performed using primers recognizing the *HSP104* promoter or the *PHO5* promoter (nonresponsive to heat shock). As can be seen, under optimal growth conditions (-), histone H4 is fully acetylated on the *HSP104* promoter (left upper panel) indicating that both antibodies immunoprecipitated the proteins.

4. Following heat shock, the levels of total and acetylated H4 decrease. As both total H4 and acetylated H4 are no longer associated with the *HSP104* promoter after heat shock (Fig. 3), it seems that H4 undergoes "removal" or dissociation of histones from the nucleosomal structures on the promoter and not by deacetylation. But how can one tell that the result is specific? Perhaps, in response to heat shock all H4 proteins dissociate from all yeast promoters.

5. By using the same ChIP product, we performed PCR with primers recognizing a region of a control gene such as *PHO5* (right panels, DNA from the same ChIP preparation). We observed that levels of acetylated H4 as well as total H4 remain mostly constant throughout the process (upper and middle panels). Therefore, we can inevitably conclude that the results presented for the *HSP104* promoter regarding histone H4 dissociation are specific to this promoter.

For various explanations and ideas on how to analyze and interpret ChIP results, we recommend consulting with the following website:(http://images.cell.com/images/EdImages/etbr/STRUHL.PDF).

4. Notes

1. In order to prepare phenol/chloroform in LETS buffer, mix phenol, chloroform, and LETS buffer at a ratio of 1:1:2, stir o/n at room temp and then store at 4°C.
2. In our hands other machines work poorly, sometimes not providing a PCR product at all.
3. Because of the large volumes required for each time points, only a few strains can be assayed in one experiment. In the particular experiment described here for example, only two strains could be tested in a single experiment, or four strains per one investigator per day.
4. If heat shock is applied for example we divide each culture to four flasks, each containing 200 mL of the culture. One control flask goes back to the 30°C shaker, while the others are incubated in a 39°C bath with shaking.
5. Another option is to manually shake the flasks every other minute; either way works well.
6. From here on all steps must proceed on ice or in a 4°C room.
7. We regularly prepare the buffer in a 5 L container and store it in a 4°C room, so that it is always ready for use.
8. We recommend at this part of the protocol to work with no more than eight glass tubes. If using two vortex instruments simultaneously (with both hands), a single worker may be able to work with two tubes at a time, changing them with two other tubes every 30 s. In this way, cells in eight tubes are broken within 1 h.
9. Efficient and reproducible nature of the cell breaking step is important for the quality of the entire experiment. If using one of the various instruments now claimed to mimic the vortex effect, it should be first compared to the above-described manual man-operated vortex.
10. This step should be equilibrated depending on the type of sonicator used; we use a continuous pulsing at 37% amplitude.
11. The efficiency of the shearing will influence the degree of background observed when analyzing the ChIP data.
12. The amount of antibodies used for immunoprecipitation depends on properties of each antibody. We have used for

example the following amounts: 1 μg of anti-HA 3F10, Roche; 1 μL of antiacetyl histone H3, Upstate; 2 μL of antiacetyl histone H4, Upstate; 2 μg of antitotal histone H3, Abcam and 8 μL of anti-nonacetylated histone H4, Serotec.

13. If the PCR is performed in a quantitative real-time PCR (RT-PCR) machine with fluorescent nucleotides, there is no need of course for the radio labeled nucleotide (see ref (2)) for protocol and quantification of RT-PCR results). However, the use of radiolabeled nucleotide allows highly sensitive and quantitative reaction that is readily applied in any laboratory at any time. The expensive RT-PCR machine may not be readily available to many laboratories. In addition, the reagents for RT-PCR are very expensive. It should be reminded that PCR products cannot be quantified when monitored on standard agarose gel stained with ethidium bromide.

14. We found the PCR step rather not trivial most probably because of the extremely low quantities of DNA recovered through the ChIP protocol (and are difficult to amplify) and, on the other hand, because of undesired nonspecific DNA that is also precipitated and gives rise to nonspecific PCR products. PCR parameters should be therefore screened and calibrated (especially regarding annealing temperature. This should be done using a gradient PCR machine) according to the properties of the specific primers used.

15. The PCR machine used is quite critical; one that enables thin wall tubes as well as a gold/silver plated block which allows for rapid temperature changes is a must.

16. Drying the gel is important because diffusion of the bands occurs if the gel is not dried.

17. If using a phosphoImager, exposure time is easily adapted to the sensitivity of the instrument.

18. In many cases, the above negative controls do results in PCR products because DNA could be nonspecifically absorbed to antibodies, sepharose, or simply precipitated by some of the reagents. These nonspecific products, observed by many are clearly one of the big problems when performing ChIP. In most cases however, if nonspecific bands remain constant throughout the experiment, there is no need to be alarmed.

References

1. Bhaumik SR, Green MR (2002) Differential requirement of SAGA components for recruitment of TATA-box-binding protein to promoters in vivo. Mol Cell Biol 22:7365–7371
2. Geisberg JV, Struhl K (2004) Cellular stress alters the transcriptional properties of promoter-bound Mot1-TBP complexes. Mol Cell 14:479–489
3. Jasiak AJ, Hartmann H, Karakasili E, Kalocsay M, Flatley A, Kremmer E, Strasser K, Martin DE, Soding J, Cramer P (2008) Genome-associated RNA polymerase II includes the

dissociable Rpb4/7 subcomplex. J Biol Chem 283:26423–26427
4. Krogan NJ, Kim M, Ahn SH, Zhong G, Kobor MS, Cagney G, Emili A, Shilatifard A, Buratowski S, Greenblatt JF (2002) RNA polymerase II elongation factors of Saccharomyces cerevisiae: a targeted proteomics approach. Mol Cell Biol 22:6979–6992
5. Li J, Lin Q, Wang W, Wade P, Wong J (2002) Specific targeting and constitutive association of histone deacetylase complexes during transcriptional repression. Genes Dev 16:687–692
6. Moqtaderi Z, Struhl K (2004) Genome-wide occupancy profile of the RNA polymerase III machinery in Saccharomyces cerevisiae reveals loci with incomplete transcription complexes. Mol Cell Biol 24:4118–4127
7. Radonjic M, Andrau JC, Lijnzaad P, Kemmeren P, Kockelkorn TT, van Leenen D, van Berkum NL, Holstege FC (2005) Genome-wide analyses reveal RNA polymerase II located upstream of genes poised for rapid response upon S. cerevisiae stationary phase exit. Mol Cell 18:171–183
8. Robyr D, Suka Y, Xenarios I, Kurdistani SK, Wang A, Suka N, Grunstein M (2002) Microarray deacetylation maps determine genome-wide functions for yeast histone deacetylases. Cell 109:437–446
9. Simic R, Lindstrom DL, Tran HG, Roinick KL, Costa PJ, Johnson AD, Hartzog GA, Arndt KM (2003) Chromatin remodeling protein Chd1 interacts with transcription elongation factors and localizes to transcribed genes. EMBO J 22:1846–1856
10. Vogelauer M, Wu J, Suka N, Grunstein M (2000) Global histone acetylation and deacetylation in yeast. Nature 408:495–498
11. Wong MM, Cox LK, Chrivia JC (2007) The chromatin remodeling protein, SRCAP, is critical for deposition of the histone variant H2A.Z at promoters. J Biol Chem 282:26132–26139
12. Krebs JE, Kuo MH, Allis CD, Peterson CL (1999) Cell cycle-regulated histone acetylation required for expression of the yeast HO gene. Genes Dev 13:1412–1421
13. Kuras L, Borggrefe T, Kornberg RD (2003) Association of the mediator complex with enhancers of active genes. Proc Natl Acad Sci U S A 100:13887–13891
14. Larschan E, Winston F (2001) The S. cerevisiae SAGA complex functions in vivo as a coactivator for transcriptional activation by Gal4. Genes Dev 15:1946–1956
15. Swanson MJ, Qiu H, Sumibcay L, Krueger A, Kim SJ, Natarajan K, Yoon S, Hinnebusch AG (2003) A multiplicity of coactivators is required by Gcn4p at individual promoters in vivo. Mol Cell Biol 23:2800–2820
16. Deckert J, Struhl K (2001) Histone acetylation at promoters is differentially affected by specific activators and repressors. Mol Cell Biol 21:2726–2735
17. Kurdistani SK, Robyr D, Tavazoie S, Grunstein M (2002) Genome-wide binding map of the histone deacetylase Rpd3 in yeast. Nat Genet 31:248–254
18. Pokholok DK, Harbison CT, Levine S, Cole M, Hannett NM, Lee TI, Bell GW, Walker K, Rolfe PA, Herbolsheimer E, Zeitlinger J, Lewitter F, Gifford DK, Young RA (2005) Genome-wide map of nucleosome acetylation and methylation in yeast. Cell 122:517–527
19. Roh TY, Ngau WC, Cui K, Landsman D, Zhao K (2004) High-resolution genome-wide mapping of histone modifications. Nat Biotechnol 22:1013–1016
20. Bernstein BE, Liu CL, Humphrey EL, Perlstein EO, Schreiber SL (2004) Global nucleosome occupancy in yeast. Genome Biol 5:R62
21. Alepuz PM, de Nadal E, Zapater M, Ammerer G, Posas F (2003) Osmostress-induced transcription by Hot1 depends on a Hog1-mediated recruitment of the RNA Pol II. EMBO J 22:2433–2442
22. Pokholok DK, Zeitlinger J, Hannett NM, Reynolds DB, Young RA (2006) Activated signal transduction kinases frequently occupy target genes. Science 313:533–536
23. Aparicio O, Geisberg JV, Sekinger E, Yang A, Moqtaderi Z, Struhl K (2005) Chromatin immunoprecipitation for determining the association of proteins with specific genomic sequences in vivo. In: Ausubel FM, Brent R, Kington RE, Moore DD, Seidman JG, Smith JA, Struhl KE (eds) Current protocols in molecular biology. Wiley, New York, pp 21.3.1–21.3.33
24. Ezhkova E, Tansey WP (2004) Chromatin immunoprecipitation to study Protein-DNA Interactions in Budding Yeast, vol 313, 2nd edn, Methods in molecular biology. Humana Press Inc, Totawa, NJ, pp 225–244
25. Nelson JD, Denisenko O, Sova P, Bomsztyk K (2006) Fast chromatin immunoprecipitation assay. Nucleic Acids Res 34:e2
26. Ren B, Dynlacht BD (2003) Use of chromatin immunoprecipitation assays in genome-wide location analysis of mammalian transcription factors. Methods Enzymol 376:304–315

Chapter 17

A Method to Visualize the Actin and Microtubule Cytoskeleton by Indirect Immunofluorescence

Flora Banuett

Abstract

The cytoskeleton provides the basic architectural organization and shape of the eukaryotic cell, and plays a key role in segregation of the genetic material. A method to visualize the actin and microtubule cytoskeleton in the fungus *Ustilago maydis* by indirect immunofluorescence is described here. The method entails growth of cells to early logarithmic phase, fixation with a cross-linking agent or organic solvent, partial digestion of the cell wall and permeabilization of cells with a detergent to allow entry of antibodies, exposure to primary antibody, followed by treatment with secondary antibody conjugated to a fluorophore to allow visualization with fluorescence microscopy.

Key words: Microtubules, Fungi, Ustilago, Actin patches, Actin cables, Nuclear stain, Immunofluorescence, Secondary antibodies

1. Introduction

The actin and microtubule cytoskeletons are dynamic filamentous structures that provide the scaffold for transport of vesicles and mRNA, distribution of organelles (nucleus, ER, Golgi, mitochondria), polarized secretion, endocytosis, cytokinesis, spindle assembly, and chromosome movement. Work in yeasts and mammalian cells has greatly contributed to our understanding of the dynamics of the cytoskeleton and its role in the above-mentioned processes.

Microtubules (MTs) are hollow tubes that consist of polymers of α and β tubulin heterodimers, and can be detected with commercially available monoclonal antibodies against α or β tubulin by immunofluorescence (see below). The ends of microtubules exhibit different properties: in living cells, MT-plus ends are the ends that exhibit dynamic instability characterized by phases of growth (elongation) and shortening, whereas MT-minus ends are often stabilized by interaction with the microtubule organizing center (MTOC) (1). Several proteins are known to associate with MT-plus ends, and regulate their dynamics or interaction with the cell cortex. In filamentous fungi, in addition to its role in spindle assembly and chromosome segregation, the microtubule cytoskeleton is proposed to be involved in long range transport of vesicles to the Spitzenkörper, a dark-phase contrast body located at the hyphal apex, from where vesicles are proposed to be delivered to the cell cortex on actin microfilaments (2).

Filamentous actin or F-actin is a polymer consisting of G-actin monomers, and can be detected with commercially available monoclonal antibodies against actin by immunofluorescence (see below). Actin filaments are also polar with a fast-growing end, the plus or "barbed" end, and a slow-growing end, the minus or "pointed end." Various proteins promote nucleation, severing, cross-linking, capping, and bundling of actin filaments. In filamentous fungi, filamentous actin is organized as patches that localize to hyphal tips and as a ring at the site of cytokinesis and septum formation (3). Actin cables, observed in *S. cerevisiae* (4), have not been detected in *A. nidulans* and *N. crassa* (5), but have been described, for example, in *Ashbya gossypii* (6), and *U. maydis* (7). Studies in filamentous fungi have shown that actin is necessary for integrity of the hyphal apex, polarized growth, cytokinesis, and endo- and exocytosis (2, 3).

U. maydis is a dimorphic basidiomycete fungus and has a unicellular, haploid, yeast-like form that is nonpathogenic, and a filamentous form that is dikaryotic and pathogenic (8). The repertoire of morphologies is increased by interaction with the plant (8). The morphogenetic changes are hypothesized to involve rearrangements of the cytoskeleton, which likely polarizes the secretory machinery (8). The *U. maydis* microtubule cytoskeleton has been extensively characterized using monoclonal anti-α antibodies and indirect immunofluorescence, and by live-imaging of green-fluorescent protein (GFP) fused to tubulin or MT-plus end associated proteins (7, 9, 10). It has been shown to be involved in cell polarity and morphogenesis, retrograde endosomal transport, organization of the endoplasmic reticulum, endocytosis, and spindle dynamics (11). In unbudded cells, MTs are arranged parallel to the long axis of the cell in antiparallel manner (plus and minus ends are located at both cell poles). During bud formation, paired tubulin structures (PTS), located at the base of the bud, nucleate MTs toward the mother and the bud (9). PTS contain γ-tubulin (10), a type of tubulin present in MTOCs in diverse eukaryotes. In large-budded cells in late G2, the nucleus migrates into the bud where nuclear division takes place. The cytoplasmic array disassembles when the nucleus is near the neck and a short intranuclear spindle is formed. Astral microtubules arise from the spindle pole bodies (SPBs; the equivalent of the mammalian centrosome) and appear to contact the cell cortex, and may play a role in movement of the nucleus into the bud (7). Astral microtubules play a key role during anaphase movements (12). After nuclear division is complete, the cytoplasmic array appears to be nucleated from cytoplasmic MTOCs (10), although the SPBs may also play a role in MT nucleation at this stage (7). The actin cytoskeleton in *U. maydis* consists of actin patches, actin cables, and an actin ring at the site of cytokinesis. Actin patches concentrate at the presumptive bud site and at the tip of the bud during bud morphogenesis. Actin cables polarize toward the patches. During cytokinesis, actin concentrates at the neck region as a broad band and as a ring (7). Little is known of the processes controlled by the actin cytoskeleton in *U. maydis*.

Indirect immunofluorescence is a method of immunofluorescence, in which visualization of a specific protein or antigen in fixed cells is done indirectly, by first using a primary unlabeled antibody that recognizes the protein or antigen of interest, and then using a fluorophore-conjugated secondary antibody that recognizes the primary antibody.

The fluorescence emitted by the excited fluorophore is indicative of the presence of the protein or antigen. The procedure involves fixation of cells with organic solvents (alcohols and acetone) or cross-linking agents (paraformaldehyde, formaldehyde). The cell wall of fungi is a formidable barrier to the entry of antibodies, therefore, it has to be partially removed with cell-wall degrading enzymes after cell fixation. The plasma membrane is impermeable to antibodies and is permeabilized by treatment with mild detergents (Triton X-100, SDS, saponin, NP-40 substitute) to allow the entry of antibodies. Because the cytoskeleton is very dynamic, it is also sensitive to disruption by chemical (buffers, detergents, fixatives) or mechanical means (handling of cells during fixation and subsequent steps). Thus, care must be taken in handling the material throughout the procedure.

The method described here has been successfully used to visualize actin patches and the elusive actin cables in both the yeast-like and filamentous form of *U. maydis*, as well as features of microtubule organization not observed using GFP-fusion proteins (7). This method can also be applied to visualization of other proteins in the cell, and is likely to be applicable to the detection of the cytoskeleton in other fungi.

2. Materials

2.1. Buffers and Reagents

1. 2× PEM-formaldehyde buffer for fixation of cells: 100 mM PIPES (Sigma), pH 6.7, 50 mM EGTA (Sigma), pH 7.0,

100 mM MgSO$_4$ (Sigma or Fisher Scientific), and 7.4% formaldehyde. The final concentration of formaldehyde after addition to the culture will be 3.7%. Prepare just before use. Stocks of 500 mM PIPES pH, 6.7, 250 mM EGTA, pH 7.0, and 1 M MgSO$_4$ can be prepared in advance, and stored at 4°C or room temperature (RT). Do not use the stock of formaldehyde or PIPES if it has turned a yellow color. Formaldehyde is toxic and should be handled in a hood.

2. PEM buffer (13): 50 mM PIPES buffer, pH 6.7, 25 mM EGTA, pH 7.0, and 50 mM MgSO$_4$. We usually prepare PEM buffer on the day of use, but it can be prepared in advance (1 month) and stored at 4°C.

3. Cell-wall-degrading enzyme solution: 10 mg/ml Novozyme (no longer available) or lysing enzymes (Sigma) in 1 M sorbitol, 50 mM sodium citrate pH, 5.7, and 5 mM EGTA. Filter sterilize. Prepare before use. Because lysing enzymes contain proteases, a protease inhibitor cocktail can be added (PMSF 1 mM, TAME 10 μg/ml, trypsin inhibitor 10 μg/ml, aprotinin 1 μg/ml, pepstatin 1 μg/ml). Alternatively, can add 50% egg whites (Sigma) solution. Neither is absolutely necessary.

4. Extraction buffer for permeabilization: 100 mM PIPES, pH 6.7, 25 mM EGTA, pH 7.0, and 0.1% Nonidet P-40 substitute (USB) or 0.1% Triton X-100 (Sigma).

5. Blocking buffer (PBS/BSA buffer): 1× PBS and 1% BSA. Prepared from 20× PBS and 10% BSA stocks. 20× PBS (100 ml): 16 g NaCl, 0.4 KCl, 2.3 g Na$_2$HPO$_4$, 0.4 g KH$_2$PO$_4$, and water to 100 ml. Filter sterilize or autoclave. Store at RT or at 4°C. Addition of 0.1% Triton X-100 may be useful, but is not essential.

6. Poly-L-lysine solution (10 mg/ml; Sigma).

7. Mounting medium. Any type of commercially available mounting medium can be used. I use Fluormount G (Southern Biotech) according to the manufacturer's instruction.

2.2. Antibodies

1. Primary antibodies. Tubulin staining: anti-α-tubulin clone DM1A (Sigma catalog no. T9026). Dilute 1:200 in 1× PBS/1% BSA buffer. Actin staining: anti-actin clone AC40 (Sigma catalog no. A4700). Dilute 1:250 in 1× PBS/1% BSA buffer. Prepare before use.

2. Secondary antibodies. A variety of secondary antibodies can be used, for example, sheep anti-mouse IgG conjugate (Sigma catalog F2883), goat antimouse (X+Y) Rhodamine (TRITC) conjugate (Jackson ImmunoResearch; catalog no. 115-026-044), goat antimouse (X+Y) DyLight conjugate (Jackson ImmunoResearch, catalog no. 115-506-003).

Dilute 1:125, 1:200, and 1:250, respectively, with 1× PBS/1% BSA in a light-tight Eppendorf tube. Fluorophores are light sensitive. Can also use Alexa-conjugated secondary antibodies (Molecular Probes/InVitrogen).

2.3. Nuclear Stain

The nucleus can be detected with a variety of dyes, for example, DAPI, Hoechst, and Sytox green. All are light sensitive. Prepare working stocks and dilutions in light-tight Eppendorf tubes. Store stocks at −20°C. DAPI (Sigma) stock: 1 mg/ml in water. Add to a final concentration of 1–2 µg/ml. Hoechst 33258 (Sigma) stock: 1 mg/ml in water. Add to a final concentration of 1 µg/ml. Sytox green (Molecular Probes/InVitrogen) is sold as a 5 mM stock solution in DMSO. Use 1–5 µl/ml.

2.4. Cell Wall Staining (Optional)

The cell wall and septa can be stained with Calcofluor White (Fluorescent Brightener 28; Sigma). Stock: 20 mg/ml in water. Light sensitive. Prepare in a light-tight Eppendorf tube. Store at −20°C. Add to a final concentration of 20 µg/ml.

3. Methods

3.1. Fixation of Cells

1. Use a fresh culture at OD_{600} 0.4–0.6 for best results. To determine OD_{600}, remove aliquots without interrupting shaking of culture (see Note 1).

2. Add an equal volume of 2× PEM-formaldehyde to the culture while it is shaking. Avoid centrifugation of the cells prior to fixation to preserve integrity of subcellular structures. Time of fixation for wild type cells: 40 min for tubulin staining; 20–30 min for actin staining (see Note 2).

3. Transfer culture to a conical 15 or 50 ml tube and spin at 2,000–3,000 × g for 5 min. Decant supernatant and immediately and gently add 10 ml of 1× PEM buffer. Resuspend the cells gently. Spin as above and wash again. Repeat three times.

4. Resuspend the pellet in 1–2 ml of 1× PEM buffer. Fixed cells can be stored at 4°C for up to 1 week before processing for indirect immunofluorescence.

3.2. Processing Fixed Cells

1. Fixed cells are affixed to a solid substrate, which can be cover slips or Teflon-coated multiwell slides (Polysciences or Erie Scientific). Prepare slides or cover slips as follows: apply poly-L-lysine for 5–10 min, aspirate liquid, allow to dry at room temperature, wash with distilled water and then with double-distilled water, place on paper towels, and dry at 37°C. Prepare before use. Do not store.

2. Gently transfer an aliquot of fixed cells (15–20 µl) to the wells of a multiwell slide or a cover slip. Care must be taken to deposit the cells gently on the surface of the slide or cover slip. Rough handling can alter cellular structure (see Note 3).
3. Allow cells to settle for 10–15 s.
4. Gently aspirate excess liquid but do not allow cells to dry.

3.3. Treatment with Cell-Wall-Degrading Enzymes

1. Gently add one drop of cell-wall-lysing solution to the cells and incubate for different times to determine the optimal digestion time needed to preserve cell integrity. The amount of enzyme and time of treatment to be used has to be determined empirically for each strain. I use from 4 to 12 min, with 2 min intervals. The best results with wild type cells are obtained with 8 min of incubation. Cell wall digestion can be monitored using phase contrast microscopy (do not cover the cells) (see Note 4).
2. Gently aspirate the liquid and add 1× PEM buffer. Repeat three times. I use a Pasteur pipette to dispense the buffer, particularly if processing several slides or cover slips. Alternatively, one can use a repeat pipettor and dispense ~30–40 µl of buffer.

3.4. Permeabilization of Cells

1. Gently aspirate PEM buffer and add one drop of extraction buffer. Incubate for 5 min at RT.
2. Gently aspirate extraction buffer and wash twice with 1× PEM buffer.

3.5. Methanol Treatment

Methanol will also contribute to permeabilization of cells.

1. Gently aspirate buffer and place slides in a couplin jar (staining jar) or the cover slips in a cover-slip reservoir containing methanol at −20°C for 10 min. It is important that methanol be at −20°C before this treatment is initiated. Keep a jar or a reservoir with methanol in the −20°C freezer at all times.
2. Remove slides or cover slips and place on a paper towel (sample face up!) and allow to air dry briefly.
3. Wash once with 1× PEM buffer.

3.6. Treatment with Blocking Buffer

1. Place slides or cover slips in a humidified chamber, which can be a plastic box with wet paper towels on the bottom.
2. Gently aspirate PEM buffer, and add blocking buffer (PBS/BSA solution) for 10–60 min.

3.7. Exposure to Antibodies

1. Gently aspirate PBS/BSA buffer and add diluted primary antibody. Incubate at room temperature or 4°C for 8 h or overnight (for convenience) in a humidified chamber sealed with Parafilm (see Note 5).

2. Gently aspirate primary antibody and wash 3× with 1× PBS/1% BSA.

3. Gently aspirate PBS/BSA and add diluted secondary antibody.

4. Incubate for 1–2 h in a humidified chamber, seal with Parafilm, and cover with Aluminum foil to avoid bleaching the fluorophore.

5. Gently aspirate secondary antibody and wash 3× with PBS/BSA.

6. Gently aspirate PBS/BSA and add a drop of mounting medium (Fluormount G) containing either 1–2 µg/ml DAPI, Hoechst 33258 (1 µg/ml), or Sytox Green (1 µl/ml). If using slides, add a cover slip. If using cover slips, invert the cover slip over a slide. Gently squeeze out any bubbles.

7. Place slides in between paper towels and press to remove excess mounting medium. Care must be taken not to introduce bubbles when excess mounting medium is removed.

8. Place a weight on top of the paper towels and let sit for 20–60 min.

9. Add commercial clear nail polish around the cover slip to seal the slide. Let dry for ~20 min.

The slides are ready to be viewed using a standard epifluorescence microscope containing the appropriate filter sets or a confocal microscope (see Note 6).

4. Notes

1. Growth media for *U. maydis* strains. YEPS broth (1% yeast extract, 2% peptone, 2% sucrose; see ref. (7), and references therein) is used routinely for growth of haploid and diploid strains as yeast-like cells. Sucrose is added after the broth has been autoclaved. For induction of filamentous growth, charcoal agar or charcoal broth is used (14). Growth of the culture is monitored spectrophotometrically (OD_{600}) and microscopically; the latter is to insure that cells are in all stages of budding.

2. Time of fixation and fixative that give optimum preservation of cellular structures is empirically determined for a given strain. Time can vary from 20 to 45 min, depending on whether actin or microtubules are to be visualized. Actin staining is more sensitive to a number of steps in the procedure. The conditions described here are specific for *U. maydis*, but similar procedures are used in other filamentous fungi.

Fixation can also be done with methanol at −20°C for 10–15 min, although results obtained are not as good as with formaldehyde.

3. If handling many samples simultaneously, it might be easier to use a Pasteur pipette to dispense solutions.

4. The concentration of lysing enzyme used is 5–10× lower than that used to generate protoplasts for DNA-mediated transformation. The incubation time in the enzyme solution has to be determined empirically, and is strain-dependent. Some mutants require longer incubation times because of altered cell wall composition; others with weak cell walls require less time.

5. Rhodamine phalloidin (tetramethyl rhodamine isothiocyanate-conjugate; Molecular Probes/InVitrogen), a reagent of choice for actin staining in budding and fission yeast, does not work in *U. maydis* and many other filamentous fungi (for example, *Aspergillus*, *Neurospora*) but see (15). I tried innumerable modifications to no avail. The reasons for this are not known.

6. TRITC and DyLight do not bleach as readily as FITC. The signal from DyLight lasts longer than that from TRITC.

Acknowledgments

I would like to thank Sylvia Sanders, who provided protocols used in her work with yeast during our tenure in the lab of Ira Herskowitz (deceased 2003) at UCSF, where I developed the method herein described. My work is funded by NIGMS grant 2S06 GM063119.

References

1. Desai A, Mitchison TJ (1997) Microtubule polymerization dynamics. Annu Rev Cell Dev Biol 13:83–117
2. Harris SD (2006) Cell polarity in filamentous fungi: shaping the mold. Int Rev Cytol 251:41–77
3. Heath IB (2000) Organization and function of actin in hyphal tip growth. In: Steiger C, Baluska F, Volkmann D, Barlow P (eds) Actin: a dynamic framework for mulitple plant cell functions. Kluwer, Dordrecht, Boston & London, pp 275–300
4. Karpova TS, McNally JG, Moltz SL, Cooper JA (1998) Assembly and function of the actin cytoskeleton of yeast: relationships between cables and patches. J Cell Biol 142:1501–1517
5. Xiang X, Plamann M (2003) Cytoskeleton and motor proteins in filamentous fungi. Curr Opin Microbiol 6:626–633
6. Schmitz HP, Kaufmann A, Köhli M, Laissue PP, Philippsen P (2006) From function to shape: a novel role of a formin in morphogenesis of the fungus *Ashbya gossypii*. Mol Biol Cell 17:130–145
7. Banuett F, Herskowitz I (2002) Bud morphogenesis and the actin and microtubule cytoskeleton during budding in the corn smut fungus *Ustilago maydis*. Fungal Genet Biol 37:149–170

8. Banuett F (2002) Pathogenic development in *Ustilago maydis*: a progression of morphological transitions that results in tumor formation and teliospore production. In: Osiewacz HD (ed) Molecular biology of fungal development. Marcel Dekker, New York, Basel, pp 349–398
9. Steinberg G, Wedlich-Söldner R, Brill M, Schulz I (2001) Microtubules in the fungal pathogen *Ustilago maydis* are highly dynamic and determine cell polarity. J Cell Sci 114:609–622
10. Straube A, Brill M, Oakley BR, Horio T, Steinberg G (2003) Microtubule organization requires cell cycle-dependent nucleation at dispersed cytosplasmic sites: polar and perinuclear microtubule organiziang centers in the plant pathogen *Ustilago maydis*. Mol Biol Cell 14:642–657
11. Steinberg G, Fuchs U (2004) The role of microtubules in cellular organization and endocytosis in the plant pathogen *Ustilago maydis*. J Microsc 214:114–123
12. Fink G, Schuchardt I, Colombelli J, Stelzer E, Steinberg G (2006) Dynein-mediated pulling forces drive rapid mitotic spindle elongation in *Ustilago maydis*. EMBO J 25:4897–4908
13. Hagan IM, Hyams JS (1988) The use of cell division cycle mutants to investigate the control of microtubule distribution in the fission yeast *Schizosaccharomyces pombe*. J Cell Sci 89:343–357
14. Banuett F, Herskowitz I (1994) Morphological transitions in the life cycle of *Ustilago maydis* and their genetic control by the *a* and *b* loci. Exp Mycol 18:247–266
15. Hoch HC, Staples RC (1983) Visualization of actin *in situ* by rhodamine-conjugated phalloidin in the fungus *Uromyces phaseoli*. Eur J Cell Biol 32:52–58

Chapter 18

Fluorescence In Situ Hybridization for Molecular Cytogenetic Analysis in Filamentous Fungi

Dai Tsuchiya and Masatoki Taga

Abstract

Fluorescence in situ hybridization (FISH) is a powerful technology for studying eukaryotic chromosomes and genomes from the combined view of cytogenetics and molecular biology, but its use in filamentous fungi has been limited. In this chapter, we describe protocols to perform three basic FISH techniques in filamentous fungi: (a) FISH mapping of unique sequences on the somatic chromosomes and interphase nuclei, (b) chromosome painting to detect a specific chromosome in the genome by fluorescent painting of the whole chromosome, and (c) fiber FISH on the stretched DNA fibers for physical mapping. The ways of preparing target specimens unique to filamentous fungi are included in the protocols.

Key words: FISH, Fluorescence in situ hybridization, Filamentous fungi, Chromosome, Nuclei, Genome

1. Introduction

In situ hybridization (ISH) is a method to detect defined DNA or RNA sequences in cytological preparations by hybridizing complementary probe sequences. Until 1980s ISH had been performed with radioisotope-labeled probes. In the late 1980s, a new ISH technique called fluorescence in situ hybridization (FISH) that utilizes fluorochromes instead of isotopes was introduced. Since FISH has significant merits in safe handling, rapid results, high resolution, and ability to simultaneously detect multiple targets using multiple colors, it has soon become popular in plant and animal cytogenetics as well as for clinical diagnostic purposes.

In contrast to the situation in higher plants and animals, application of FISH to filamentous fungi has been limited. As long as DNA FISH is concerned, only *Coprinus cinereus* (1), *Botrytis cinerea* (2), *Alternaria alternata* (2, 3), *Nectria haematococca* (4–7), *Cochliobolus heterostrophus* (6, 8), *Ustilago maydis* (9), and some glomalean species of mycorrhiza (10) have been subjected to FISH in the published papers. Among these, meiotic pachytene chromosomes were targeted in *C. cinereus*, while mitotic chromosomes, interphase nuclei, and DNA fibers from somatic cells were used in other species. FISH is admittedly more difficult to perform in fungi than in higher plants and animals due to much smaller dimension of fungal specimens. However, the scarce application of FISH to filamentous fungi may also be attributable to the insufficient recognition of the feasibility of FISH in filamentous fungi among the researchers.

In this chapter, FISH protocols that we have developed to suit to filamentous fungi are described. Procedures unique to filamentous fungi are mainly related to the preparation of target specimens, and the rest parts are similar to those used for plants and animals. Therefore, readers also should refer to the protocols of FISH for plant and animals (for examples, see refs. (11–15)).

1.1. FISH with Unique DNA Sequences on Mitotic Chromosomes and Nuclei

This is the most basic FISH technique for localizing specific DNA sequences on the metaphase chromosomes and interphase nuclei. Various DNA sequences from single genes to site-specific repetitive sequences that are cloned in plasmid, cosmid, BAC, and YAC vectors have been used as probes in plants and animals. In filamentous fungi, plasmid and cosmid clones have been proven useful as probes, while the usability of BAC and YAC clones has not been testified yet in this type of FISH. Thus, the protocol described here is for plasmid and cosmid clones. In using these probes, amplification of hybridization signals with secondary antibody is usually necessary for plasmid clones but not for cosmid clones. Exceptionally, highly repetitive targets such as ribosomal DNA can be easily detected using a repeat unit cloned in a plasmid as a probe without signal amplification (Fig. 1a).

Regarding specimens, metaphase chromosomes and interphase nuclei can be prepared by the germ tube burst method (GTBM) (2, 16). The point to notice is that specimens should have minimum cytoplasm, as residual cytoplasm hampers both efficient probe hybridization and detection of hybridized probe, resulting in reduced FISH signals and higher background. In our hands, GTBM is the preferred method for mitotic specimens in terms of the residual cytoplasm.

1.2. Chromosome Painting

Chromosome painting implies highlighting of whole chromosome or chromosome segment in metaphase or interphase specimens by FISH. It has various applications including detection of structural and numerical chromosome rearrangement, identification of specific chromosomes in a genome, and comparative genomics. In filamentous fungi, chromosome painting has been

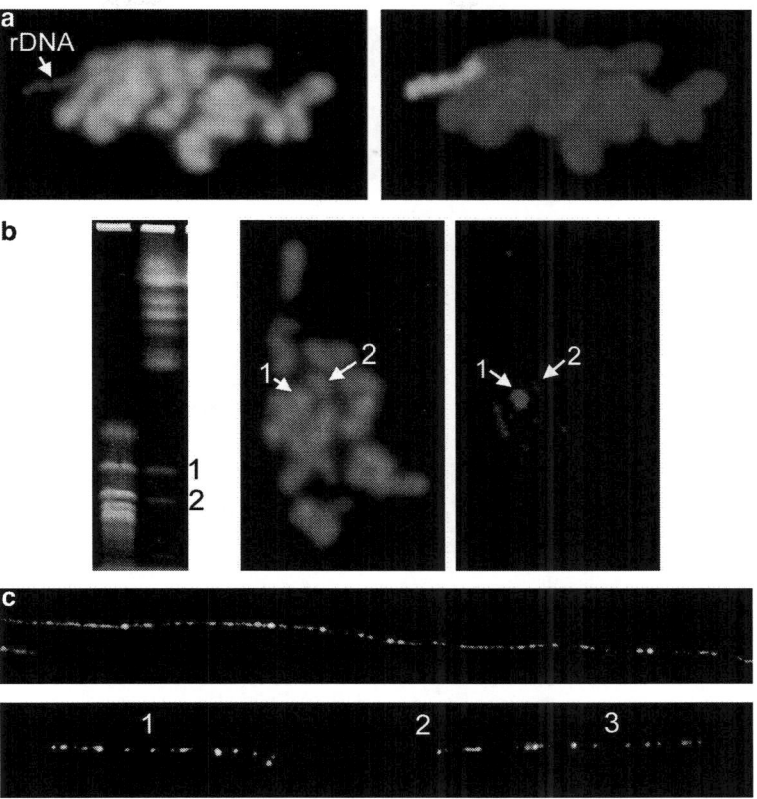

Fig. 1. Examples of FISH in the plant pathogenic ascomycete *Nectria haematococca*. (**a**) Detection of specific DNA sequences on the mitotic metaphase chromosome. *Left*: Chromosomes stained with both DAPI and PI. Nucleolar organizing region (rDNA) appears as a thread-like protrusion. *Right*: Detection of rDNA with the cloned repeat unit of rDNA cluster as a probe. Biotin:avidin-FITC system was used. (**b**) Chromosome painting. Chromosomes denoted by the numbers 1 and 2 were simultaneously detected by chromosome painting. Chromosome 1 is a special type of supernumerary chromosome called conditionally dispensable chromosome. *Left*: Separation of chromosomes by pulsed field gel electrophoresis (PFGE). *Middle*: Mitotic metaphase chromosomes stained with DAPI. *Right*: Painting with different fluorescence colors. Biotin:avidin-FITC (*green*) and DIG:anti-DIG-rhodamine systems (*red*) were used. Probes were prepared by DOP-PCR. (**c**) Fiber FISH. *Upper*: Chromosome-specific DNA fiber visualized by staining with YOYO-1. *Lower*: Simultaneous mapping of three cosmid clones (denoted by 1, 2, and 3) on a chromosome-specific DNA fiber. Green and red hybridization signals were yielded by Alexa 488 and Alexa 555, which were conjugated to the secondary antibody.

used to visualize a supernumerary chromosome called conditionally dispensable (CD) chromosome in *N. haematococca* (Fig. 1b) (4, 7). Chromosome painting is unique in using composite probes that hybridize to numerous sites to collectively represent most part of a chromosome or a segment thereof. The straightforward way of preparing probes is the isolation of whole DNA from the target chromosome. Although this is achieved by flow-sorting or micro-dissecting of chromosomes in higher plants and animals, fungal chromosomes are too small to apply such methods. Instead, pulsed field gel electrophoresis (PFGE) is applicable to filamentous fungi. With PFGE, individual chromosomes can be separated on an agarose gel, and the chromosomal DNA isolated from the excised bands is amplified by degenerate oligonucleotide-primed PCR (DOP-PCR) (17) or multiple displacement amplification (MDA) (18) to the sufficient amount for labeling. An alternative approach to prepare probes for chromosome painting is the use of a composite of chromosome-specific large insert clones that collectively represent most part of the chromosome. This was successfully applied to yeast meiotic chromosomes (19).

A variation of chromosome painting is genomic in situ hybridization (GISH) that employs total genomic DNA as a probe. Using GISH, chromosomes of different parental origin in hybrids or alien chromosomes incorporated in a genome can be distinguished. In filamentous fungi, a CD chromosome was detected with GISH in the genome of a specific strain of *N. haematococca* (4).

1.3. Fiber FISH

Fiber FISH refers to the approaches that perform FISH on a stretched DNA. Since the specimens are prepared in a stretched form, resolution of fiber FISH is much higher than other FISH techniques that are performed on the chromosomes or interphase nuclei. Owing to this resolution, fiber FISH has been appreciated as a tool of physical mapping. In filamentous fungi, mapping of rRNA genes in the rDNA region (8), estimation of physical distance between the two toxin-producing genes (6), and contig mapping with cosmid clones (6) were conducted with fiber FISH in *C. heterostrophus* and *N. haematococca* (Fig. 1c).

DNA fibers of filamentous fungi for performing fiber FISH are prepared with protoplasts or PFGE-separated chromosomal DNAs, both of which are embedded in low melting agarose block. DNA fibers are released in the melted agarose and the stretched fibers spanning up to 900 kb can be generated on the slides without difficulty (8). Cosmid, BAC, and YAC are also usable as a source of DNA fibers. As probes, DNA fragments amplified by PCR or cloned in vectors such as plasmids and cosmids are employed.

2. Materials

2.1. General Reagents, Buffers, and Supplies

The following list includes details of reagents, buffers, and supplies that are common to all or most parts. Specific items will be listed again without detailed description in appropriate places of the protocols.

1. Paper cement (rubber cement) tubes.
2. Moist chamber: Airtight plastic container with blotting papers soaked with water and two glass rod to hold the glass slides horizontally.
3. Cover slips (18 × 32 mm).
4. Glass microscope slides.
5. Parafilm slips: Rectangle pieces (18 × 32 mm) cut from a rolled sheet.
6. TE buffer: 10 mM Tris–HCl, 1 mM EDTA, pH 8.0.
7. Sodium acetate buffer: 3 M sodium acetate, pH 5.5.
8. 20× SSC: 3 M NaCl, 0.3 M sodium citrate, pH 7.4. Store at room temperature after autoclaving.
9. Vectashield (Vector Laboratories, Burlingame, CA): Antifade and mounting solution.
10. 4′,6-diamidino-2phenylindole (DAPI): 1 µg/mL dissolved in Vectashield (Vector Laboratories). Prepare 100 µg/mL stock solution in distilled water, aliquot and store at −20°C.
11. 10× dNTP mix: 2 mM of each dATP, dCTP, dGTP, dTTP.
12. Blocking solution: 4% (w/v) Block Ace (AbD Serotec, Oxford, UK) in 4× SSC. Dissolve 4 g of Block Ace powder in 100 mL of 4× SSC, aliquot and keep in freezer at −20°C.
13. Antibody diluent buffer: 1% Block Ace in 4× SSC. Dilute from 4% Block Ace with 4× SSC before use.
14. Fluorescence microscope equipped with oil immersion 40× and 100× lenses and a CCD camera.
15. RNase A: 100 µg/mL in 2× SSC. Dilute from a DNase-free 10 mg/mL stock solution (20). Stock and diluted solution can be stored at −20°C.

2.2. FISH with Unique DNA Sequences on Mitotic Chromosomes and Nuclei

2.2.1. Poly-L-Lysine-Coated Slides

1. Glass microscope slides (pre-cleaned). One end frosted is convenient for marking with pencil.
2. Staining jar.
3. Acid alcohol (e.g., 1% HCl in 70% ethanol).
4. Poly-L-lysine solution (Sigma, St. Louis, MO).
5. Drying oven.

2.2.2. Preparation of Mitotic Chromosomes and Nuclei

1. Poly-L-lysine-coated glass microscope slides.
2. Paper cement tubes.
3. Glass Coplin jars.
4. Moist chamber.
5. Cover slips (18 × 32 mm).

6. Diamond glass cutter.
7. Parafilm slips.
8. Hemacytometer.
9. Incubator.
10. Low speed centrifuge.
11. Fixative: 99% methanol: glacial acetic acid (17:3 (v/v) methanol/glacial acetic acid).
12. DAPI.
13. Thiabendazole (TBZ): Freshly prepare 50 μg/mL solution by diluting with liquid culture medium from stock solution (10 mg/mL in dimethyl sulfoxide). Stock solution can be stored at −20°C.
14. RNase A.
15. Ethanol: 70%, 85%, and 95% in water.
16. Bright filed light microscope.
17. Fluorescent microscope.

2.2.3. Probe Preparation

1. Plasmid preparation kit (see Note 1).
2. Mini agarose gel electrophoresis apparatus.
3. Spectrophotomer: Equipped with a microcuvette (7–10 μL capacity).
4. Centrifuge.

2.2.4. Probe Labeling

1. Sample DNA.
2. Labeling kit: Bio-Nick labeling system (Invitrogen, Carlsbad, CA) for labeling with biotin-14-dATP or DIG-nick translation mix (Roche Diagnostics, Mannheim, Germany) for labeling with digoxigenin (DIG)-11-dUTP.
3. Incubator or water bath (15°C).
4. Agarose gel electrophoresis apparatus.
5. TE buffer.
6. Sodium acetate buffer.
7. Ethanol (70%, 100%).
8. Heating block (65°C).
9. Microcentrifuge.

2.2.5. Hybridization

1. Labeled probe DNA.
2. Formamide: Molecular biology grade. Deionization is not necessary if immediately stored at −20°C in small aliquots after purchase.

3. Dextran sulfate: Prepare a 50% (w/v) solution in water by heating to 65°C. Filter sterilize (0.22 μm) and store at −20°C in small aliquots.
4. Sonicated salmon sperm DNA: Commercially available DNase-free solution (10 mg/mL). Store at −20°C in small aliquots.
5. 20× SSC.
6. Moist chamber.
7. Glass cover slips.
8. Paper cement tubes.
9. Heating block (80°C).
10. Incubator (37°C).

2.2.6. Washing and Signal Detection

1. 2× SSC: dilute from 20× SSC.
2. 4× SSC: dilute from 20× SSC.
3. 50% (v/v) formamide/2× SSC: Freshly prepare by diluting 25 mL formamide of a good grade formamide with an equal volume of 4× SSC.
4. Blocking solution.
5. Antibody diluent buffer.
6. Hybridization detection agents without signal amplification: Avidin or streptavidin conjugated to fluorescein (FITC), rhodamine (TRITC), Alexa fluorophores, cyanine dyes, or others for biotin-labeled DNA probes. Anti-DIG (fab fragment) conjugated to FITC, TRITC, Alexa fluorophore, or others for DIG-labeled DNA probes (see Note 2).
7. Hybridization detection agents with signal amplification:
   (a) Primary antibody: Goat anti-biotin (Vector Laboratories) for biotin-labeled DNA probes. Mouse anti-DIG (Roche) for DIG-labeled DNA probes. Both antibodies are supplied as lyophilized powder. Dissolve each antibody in 1 mL water to make stock solution (1 mg/mL goat anti-biotin; 0.1 mg/mL mouse anti-DIG), aliquot and keep at −20°C.
   (b) Secondary antibody-fluorophore conjugate: Rabbit anti-goat IgG conjugated to fluorescein (Vector Laboratories), Alexa Fluor 488 (Molecular Probes, Carlsbad, CA), or others for biotin-labeled DNA probes. Sheep anti-mouse IgG conjugated to TRITC (Chemicon international, Temecula, CA), Alexa Fluor 555 (Molecular Probes), or others for DIG-labeled DNA probes.
8. DAPI.
9. Propidium iodide (PI): Prepare stock solution of 100 μg/mL in water, store at −20°C.

10. Vectashield.
11. Water bath (37°C).
12. Coplin jars.
13. Glass cover slips.
14. Parafilm slips.
15. Moist chamber.
16. Shaker.
17. Nail polisher.
18. Fluorescent microscope.

2.3. Chromosome Painting

2.3.1. Preparation of Total DNA from a Specific Chromosome

DOP-PCR:

1. PFGE apparatus (e.g., CHEF-DR II system, Bio-Rad, Hercules, CA).
2. Mini agarose gel electrophoresis apparatus.
3. UV transilluminator.
4. Agarose gel DNA extraction kit.
5. Thermocycler.
6. Primer 6-MW (50 µM): 5′-CCGACTCGAGNNNNNNATGTGG-3′ (see (17)).
7. Taq polymerase and buffer.
8. 10× dNTP mix.
9. Centrifuge.

MDA

1. REPLI-g kit (QIAGEN, Valencia, CA).
2. PFGE apparatus (e.g., CHEF-DR II system, Bio-Rad).
3. Mini agarose gel electrophoresis apparatus.
4. Agarose gel DNA extraction kit (e.g., GENECLEAN II Kit, Qbiogene).
5. UV transilluminator.
6. Centrifuge.

2.3.2. Probe Labeling

DOP-PCR:

1. Product of DOP-PCR.
2. Biotin-16-2′-deoxy-uridine-5′-triphosphate (biotin-16-dUTP) (Roche Diagnostic).
3. 10× dNTP mix minus dTTP.
4. dTTP: 2 mM stock, final concentration 80 µM.
5. Taq polymerase and buffer (Takara Taq, Takara Bio, Otsu, Japan).
6. Thermocycler.

7. 10× dNTPs mix.
8. Centrifuge.
9. Mini agarose gel electrophoresis apparatus.

MDA:

1. Product of MDA.
2. Bio-Nick labeling system (Invitrogen) for labeling with biotin-14-dATP.
3. Incubator or water bath (15°C).
4. TE buffer.
5. Sodium acetate buffer.
6. Ethanol (70%, 100%).
7. Heating block (65°C).
8. Microcentrifuge.

2.3.3. Hybridization

1. Labeled DNA probe.
2. Blocking DNA (option): purified genomic DNA in TE or water.
3. Dextran sulfate (50% (w/v)).
4. Salmon sperm DNA (10 mg/mL).
5. 20× SSC.
6. Moist chamber.
7. Glass cover slips.
8. Paper cement tubes.
9. Heating block (80°C).
10. Incubator (37°C).

2.3.4. Washing and Signal Detection

1. 2× SSC.
2. 4× SSC.
3. 4× SSC/0.1% (w/v) Triton X: Dissolve Triton X-100 into 4× SSC.
4. 50% (v/v) formamide/2× SSC.
5. Blocking solution.
6. Antibody diluent buffer.
7. Hybridization detection agents: Avidin or streptavidin conjugated to FITC, TRITC, Alexa fluorophores, or others for biotin-labeled DNA probes. Anti-DIG (fab fragment) conjugated to FITC or TRITC for DIG-labeled DNA probes.
8. Mounting and staining solution: Vectashield containing 1 μg/mL each of DAPI and PI.
9. Water bath (37°C).

10. Coplin jars.
11. Glass cover slips.
12. Parafilm slips.
13. Moist chamber.
14. Shaker.
15. Fluorescence microscope.

2.4. Fiber FISH

2.4.1. Preparation of DNA Fibers

1. Glass microscope slides.
2. Glass cover slips.
3. Poly-L-lysine.
4. Whatman no. 4 filter paper.
5. Enzyme solution (for protoplasting *C. heterostrophus*). Dissolve the following enzymes in 0.7 M NaCl: 2 mg/mL Lysing enzymes (L1412, Sigma), 3 mg/mL of Kitalase (Wako Pure Chemical, Osaka, Japan), 3 mg/mL of β-glucuronidase (type HA-4, Sigma). Filter-sterilized and use fresh (see Note 5).
6. SE: 1 M sorbitol, 50 mM EDTA, pH 8.0.
7. Low melting agarose: Dissolve low melting agarose (Bio-Rad) in SE at a concentration of 1% (w/v).
8. NDS: 0.5 M EDTA, 10 mM Tris-HCl, pH 8.0, 1% (w/v) sodium lauroyl sarcosinate.
9. RNase.
10. Ethanol (70%, 85%, 95%).
11. 1% (w/v) pepsin solution: Dissolve 10 mg of pepsin (Sigma) in 1 mL of 0.9% (w/v) NaCl, pH 1.8.
12. Plug mold: For making protoplasts-embedded agarose plugs (e.g., 170-3713, Bio-Rad).
13. Razor blade (oil-free).

2.4.2. Probe Labeling

Probe DNAs are labeled with biotin-14-dATP and DIG-11-dUTP by nick translation.

1. Labeling kit: Bio-Nick labeling system (Invitrogen) for labeling with biotin. DIG-nick translation mix (Roche Diagnostic) for labeling with DIG.
2. Ultra pure distilled water (Gibco, Rockville, MD).

2.4.3. Hybridization

1. Labeled probe DNA.
2. Hybridization mixture: 50% (v/v) formamide, 10% (v/v) dextran sulfate, 100 ng/μL salmon sperm DNA, and 0.1–0.2 μg labeled probe DNA in 2× SSC. Fifteen microliter is used per slide.
3. Paper cement tubes.

2.4.4. Washing and Signal Detection	1. 2× SSC. 2. 4× SSC. 3. Formamide. 4. Blocking solution. 5. Detection buffer: 1% Block Ace in 4× SSC. 6. Primary antibody: Goat anti-biotin (Vector Laboratories) for biotin-labeled probe. Mouse anti-DIG (Roche Diagnostics) for DIG-labeled probe. 7. Secondary antibody: Alexa Fluor 488 rabbit anti-goat IgG antibody (Molecular Probes, 2 mg/mL). Alexa Fluor 594 rabbit anti-mouse IgG antibody (Molecular Probes, 2 mg/mL). 8. Primary detection solution: For biotin-labeled DNA probes, use goat anti-biotin in detection buffer at a final concentration of 5 µg/mL. For DIG-labeled DNA probes, use mouse anti-DIG in detection buffer at a final concentration of 0.8 µg/mL. 9. Secondary detection solution: For biotin-labeled DNA probes, use Alexa Fluor 488 rabbit anti-goat antibody in detection buffer at a concentration of 20 µg/mL. For DIG-labeled DNA probes, use Alexa Fluor 594 rabbit anti-mouse antibody in detection buffer at a concentration of 20 µg/mL (see Note 6). 10. Vectashield. 11. DAPI staining solution.

3. Methods

3.1. FISH with Unique DNA Sequences on Mitotic Chromosomes and Nuclei	Wear latex gloves and use forceps to handle slide glasses.
3.1.1. Preparation of Poly-L-Lysine-Coated Microscope Slides	1. Immerse slides in acid alcohol for 1 h at room temperature. 2. Wash slides in running tap water. 3. Rinse slide thoroughly in distilled water. 4. Stand slides in a rack and air-dry in a dust-free condition. 5. Coat slide with diluted poly-L-lysine solution in a plastic staining jar according to the manufacturer's protocol supplied with the reagent. 6. Drain slides and dry on the rack in an oven at 60°C for 1 h or at room temperature over night. 7. Use slide within 1 week or so.
3.1.2. Preparation of Mitotic Chromosomes and Nuclei	The method described below uses germling cells from conidia (see Note 7). For producing conidia and preparing suspension of conidia, follow the standard protocol developed for individual

species. Aseptic handling in steps 1–13 is not needed if the conidia can germinate within several hours after the initiation of slide culture. When overnight or longer incubation of conidia is necessary, care must be taken in avoiding the overgrowth of contaminated bacteria.

1. Squeezing paper cement from a tube, draw a rectangle frame (ca. 18×32 mm) at the center of a poly-L-lysine-coated slide glass.
2. Allow paper cement to dry completely at room temperature. This may take 20–30 min.
3. Harvest fresh conidia from cultures and remove mycelia debris by filtration.
4. Pellet conidia by centrifugation, and then resuspend in a liquid medium suitable for conidia germination such as PDB.
5. Repeat step 4.
6. Count and calculate conidia density using a hemacytometer and adjust to an appropriate concentration ($3-5 \times 10^5$/mL in most cases) by diluting with the liquid medium.
7. Drop and spread ca. 100 μL conidial suspension inside the paper cement-made frame (see Note 8). Paper cement serves as a dike to enclose the suspension inside the frame.
8. Incubate slides in a humid chamber at a temperature suitable for the germination of conidia and check germination intermittently under a microscope without putting a cover slip.
9. Continue incubation until germ tubes grow up to around 100 μm in length or until they begin to branch. Avoid overlapping of germ tubes (dilute conidia if necessary). Skip steps 10–12 if only interphase nuclei are analyzed or species inherently resistant to TBZ is used (see Note 9).
10. Tilting the slide, gently aspirate liquid medium with a micropipette. A certain percentage of the germinated conidia remain on the slide with the germ tubes adhering to the surface of slide.
11. Apply ca. 100 μL of fresh medium containing 50 μg/mL TBZ onto the frame. This medium should be freshly prepared prior to step 10 by diluting stock solution of TBZ with liquid medium.
12. Resume incubation for additional 2–3 h.
13. After incubation, peel off paper cement from the slide with a forceps.
14. Dip slides in a staining jar containing distilled water for a few minutes to wash off liquid medium.
15. Take out slides and blot with a filter paper except the area containing the conidia. It is important to keep the specimens wet (covered with a small amount of water).

16. Slip and place slides horizontally in fixative for at least 15 min (see Note 10).
17. Take out slides, briefly drain leaving a small amount of fixative on slides and horizontally pass through a flame of alcohol lamp. Fixative on the slide catches fire, but let it burn out.
18. Leave slides to dry at room temperature. Mark the area of specimens on the slide with a diamond glass cutter, and store in dry condition until use.
19. Stain one slide with DAPI, and check the quality of the specimen under a fluorescent microscope. Do not proceed with the procedure if the specimens are not good e.g., low bursting frequency or destroyed morphology.
20. Apply 70 µL of RNase to each of the remaining specimens, cover with a piece of parafilm, and incubate for 1 h at 37°C in a humid chamber.
21. Peel off the parafilm, and put the slides into a Coplin jar containing 2× SSC for 5 min.
22. Drain excessive fluid and dehydrate by successive transfers to 70%, 85%, and 95% ethanol, for 5 min each in Coplin jar.
23. Following the final ethanol dehydration, drain slides and dry on a rack.

3.1.3. Probe Preparation

1. Isolate plasmid or cosmid from bacteria by a common method such as the standard alkaline lysis or by a commercially available extraction kit.
2. Check the quality of DNA by agarose gel electrophoresis. When DNA is contaminated with RNA that is seen as a diffuse bright smudge below DNA bands, treat with RNase followed by phenol-chloroform extraction and ethanol precipitation.
3. Measure the amount and purity of DNA with a spectrophotometer. Proceed only if DNA purity is 80% or higher.

3.1.4. Probe Labeling

When plasmid- or cosmid-cloned DNA is used as a probe, usually elimination of vector DNA is not necessary. Bio-Nick labeling system (Invitrogen) and DIG-Nick translation mix (Roche Diagnostic) are used for labeling with biotin and DIG, respectively.

1. Using 1 µg of probe DNA and reagents supplied in a kit, prepare reaction mixture in a total volume of 50 µL for labeling with biotin or 20 µL for DIG.
2. Incubate for 90 min at 15°C.
3. Stop reaction by heating to 65°C for 10 min.
4. To remove unincorporated nucleotides, add 1/10 volume of sodium acetate buffer and 2 volumes of 100% ethanol and leave for 15 min at −80°C or 2 h at −20°C.

5. Centrifuge at 15,000 × g for 15 min, remove supernatant, and wash the pellet with 70% cold ethanol.
6. Discard the supernatant and dry the pellet.
7. Dissolve pellet in 50 µL water and repeat the steps 4–6.
8. Dissolve the pellet in 10–20 µL TE and store at −20°C until use.

3.1.5. Hybridization

This protocol for hybridization follows the method of ref. 21, in which denaturation of both probe and target DNA and the subsequent hybridization are consecutively carried out on the slide in the steps 5 and 6.

1. In a 1.5-mL microtube, prepare hybridization mixture in a total volume of ca. 50 µL. The mixture is made up of 6–10 µL of the labeled probe (see Note 11), 25 µL formamide, 10 µL 50% dextran sulfate, 5 µL 20× SCC and 0.5 µL salmon sperm DNA (10 mg/mL) and water to fill up to 50 µL. Prior to measuring dextran sulfate and salmon sperm DNA with a micropipette, heat to 70–80°C to decrease their viscosity.
2. Mix thoroughly by repeated pipetting with a micropipette and vigorous vortexing, and then spin down briefly in a microcentrifuge.
3. Put 15 µL of the hybridization mixture on the slide and cover with a cover slip. If air bubbles are seen on the mounted specimen, press them out. Blot the excessive hybridization mixture spilling out from the edge of the cover slip.
4. Seal with paper cement, and wait till the cement is completely dry.
5. Place the slide on a heat block at 80°C, and leave for 2 min.
6. Incubate the slide at 37°C in a moist chamber overnight (more than 16 h).

3.1.6. Washing and Signal Detection

Do not allow the specimens to dry out at anytime during the wash and detection procedures. Coplin jars are used as container for solutions. To change solution in the steps 4 and 5, pour liquid off carefully and replace slowly with next solution.

Detection without Signal Amplification:

1. Warm 50% formamide/2× SSC to 37°C in a water bath.
2. Take slides from a moist chamber, remove the paper cement using forceps, and immerse slides in 2× SSC to allow the cover slip to come off.
3. After wiping slides except the area of specimens, immerse in pre-warmed 50% formamide/2× SSC for 15 min at 37°C. During the incubation, gently agitate slides several times.
4. Immerse the slides in 2× SSC twice for 10 min and in 4× SSC once for 5 min.

5. Drain slides and blot excess fluid from the edge of area of specimens with a filter paper.
6. Apply 70 µL of blocking solution, cover with a parafilm slip, and incubate for 15 min at 37°C in a humid chamber.
7. Prepare detection solution by diluting stock detection reagent in 1% Block Ace/4× SSC at a proper dilution ratio (e.g., 1:100 for avidin-FITC (Roche Diagnostics)) (see Note 12).
8. Remove parafilm and wipe excess blocking solution carefully by tilting the slide. Apply 70 µL of detection solution, cover with a parafilm slip, and incubate for 1 h at 37°C in a moist chamber.
9. Remove parafilm, immerse in 4× SSC for 10 min, 4× SSC containing 0.1% Triton X-100 for 10 min, and 4× SSC for 10 min with gentle shaking on a shaker.
10. Immerse slides in 2× SSC for 5 min, wipe excess fluid.
11. Mount each slide with 15 µL of antifade solution, cover with a cover slip, and incubate for at least 10 min protected from light (see Note 13). Seal with nail polisher.
12. View specimens under fluorescent microscope using appropriate filters. Look for good specimens with ×40 and ×100 immersion objective lens under DAPI-excitation, and then switch filter for FITC- or TRITC-excitation.

Detection with Signal Amplification:

1. Perform steps 1–6 of Detection without Signal Amplification in Subheading 3.1.6.
2. Prepare the necessary volume of primary antibody solution and fluorescent dye-conjugated secondary antibody solution in 1% Block Ace/4× SSC at a proper dilution ratio (e.g., 1:100 for avidin-FITC (Roche Diagnostics)) (see Note 12). For each slide, 70 µL is needed.
3. Remove parafilm and wipe excess blocking solution carefully by tilting slides. Apply 70 µL of primary antibody, cover with parafilm, and incubate for 45 min at 37°C in a moist chamber.
4. Immerse in 4× SSC for 10 min, 4× SSC/0.1% Triton X-100 for 10 min, and 4× SSC for 10 min with gentle shaking on a shaker.
5. Wipe excess fluid, apply 70 µL of fluorescent dye-conjugated secondary antibody, cover with a parafilm slip, and incubate for 45 min at 37°C in a moist chamber.
6. Immerse in 4×SSC for 10 min, 4× SSC/0.1% Triton X-100 for 10 min, and 4× SSC for 10 min with gentle shaking on a shaker. Leave the slides in 2× SSC for 5 min, wipe excess fluid carefully.

7. Mount each slide with 15 μL of antifade solution, cover with a cover slip, and incubate for at least 10 min protected from light (see Note 13). Seal with nail polisher.
8. Observe under fluorescent microscope.

3.2. Chromosome Painting

3.2.1. Preparation of Total DNA from Specific Chromosome

DOP-PCR:

1. Separate chromosomes using all lanes of a gel by PFGE (see Note 14).
2. Take out the gel from the buffer tank, place it on a glass plate, and cut the leftmost and rightmost lanes into strips with a sharp scalpel.
3. Stain the strips with ethidium bromide and mark the position of the chromosome to be analyzed on each strip under UV illumination.
4. Place the two strips back to the original gel, and mark the estimated position of the chromosome in the remaining unstained lanes using the chromosome position on two strips as a guide.
5. Cut out a slice containing the chromosome from each lane, combine gel slices, and measure weight.
6. Extract DNA from gel using a commercial DNA extraction kit.
7. Carry out ethanol precipitation, wash DNA with 70% ethanol, and dry. Finally dissolve DNA in 10 μL TE.
8. Using the primer 6-MW, perform DOP-PCR in 50 μL. The reaction mixture contains 200 μM of each dNTP, 2 μL of template DNA (prepared at step 6), and 1.25 U of Taq polymerase (Takara) in 1× buffer. Reaction conditions are 10 min at 93°C, followed by five cycles of 1 min at 94°C, 1.5 min at 30°C, 3 min transition 30–72°C, and 3 min extension at 72°C. This is followed by 30 cycles of 1 min at 94°C, 1 min at 62°C, and 3 min at 72°C. The final extension is 10 min at 72°C.
9. Using 4 μL of PCR product, check the quality of amplified DNA by agarose gel electrophoresis. A smear of DNA in a broad size range below 20 kb should be visible.
10. Store in a freezer until use.

MDA:

1. Prepare chromosomal DNA from PFGE gel as in the steps 1–6 of DOP-PCR in Subheading 3.2.1.
2. Place DNA into a microcuvette, and measure the concentration with a spectrophotometer. After the measurement, recover DNA to use as the template for MDA.

3. Amplify DNA with REPLI-g kit (QIAGEN) in 50 µL of reaction mixture according to the manufacturer's protocol optimized for purified genomic DNA (see ref. 22).

4. Check amplification of DNA by agarose gel electrophoresis.

5. Ethanol precipitate DNA, wash with 70% ethanol, and dry. Dissolve DNA in 10 µL H_2O or TE.

6. Store in a freezer until use.

3.2.2. Probe Labeling

DOP-PCR:

1. Prepare 50-µL PCR reaction mixture containing 1 µL of DOP-PCR product, 160 µM biotin-16-dUTP, 80 µM dTTP, 200 µM each of the other dNTPs, 1.25 U of Taq polymerase in 1× buffer supplied by the polymerase manufacture.

2. Perform PCR with the thermal conditions: 5 min at 94°C, followed by 25 cycles of 1 min at 94°C, 1 min at 62°C, 3 min at 72°C, and final extension of 10 min at 72°C.

3. Check if DNA is successfully labeled by agarose gel electrophoresis with 2 µL each of labeled product and DOP-PCR product. Mobility of the labeled DNA on the gel should be smaller than the DOP-PCR product.

4. Ethanol precipitate the labeled DNA and resuspend in 10 µL TE.

MDA:

The probe DNA prepared by the amplification of MDA can be labeled with biotin by nick translation (see Subheading 3.1.4).

3.2.3. Hybridization

1. In a 1.5-mL microtube, prepare hybridization mixture in a total volume of ca. 50 µL. The mixture is made up of 5 µL of the labeled probe, 25 µL formamide, 10 µL 50% dextran sulfate, 5 µL 20× SCC, 0.5 µL salmon sperm DNA, and 5 µL water. For painting supernumerary chromosomes such as CD or B chromosome, add 4 µg of unlabeled genomic DNA as blocking DNA that is isolated from the strain without supernumerary chromosome. Prior to measuring dextran sulfate and salmon sperm DNA with a micropipette, heat to 70–80°C to decrease their viscosity.

2. Mix thoroughly by repeated pipetting with a micropipette and vigorous vortexing, and then spin down briefly in a microcentrifuge.

3. Put 15 µL of the hybridization mixture on the slide and cover with a cover slip. If air bubbles are seen on the mounted specimen, press them out. Blot the excessive hybridization mixture spilling out from the edge of the cover slip.

4. Seal with paper cement, and wait till the cement is completely dry.
5. Place the slide on a heat block at 80°C, and leave for 2 min.
6. Incubate the slide at 37°C in a moist chamber overnight (more than 16 h).

For washing and signal detection, follow the procedures in steps 1–12 of Detection without Signal Amplification in Subheading 3.1.6.

3.3. Fiber FISH

3.3.1. Preparation of Extended DNA Fibers

Genomic DNA Fiber: Once protoplasts are prepared, any species should be applicable for preparation of extended genomic DNA fibers (steps 4–12). Here, protoplasting method for *Cochliobolus* spp. (steps 1–3) is described as an example.

1. Harvest conidia from 1-week-old cultures by washing plates with sterile distilled water containing 0.05% Tween 20. Filter through one layer of Kimwipe to remove hyphal debris.
2. Add conidial suspension to 100 mL liquid culture medium in a 300-mL flask (final concentration 1×10^6 conidia/mL), and incubate on a reciprocal shaker (110 strokes/min) for 12–15 h at 27°C.
3. Collect the germlings by vacuum-filtration on a Whatman no. 4 filter paper, wash with 0.7 M NaCl, and then suspend in 6–10 mL of filter-sterilized enzyme solution. Incubate on a shaker (65–75 strokes/min) for 2–4 h at 30°C for protoplasting.
4. Filter the protoplasts through four layers of sterile Kimwipe, centrifuge the protoplast suspension at $550 \times g$ for 10 min, and wash the pellet twice with 0.7 M NaCl.
5. Suspend the final protoplast pellet in SE, and adjust the density to ca. 2×10^8 cells/mL. Mix the protoplasts with an equal volume of 1% melted (50°C) low melting agarose, with caution not to break protoplasts, and then transfer the mixture to molds for casting into plugs. Allow agarose to solidify at 4°C for 15 min, and then incubate in NDS for 14 h at 37°C in order to lyse cells. Rinsing plugs three times in 50 mM EDTA for 30 min each at 37°C and store at 4°C.
6. Cut a tiny piece (ca. 1 mm³) from a plug with a razor blade, place on a poly-L-lysine-coated slide, and mount with 40 μL sterile water.
7. Place slide on a heat block at 85°C for 20–30 s to melt the agarose. Mechanically extend the drop of liquified agarose using the edge of a cover slip and air-dry.
8. Add 100 μL of RNase to the target area, cover with parafilm, and incubate for 1 h at 37°C in a moist chamber.

9. Remove parafilm, and wash in 2× SSC for 5 min.
10. Wipe excess fluid, add 100 µL of 1% pepsin solution, cover with parafilm, incubate for 5 min at room temperature in a moist chamber, immerse in water for 1 min, and then in 2× SSC for 5 min (see Note 15).
11. Dehydrate by successive transfers to 70%, 85%, and 95% ethanol each for 5 min and air-dry.

Specific Chromosomal DNA Fibers:

1. Separate individual chromosomal DNA on low melting agarose by PFGE and excised the chromosomal band containing target sequence using a razor blade. Store the excised chromosomal DNA embedded agarose blocks (DAB) in TE or 50 mM EDTA at 4°C. For detailed procedures see ref. (4).
2. Cut the DAB into a tiny piece (ca. 1 mm^3) with a razor blade, place it on a poly-L-lysine-coated slide, and mount with 40 µL of sterile water.
3. Place slide on a heat block at 85°C for 20–30 s to melt the agarose. Extend the liquified agarose drop using a cover slip, and air-dry.
4. Add 100 µL RNase to the target area, cover with parafilm, and incubate for 1 h at 37°C in a moist chamber (see Note 16).
5. Remove parafilm, and wash in 2× SSC for 5 min.
6. Dehydrate by successive transfers to 70%, 85%, and 95% ethanol each for 5 min and air-dry.

Cosmid DNA Fibers:

1. Purify cosmid DNA by phenol-chloroform extraction to remove all residual enzymes and proteins (see Note 17).
2. Add 0.5 µg cosmid DNA to distilled water, and end up with a final volume of 10 µL.
3. Apply the DNA solution onto a poly-L-lysine-coated slide.
4. Add 2 µL of NDS to the DNA solution, and mix gently (using a tip).
5. Leave for 5 min, and add 10 µL water.
6. Put the slide on a heat block at 65°C, and allow to dry (see Note 18).
7. Put in 100% ethanol for 5 min and air-dry.

3.3.2. Probe Labeling

Probe DNAs are labeled with either biotin-14-dATP or DIG-11-dUTP by nick translation.

1. Biotin labeling and DIG labeling are performed with the Bio-Nick labeling system and DIG-Nick Translation Mix, respectively.

2. After labeling and ethanol precipitation by following the manufacture's instruction, labeled probe DNAs are dissolved in 20 μL of ultra pure distilled water to end up with the concentration of 50 ng/μL.

3.3.3. Hybridization

1. Prepare hybridization mixture.
2. Apply 15 μL of hybridization mixture directly to the prepared area on the slide, cover with a cover slip, and seal with paper cement.
3. After paper cement is completely dry, place the slide on a heat block at 80°C and leave for 5 min.
4. Incubate the slide at 37°C in a moist chamber for at least 72 h.

3.3.4. Washing and Signal Detection

1. Remove paper cement carefully with forceps, and immerse in 2× SSC in a Coplin jar to float off the cover slip. After the cover slip is removed, transfer the slide into fresh 2× SSC and incubate for 5 min.
2. Transfer slides into pre-warmed 50% formamide/2× SSC, and leave for 15 min at 37°C.
3. Wash slides in 2× SSC twice for 10 min, and in 4× SSC for 5 min.
4. After wiping excess fluid, apply 80 μL of blocking solution, cover with parafilm, and incubate for 15 min at 37°C in a moist chamber.
5. Remove parafilm and wipe excess blocking solution carefully by tilting the slides. Apply 80 μL of primary antibody in 1% Block ace/4× SSC, cover with parafilm, and incubate for 1 h at 37°C in a moist chamber.
6. Wash in 4× SSC for 10 min, 4× SSC containing 0.1% Triton X-100 for 10 min, and 4× SSC for 10 min with gently shaking on an orbital shaker.
7. Wipe excess fluid, apply 80 μL of fluorescent dye-conjugated secondary antibody in 1% Block Ace/4× SSC, cover with parafilm, and incubate for 1 h at 37°C in a moist chamber protected from light.
8. Wash in 4× SSC for 10 min, 4× SSC containing 0.1% Triton X-100 for 10 min, and 4× SSC for 10 min in a Coplin jar protected from light with gently shaking on an orbital shaker.
9. Leave the slides in 2× SSC for 5 min, wipe excess fluid carefully, apply 15 μL of Vectashield, and cover with a cover slip.
10. Observe sample using a fluorescent microscope.

4. Notes

1. The standard alkaline method also works well.
2. Numerous fluorescent dye-conjugated antibodies are available. Choose as required for a particular application.
3. The size of the cover slip is not critical, but the amount of solution applied to the slide should be carefully matched to the volume underneath the cover slip. Volumes described here are for a cover slip of dimensions 18 × 32 × 0.15 mm. If a different size of cover slip is used, the volume of solution used must be adjusted accordingly.
4. Poly-l-lysine ensures the adhesion of DNA fibers to slides during all procedures although they do tend to cause more background in FISH compared with uncharged slides. Therefore, positively charged slides are recommended for DNA fibers preparation.
5. Enzymes used for protoplasting varies depending on the fungal species.
6. Dilution of antibody should be optimized and will be antibody dependent.
7. Fragmented hyphae prepared with homogenizer can replace conidia.
8. We usually prepare more than six slides for one FISH experiment.
9. Time-course sampling is recommended to obtain metaphase specimens for the species inherently resistant to TBZ (see ref. (23)).
10. Optimal ratio of methanol to acetic acid may vary depending on species. Generally, high methanol rate is effective in bursting germ tube but tend to break morphology of chromosome and nuclei.
11. It is difficult to assess the concentration of probe DNA after labeling. The best test is a FISH run. Probe concentration in the mixture can be increased by making more concentrated labeled probe in Subheading 3.1.4. However, it is not helpful to use excess probe DNA because too much probe may cause unspecific signals and background.
12. Optimal ratio might vary depending on the specimens. Try the concentration recommended by the supplier first. Then, adjust the optimal ratio.
13. DAPI is a known mutagen and should be handled with care.
14. Using low melting agarose yields DNA of higher quality. See ref. 24 for the general procedures to separate fungal

15. Pepsin treatment is not essential, but it helps to clean the cytoplasm or nucleoplasm covering the target. However, longer treatment times may cause morphological artifacts.
16. RNase treatment is not essential for chromosomal DNA fiber preparation.
17. Residual enzyme and proteins cause severe background fluorescent staining.
18. During drying out the DNA solution, frequent shaking and tilting the slide would be helpful to spread DNA fibers well.

chromosomes by PFGE, and recover DNA from the separated bands.

References

1. Li L, Gerecke EE, Zolan ME (1999) Homolog pairing and meiotic progression in *Coprinus cinereus*. Chromosoma 108:384–392
2. Taga M, Murata M (1994) Visualization of mitotic chromosomes in filamentous fungi by fluorescence staining and fluorescence in situ hybridization. Chromosoma 103:408–413
3. Akamatsu H, Taga M, Kodama M, Johnson R, Otani H, Kohmoto K (1999) Molecular karyotypes for *Alternaria* plant pathogens known to produce host-specific toxins. Curr Genet 35:647–656
4. Taga M, Murata M, VanEtten HD (1999) Visualization of a conditionally dispensable chromosome in the filamentous ascomycete *Nectria haematococca* by fluorescence in situ hybridization. Fungal Genet Biol 26:1 69–177
5. Taga M, Tsuchiya D, Murata M (2003) Dynamic changes of rDNA condensation state during mitosis in filamentous fungi revealed by fluorescence in situ hybridisation. Mycol Res 107:1012–1020
6. Tsuchiya D, Matsumoto A, Covert SF, Bronson CR, Taga M (2002) Physical mapping of plasmid and cosmid clones in filamentous fungi by fiber-FISH. Fungal Genet Biol 37:22–28
7. Garmaroodi HS, Taga M (2007) Duplication of a conditionally dispensable chromosome carrying pea pathogenicity (*PEP*) gene clusters in *Nectria haematococca*. Mol Plant Microbe Interact 20:1495–1504
8. Tsuchiya D, Taga M (2001) Application of fibre-FISH (fluorescence in situ hybridization) to filamentous fungi: visualization of the rRNA gene cluster of the ascomycete *Cochliobolus heterostrophus*. Microbiology 147:1183–1187
9. Li S, Harris CP, Leong SA (1993) Comparison of fluorescence in situ hybridization and primed in situ labeling methods for detection of single-copy genes in the fungus *Ustilago maydis*. Exp Mycol 17:301–308
10. Trouvelot S, van Tuinen D, Hijri M, Gianinazzi-Pearson V (1999) Visualization of ribosomal DNA loci in spore interphasic nuclei of glomalean fungi by fluorescence in situ hyhridization. Mycorrhiza 8:203–206
11. Schwarzacher T, Heslop-Harrison P (2000) Practical in situ hybridization. BIOS Scientific, Oxford
12. Rautenstrauss BW, Liehr T (eds) (2002) FISH technology. Springer, New York
13. Fan Y (ed) (2002) Molecular cytogenetics. Humana, Totowa
14. Darby IA, Hewitson TD (eds) (2006) In situ hybridization protocol, 3rd edn. Humana Press, Totowa
15. Beatty B, Mai S, Squire J (2002) FISH – a practical approach. Oxford University Press, Oxford
16. Shirane N, Masuko M, Hayashi Y (1988) Nuclear behavior and division in germinating conidia of *Botrytis cinerea*. Phytophathology 78:1627–1630
17. Telenius H, Pelmear AH, Tunnacliffe A, Carter NP, Behmel A, Ferguson-Smith MA, Nordenskjöld M, Pfragner R, Ponder BAJ (1992) Cytogenetic analysis by chromosome painting using DOP-PCR amplified flow-sorted chromosomes. Genes Chrom Cancer 4:257–263
18. Dean FB, Hosono S, Fang L, Wu X, Faruqi AF, Bray-Ward P et al (2002) Comprehensive human genome amplification using multiple displacement amplification. Proc Natl Acad Sci U S A 99:5261–5266

19. Scherthan H, Loidl J, Schuster T, Schweizer D (1992) Meiotic chromosome condensation and paring in *Saccharomyces cerevisiae* studied by chromosome painting. Chromosoma 101:590–595
20. Sambrook J, Fritsch EF, Maniatis T (1989) Molecular cloning: a laboratory manual, 2nd edn. Cold Spring Harbor Laboratory, Cold Spring Harbor
21. Murata M, Nakata N, Yasumuro Y (1992) Origin and molecular structure of a midget chromosome in a common wheat carrying rye cytoplasm. Chromosoma 102:27–31
22. QIAGEN (2005) REPLI-g handbook. Qiagen, Valencia
23. Tsuchiya D, Taga M (2001) Cytological karyotyping of three *Cochliobolus* spp. by the germ tube burst method. Phytophathology 91:354–360
24. Mills D, McCluskey K, Russell RW, Agnan J (1995) Electrophoretic karyotyping: method and applications. In: Singh RP, Singh US (eds) Molecular methods in plant pathology. CRC Press, Boca Raton, pp 81–96

Chapter 19

Live-Cell Imaging of Microtubule Dynamics in Hyphae of *Neurospora crassa*

Maho Uchida, Rosa R. Mouriño-Pérez, and Robert W. Roberson

Abstract

Due to the large number of microtubules in the wild-type strain, total internal reflection fluorescence (TIRF) microscopy was used to study cortical microtubule dynamics in leading hyphae of *Neurospora crassa* expressing β-tubulin-GFP. Detection of plus-end dynamics of individual microtubule was much improved with this approach compared to the other commonly used methods such as confocal and widefield fluorescence microscopy. In order to address the roles of motor proteins in microtubule dynamics, microtubule-motor mutant strains, Δ*nkin* and *ro-1* were examined. Unlike the wild-type strain, there were fewer microtubules in these hyphal cells; therefore, imaging was done using widefield fluorescence microscopy. We have shown that polymerization and depolymerization rates as well as hyphal extension rates were reduced by one half relative to those of wild type. Therefore, we believe that the hyphal extension rates are dependent upon the dynamic characteristics of microtubules, which are then regulated by microtubule motors in *N. crassa*.

Key words: Live-cell imaging, Green fluorescent protein, Microtubule dynamics, TIRF, *Neurospora crassa*, Kinesin, Dynein

1. Introduction

Neurospora crassa exhibits one of the fastest cellular growth rates among filamentous fungi. In order to address the roles of microtubule dynamics in polarized hyphal growth, it was our interest to study apical microtubule dynamics in leading hyphae of both *N. crassa* wild-type and microtubule-motor mutants expressing β-tubulin-GFP using live-cell imaging. Principle difficulties arose during the study of wild-type strain. Unlike microtubule-motor mutant strains, a large number of microtubules in wild-type hyphae reduced the image clarity required to monitor individual microtubule exhibiting dynamic instability using widefield fluorescence microscopy. Although other microscope methods such as confocal laser scanning microscopy and spinning disk confocal microscopy were employed, we were not able to achieve the appropriate temporal and spatial resolutions required for the study. For this reason, we used total internal reflection fluorescence (TIRF) microscopy to study microtubule dynamics in leading hyphae of wild-type strain. This optical technique uses evanescent waves to excite fluorophores within 200 nm from the interface of the specimen and the cover slip, thus only detecting the microtubules located within the cell cortex. This effectively eliminates background fluorescence from other regions of the cell and improves spatial resolution by increasing the signal to noise ratio. In addition, unlike confocal and spinning disk confocal microscopy, live-cell imaging can be achieved with high temporal resolution and low phototoxic effects. With this technique, we were able to study four elements of microtubule instability, i.e., polymerization and depolymerization rates, frequency of catastrophe, and duration of pausing in wild-type leading hyphae of *N. crassa*.

2. Materials

2.1. Strains

For list of strains see Table 1.

2.2. Growth Media and Live-Cell Imaging

1. Vogel's medium N (3) supplemented with 0.5% (w/v) yeast extract and 0.5% (w/v) casamino acids. Vogel's medium N contains 50× salt solution, 2.0% (w/v) sucrose, and 1.5% (w/v) agar.
2. 50× salt solution (see Note 1): 12.5% (w/v) Na_3 citrate·$2H_2O$, 25% (w/v) KH_2PO_4, 10% (w/v) NH_4NO_3, 1% (w/v) $MgSO_4·7H_2O$, 0.5% (w/v) $CaCl_2·2H_2O$, 0.5% (v/v) trace element solution, 0.25% (v/v) biotin solution (0.1 mg/mL).

Table 1
Neurospora crassa strains used in this study

Strain	Genotype	Reference
N2526	rid[RIP1] mat A his3+::Pccg1Bml+sgfp+	(1)
FGSC4351	mat a *ro-1*	FGSC[a]
	mat a Δ*nkin*	(2)
RL21SG150	mat a	FGSC
XMF11343	Δ*nkin* his3+::Pccg1Bml+sgfp+	(8)
XRM1779	*ro-1* his3+::Pccg1Bml+sgfp+	(8)

[a]Fungal Genetics Stock Center

3. Trace element solution (see Note 1): 5 % (w/v) citric acid·H_2O, 5% (w/v) $ZnSO_4·7H_2O$, 1% (w/v) $Fe(NH_4)_2(SO_4)_2·6H_2O$, 0.25% (w/v) $CuSO_4·5H_2O$, 0.05% (w/v) $MnSO_4·H_2O$, 0.05% (w/v) H_3BO_3, 0.05% (w/v) $Na_2MoO_4·2H_2O$.

4. Synthetic cross medium (4) (2× stock): 0.2% (w/v) KNO_3, 0.14% (w/v) K_2HPO_4, 0.1% (w/v) KH_2PO_4, 0.1% (w/v) $MgSO_4·7H_2O$, 0.02% (w/v) NaCl, 0.02% (w/v) $CaCl_2·2H_2O$ (dissolved separately), 0.01% (v/v) biotin solution, 0.02% (v/v) trace element solution.

5. Water agar: 1.5% (w/v) agar.

6. Microscope cover slips (22 × 50 mm).

3. Methods

3.1. Growth Conditions

All *N. crassa* strains used in this study were maintained at 25°C on Vogel's medium with 2% (w/v) sucrose. One could maintain the plates by excising the leading edge of actively growing hyphae and transferred to the fresh Vogel plates. However, we recommend using glycerol stock (see Note 2) to generate a new plate for each experiment.

3.2. Generation of Microtubule-Motor Mutants Expressing β-Tubulin-GFP

Crossing is a useful tool for basic genetic modeling. Inducing the sexual cycle of two different strains with complementary mating type enables to obtain a new strain expressing two parental genetic features inherited from each parent after karyogamy.

To generate microtubule-motor mutants expressing β-tubulin-GFP, *ro-1* (FGSC4351), and Δ*nkin* (2) strains were crossed separately with a mating compatible strain that was transformed with the β-tubulin-GFP construct (N2526; ref. 1) by following these steps:

1. To obtain conidia from each strain, each of the β-tubulin-GFP, *ro-1*, and Δ*nkin* strains are first grown on small Petri dishes (5 cm) containing synthetic cross medium (4) supplemented with 1% (w/v) sucrose and 2% (w/v) agar for 5 days at 25°C.

2. Once conidia are formed, conidia from both mating types are inoculated for fertilization.

3. After 14 days of incubation at 25°C in a dark chamber, perithecia develop and ascospores are produced. The ascospores are collected by adding distilled water inside the lid of Petri dish.

4. To activate the ascospores, it requires to heat shock for 2 h at 60°C. These spores are then placed on Vogel's medium with 2% (w/v) sucrose and incubated for 12 h at 28°C (see Note 3).

5. Each developing colony is transferred to 5-mL culture tubes containing Vogel's medium and incubated for 24 h at 28°C.
6. The final screening is done to select the ones with the mutant phenotype and β-tubulin-GFP expression.

3.3. Live-Cell Imaging of Apical Hyphae Expressing β-Tubulin-GFP

3.3.1. Total Internal Reflection Fluorescence Microscopy

TIRF microscopy utilizes the unique properties of an induced evanescent wave in a limited specimen region immediately adjacent to the interface between two media having different refractive indices. By confining illumination, it enables the user to image cellular and molecular interactions such as vesicle trafficking and protein–protein interactions at high temporal and spatial resolution. Although this technique has been available for last 20 years, it has not received a considerable amount of attention until recently. Much of the trend toward greater utilization of TIRF microscopy (e.g., (5–8)) is due to technological advances that facilitated its use, developments of vital fluorescent dyes, and most importantly development of strains expressing fluorescent protein.

In this study, TIRF microscopy provided the capabilities to resolve individual microtubule plus ends optically and follow their dynamics directly, which were not possible with other optical systems.

1. To obtain active hyphal growth of the wild-type strain (N2526), glycerol stock is used to make new plates. It usually takes over night before hyphal growth is observed. These plates are kept at 25°C.
2. Once hyphae are formed on the plates containing growth medium, small sections of growing hyphal tips are transferred to water agar. Using the plates containing water agar alone help eliminating any fluorescence produced from the growth medium and reduces cytoplasmic background during image acquisition. It will take a few days for hyphae to grow out.
3. For live-cell imaging, small sections of growing hyphal tips are excised and placed on a cover slip (9).
4. Growing hyphal tips expressing β-tubulin-GFP are observed using TIRF microscopy with excitation at 488 nm. In this study, we used an IX-70 inverted microscope equipped with a 100×/1.45 apochromat objective lens (Olympus America Inc., Center Valley, PA) and a krypton/argon laser (Melles Griot, Carlsbad, CA) at 488 nm excitation.
5. Images are recorded with an EMCCD camera such as cascade 512B (Photometrics, Tucson, AZ) for durations of 2–3 min at 512×512 resolution and frame rates of 50–200 ms.
6. MetaMorph 6.0/6.1 software (Universal Imaging, Downingtown, PA) is used to control the camera and capture images. TIRF microscopy of microtubule polymerization and depolymerization in leading hyphae of wild-type strain are shown in Figs. 1 and 2 as examples.

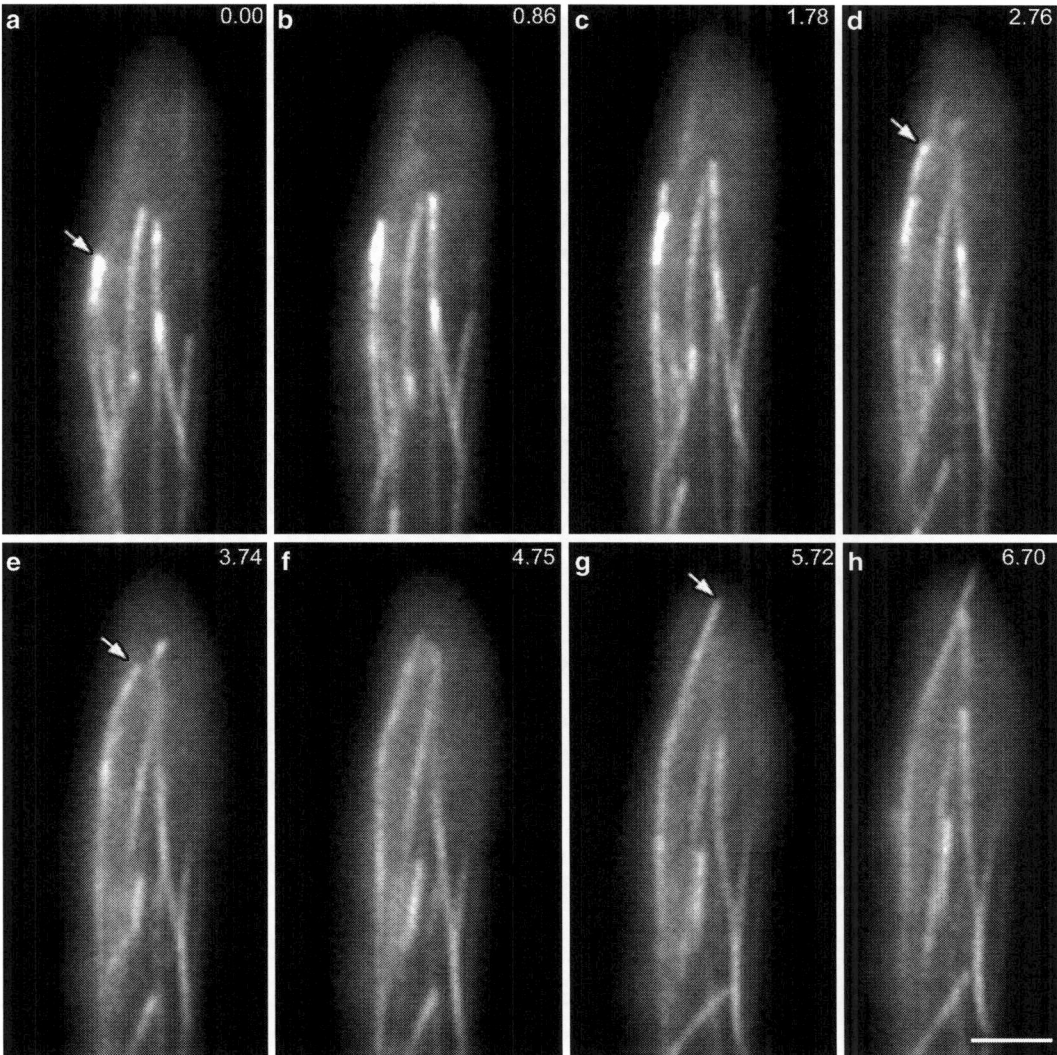

Fig. 1. TIRF microscopy of microtubule polymerization in leading hyphae of a β-tubulin-GFP strain. The plus end of a single microtubule is indicated by arrows. Elapsed time (s, ms) is shown in the *top right* corner of each panel. Scale bar, 5 μm. Reproduced from (8) with permission from Elsevier Science.

3.3.2. Widefield Fluorescence Microscopy

1. To obtain active hyphal growth of the Δ*nkin* (XMF11343) and *ro-1* (XRM1779), glycerol stocks of these strains are used to make new plates. These plates are kept at 25°C.

2. Once hyphae are formed on the plates containing growth medium, small sections of growing hyphal tips are transferred to water agar.

3. For live-cell imaging, small sections of growing hyphal tips are excised and placed on a cover slip.

4. Green fluorescence-tagged microtubules are observed using a standard inverted epifluorescence microscope such as Eclipse

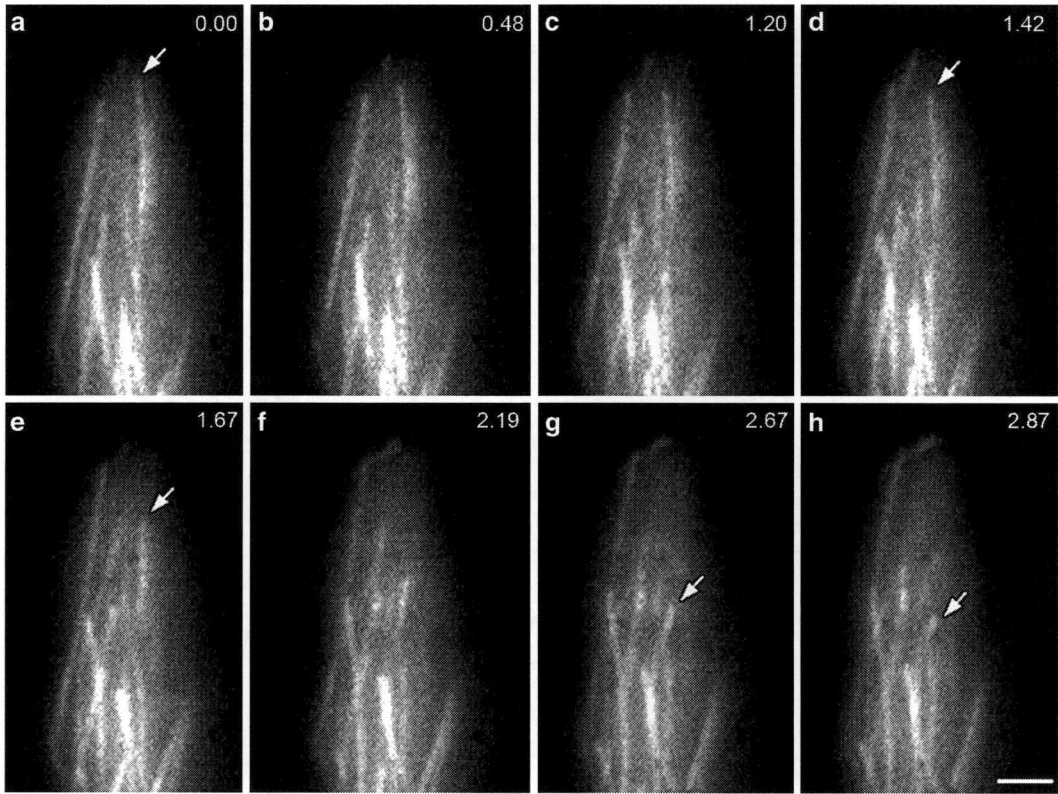

Fig. 2. TIRF microscopy of microtubule depolymerization in leading hyphae of a β-tubulin-GFP strain. The plus end of a single microtubule is indicated by arrows. Elapsed time (s, ms) is shown in the top right corner of each panel. Scale bar, 5 μm. Reproduced from (8) with permission from Elsevier Science.

TE300 (Nikon Inc., Instrument Group, Melville, NY) with a Plan Neofluar 100×/1.3 N.A. oil immersion objective lens, a 150-W Xenon bulb, and a blue excitation fluorescence filter (420–495 nm).

5. Time-lapse images are recorded with a low-light level CCD camera such as Quantix CCD camera (Roper Scientific, Tucson, AZ) for durations of approximately 2–3 min at 512 × 512 resolution and frame rates of 300–800 ms.

6. MetaMorph 6.0/6.1 software (Universal Imaging, Downingtown, PA) is used to control the camera and capture images. Widefield epifluorescence microscopy of a single microtubule polymerizing and depolymerizing in *ro-1* and Δ*nkin* strains are shown in Figs. 3 and 4 as examples.

3.4. Measurement of Microtubule Dynamic Instability

The protocol described here was based on MetaMorph Offline 6.2r6. However, analysis of microtubule dynamics was carried out in a similar way as previously described (10–12).

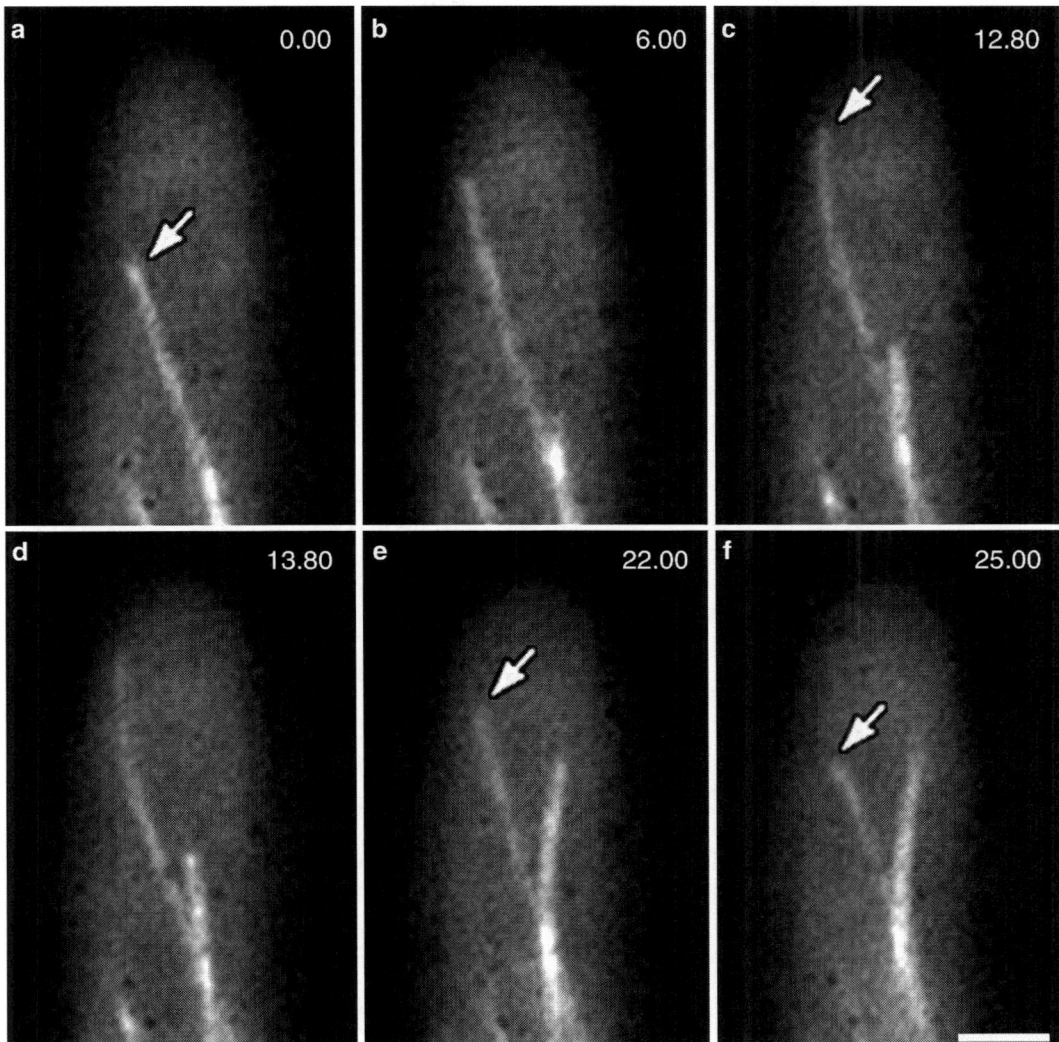

Fig. 3. Widefield epifluorescence microscopy of a single microtubule polymerizing (**a–c**), and depolymerizing (**d–f**) in a β-tubulin-GFP::*ro-1* strain. The plus end of microtubule is indicated by arrows. Elapsed time (s, ms) is shown in the *top right* corner of each panel. Scale bar, 5 μm. Reproduced from (8) with permission from Elsevier Science.

1. Add the time stamp and apply appropriate calibration.
2. Convert the time-lapse images to gray scale. This would ease the detection of individual microtubules. If necessary, use the command, i.e., inversion, which makes microtubules and cell membrane black and the background as white.
3. To measure the rates of microtubule dynamics, the original position of microtubule (+) end is first marked by drawing a line (= the start line) that is perpendicular to the end of microtubule. In addition, a time point of the current image (= the start point) is noted.

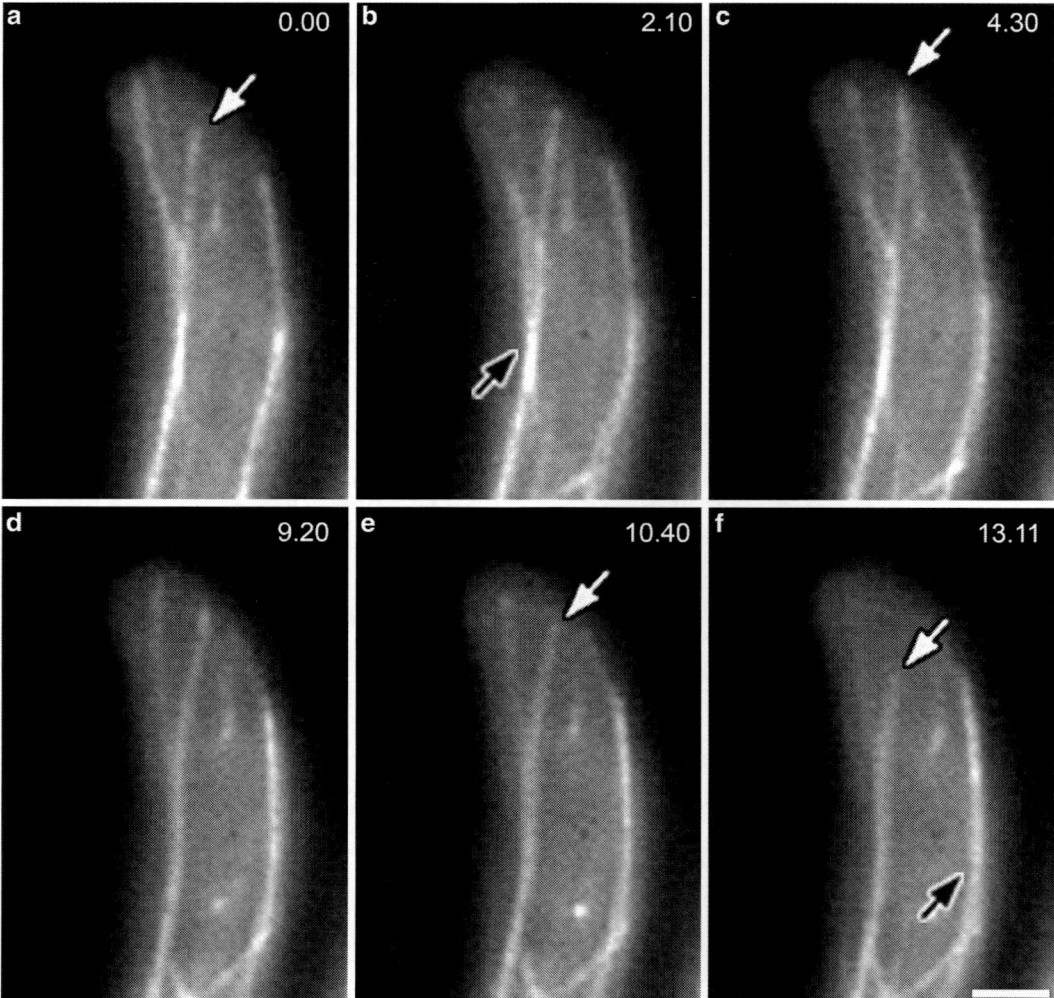

Fig. 4. Widefield epifluorescence microscopy of a single microtubule polymerizing (**a–c**) and depolymerizing (**d–f**) in β-tubulin-GFP::Δ*nkin* strain. The plus end of microtubule is indicated by arrows. Elapsed time (s, ms) is shown in the top right corner of each panel. Scale bar, 5 μm. Reproduced from (8) with permission from Elsevier Science.

4. The extension of the (+) end is then tracked through individual frames of a time-lapse sequence until the microtubule growth ends. At this point, a second line (= the finish line) is drawn at the (+) end, and a time point of the last image sequence (= the end point) is noted.

5. The distance between the start and finish lines are obtained by the command "measure distance." Duration of time for the particular microtubule dynamic event is obtained by subtracting the start point from the end point.

6. The rate calculations are done using Microsoft Excel.

Table 2
Summary of hyphal growth rates and apical microtubule dynamic instability in leading hyphae of wild type, *ro-1*, and Δ*nkin* strains

	Wild type	ro-1	Δnkin
Hyphal growth rate (μm/min)	15.31 ± 0.98 $n=20$	0.71 ± 0.43 $n=21$	0.78 ± 0.46 $n=20$
Polymerization rate (μm/min)	27.60 ± 7.20 $n=30$ (20)	13.86 ± 4.66 $n=55$ (20)	14.97 ± 4.06 $n=58$ (20)
Depolymerization rate (μm/min)	53.70 ± 20.40 $n=27$ (20)	25.44 ± 10.77 $n=45$ (20)	29.58 ± 11.55 $n=48$ (20)
Frequency of catastrophe (per second)	0.21	0.06	0.08
Frequency of rescue (per second)	Not determined	0.01	0.01
Duration of pausing (per second)	1.46	7.52	4.11

n = number of hyphae/microtubules measured; numbers in parentheses indicate the number of cells observed for microtubule dynamic instability study; ±indicates standard deviation

7. Frequencies of catastrophe and rescue are calculated by dividing the total number of catastrophe or rescue events per sequence by the total time of the sequence.

8. Pausing is recognized as a brief stationary state when microtubule rate of growth or shrinkage is equal to or less than 0.5 μm per frame. Summary of apical microtubule dynamic instability is shown in Table 2.

3.5. Measurement of Hyphal Extension

The same methodology described in Subheading 3.4 is used to measure hyphal extension rates except that the start and end lines are drawn on the apical hyphal tip. All measurements are done in the same cells used to study microtubule dynamics. Growth rates for each movie sequence are calculated and averaged. Summary of hyphal growth rates is shown in Table 2.

4. Notes

1. Add 1–2% (v/v) chloroform as preservative, and store at room temperature.

2. 20% autoclaved glycerol solution. For the glycerol stock, it is of best to use a plate on which mycelium and conidia are formed. Few agar plugs (~5 × 5 mm) are excised from the plate and placed in an Eppendorf tube containing sterile glycerol solution. Store at −80°C.

3. During the 12 h incubation at 28°C, it is best to prescreen the developing colonies expressing mutant phenotype and β-tubulin-GFP.

References

1. Freitag M, Hickey PC, Raju NB, Selker EU, Read ND (2004) GFP as a tool to analyze the organization, dynamics, and function of nuclei, and microtubules in *Neurospora crassa*. Fungal Genet Biol 41:897–910
2. Seiler S, Nargang FE, Steinberg G, Schliwa M (1997) Kinesin is essential for cell morphogenesis and polarized secretion in *Neurospora crassa*. EMBO J 16:3025–3034
3. Vogel HJ (1956) A convenient growth medium for *Neurospora* (Medium N). Microbial Genet Bull 13:42–43
4. Westergaard M, Mitchell HK (1947) Neurospora V. A synthetic medium favoring sexual reproduction. Am J Bot 34:573–577
5. Becherer U, Pasche M, Nofal S, Hof D, Matti U, Rettig J (2007) Quantifying exocytosis by combination of membrane capacitance measurements and total internal reflection fluorescence microscopy in chrofaffin cells. PLoS One 2:e505
6. Shaw RM, Fay AJ, Puthenveedu MA, Zastrow M, Jan Y, Jan LY (2007) Microtubule plus-end-tracking proteins target gap junctions directly from the cell interior to adherens junctions. Cell 128:547–560
7. Popp D, Yamamoto A, Maeda Y (2007) Crowded surfaces change annealing dynamics of actin filaments. J Mol Biol 368:365–374
8. Uchida M, Mouriño-Pérez RR, Freitag M, Bartnicki-García S (2008) Microtubule dynamics and the role of molecular motors in *Neurospora crassa*. Fungal Genet Biol 45:683–692
9. Hickey PC, Swift SR, Roca MG, Read ND (2005) Live-cell imaging of filamentous fungi using vital fluorescent dyes and confocal microscopy. In: Savidge T, Pothoulakis C (eds) Methods in microbiology, microbial imaging, vol 35. Elsevier, London, pp 63–87
10. Dhamodharan R, Jordan MA, Thrower D, Wilson L, Wadsworth P (1995) Vinblastine suppresses dynamics of individual microtubules in living interphase cells. Mol Biol Cell 6:1215–1229
11. Dhamodharan R, Wadsworth P (1995) Modulations of microtubule dynamic instability in vivo by brain microtubule associated proteins. J Cell Sci 108:1679–1689
12. Han G, Liu B, Zhang J, Zuo W, Morris NR, Xiang X (2001) The *Aspergillus* cytoplasmic dynein heavy chain and NUDF localize to microtubule ends and affect microtubule dynamics. Curr Biol 11:719–724

Chapter 20

Methods to Detect Apoptotic-Like Cell Death in Filamentous Fungi

Camile P. Semighini and Steven D. Harris

Abstract

Fungi are capable of undergoing apoptotic-like cell death, and display many of the characteristic features of apoptosis observed in multicellular organisms. These features include nuclear condensation, DNA fragmentation, translocation of phosphatidylserine from the cytoplasmic to the extracellular side of the plasma membrane, and increased levels of reactive oxygen species (ROS). Several assays can be used to detect apoptotic cells, and here we describe adaptations of assays such as TUNEL, Annexin V, and Evan's Blue for the investigation of apoptotic-like cell death in fungal hyphae. We also present approaches for monitoring nuclear condensation and production of ROS.

Key words: Apoptosis, Filamentous fungus, Nuclear staining, TUNEL, Annexin V, Evans Blue, Reactive oxygen species

1. Introduction

Apoptosis is a form of programmed cell death critical for the development and homeostasis in multicellular organisms (1). The process of apoptosis depends on numerous extra- and intracellular pro- and antiapoptotic signals that are integrated to activate apoptotic effectors only when necessary. Two separate, but overlapping, mechanisms of inducing apoptosis have been identified (2). For the first mechanism, intracellular death messengers lead to disruption of mitochondrial membrane potential. The subsequent release of cytochrome *c* from mitochondria activates the apoptotic executors; cysteine proteases that are known as caspases. Alternatively, signals from membrane-bound death receptors can lead to mitochondria-independent activation of caspases.

Morphological and biochemical changes follow activation of either apoptotic pathway. These changes include nuclear and cytoplasmic condensation, DNA fragmentation, plasma membrane alterations including the translocation of phosphatidylserine (PS) from the cytoplasmic to the extracellular side of the membrane, increased levels of reactive oxygen species (ROS) indicative of cellular oxidative stress, and partitioning of cytoplasm and nuclei into membrane bound-vesicles (apoptotic bodies). The latter structures are known to contain ribosomes, morphologically intact mitochondria, and nuclear material. In vivo, apoptotic bodies are rapidly recognized and phagocytized by either macrophages or adjacent epithelial cells (1).

Apoptotic-like cell death has been reported in several fungi, with many of its underlying features resembling apoptosis of multicellular organisms, including DNA condensation and fragmentation, exposure of PS and high levels of ROS. For example, we previously reported that the isoprenoid compound farnesol induces apoptotic-like cell death both in *Aspergillus nidulans* (3) and *Fusarium graminearum* (4). Several assays have been developed for the detection of mammalian cells undergoing apoptosis. In this chapter, we describe assays that were adapted to detect apoptotic-like cell death in fungal hyphae.

One of the first morphological changes associated with apoptosis is chromatin condensation, similar to that observed in mitosis. Nuclear condensation can be easily detected by different fluorochromes that label nuclear material, followed by visual analysis with fluorescence microscope. We prefer using Hoechst 33258 for nuclear staining since this fluorochrome stains chromatin on the nuclear DNA, whereas DAPI, for example, stains both mitochondrial and nuclear DNA. The next assay described, TUNEL (TdT-mediated dUTP nick end labeling), is the standard method for identification and quantification of apoptotic cells. This method detects DNA fragmentation by using terminal deoxynucleotidyl transferase (TdT) to incorporate dUTP tagged with biotin, DIG, or fluorescein into the blunt ends of double-stranded DNA breaks (5, 6). In order to allow the TdT and dUTP to enter fixed hyphae, their cell wall and plasma membrane have to be permeabilized prior to the enzymatic reaction. The assay described here uses FITC-conjugated dUTP, which can be directly visualized by fluorescence microscopy, thereby avoiding the second incubation step with streptavidin or anti-DIG antibody necessary to detect dUTP tagged with biotin or DIG respectively.

The Annexin V assay detects translocation of PS from the inner side of the plasma membrane to the outer layer (7). This test allows the detection of cells at early stages of apoptosis, because changes in PS asymmetry become apparent before the other morphological changes associated with apoptosis, including DNA fragmentation. Annexin V is a member of the annexin family of calcium-dependent phospholipid-binding proteins and preferentially binds to negatively charged phospholipids like PS. By conjugating FITC to Annexin V, it is possible to identify and quantify apoptotic cells by fluorescence microscopy. Annexin V does not bind cells with an intact plasma membrane, but it can falsely detect the inner membrane PS of lysed (necrotic) cells. Therefore, simultaneous staining with propidium iodide (red fluorescence), which will be excluded from intact cells, allows the discrimination of intact (FITC–, PI–), early apoptotic (FITC+, PI–) and late apoptotic or necrotic hyphae (FITC+, PI+). An extra step has to be added to the protocol in order to remove the cell wall from hyphae and allow the detection of exposed PS on the external surface of protoplasts. Late apoptotic hyphae can also be detected by Evans Blue staining. Normal hyphae with an intact plasma membrane exclude Evans blue and remain their natural color. Dying hyphae that have undergone plasma membrane lysis are unable to exclude the dye and stain deep blue. Finally, we describe the detection of increased ROS production by apoptotic cells using two staining methods. The first method uses 2′,7′-dichlorodihydrofluorescein diacetate (H_2DCFDA) which is permeable to protoplasts and undergoes intracellular conversion by nonspecific esterases to nonfluorescent 2′,7′-dichlorofluorescein (DCFH). DCFH oxidizes in the presence of H_2O_2 and other ROS to form 2′,7′-dichlorofluorescein (DCF), which emits bright green fluorescence (8). The second method uses Nitro blue Tetrazolium (NBT), a pale yellow compound that is reduced by ROS to a blue-purple formazan precipitate (9, 10). Superoxide was reported to be the major oxidant species responsible for reducing NBT to formazan (11).

2. Materials

2.1. Detection of Nuclear Condensation

1. Phosphate Buffered Saline (PBS) 10× stock solution: 1.37 M sodium chloride, 27 mM potassium chloride, 100 mM sodium dihydrogen phosphate (NaH_2PO_4), 18 mM potassium dihydrogen phosphate (KH_2PO_4). Adjust pH to 7.2 with HCl if necessary and autoclave before storing at room temperature (see Note 1).

2. Fixing Solution (see Note 2): 1× PBS, 5% dimethyl sulfoxide (DMSO), 3.7% formaldehyde. Better if prepared fresh but can be stored at room temperature for 1 week.

3. Nuclear Staining Solution: Prepare stock of 1 mg/mL Hoechst 33258 (Molecular Probes) and store at 4°C in the dark for several months. Prepare working solution fresh by diluting to 0.1–0.5 mg/mL (see Note 3).

4. Mount Solution: 1× PBS buffer, 50% glycerol, 0.1% *n*-propyl-gallate. Divide into aliquots and store at −20°C.
5. Sterile 22 mm × 22 mm square glass coverslips.
6. Staining jars for coverslips (see Note 4).
7. Precleaned microscopy slides (available from Fisher or VWR).
8. Fluorescence microscope.

2.2. TUNEL Assay

1. This method uses all materials described in the previous Subheading 2.1, however, a different fixing solution is routinely used. Substitute PBS with PEM; 50 mM piperazine-N,N'-bis(2-ethanesulfonic acid) (PIPES) pH 6.7, 25 mM ethylene glycol tetra-acetic acid (EGTA) pH 7.0, and 5 mM magnesium sulfate ($MgSO_4$).
2. Digestion Solution: Dissolve 50 mg/mL lysing enzymes (Sigma-Aldrich), 67.5 mg/mL beta-D glucanase (Interspex), and 22 mg/mL driselase (Sigma-Aldrich) in PEM and incubate on ice for 15 min. Centrifuge at 2,000 rpm ($447 \times g$) for 5 min and transfer the supernatant to a new tube and add 1 mL of egg white (aliquot previously and store at −20°C) (see Note 5).
3. Extraction Solution: 100 mM PIPES pH 6.7, 25 mM EGTA pH 7.0, and 0.01% Igepal CA-630.
4. PEM/BSA: Dissolve 0.1% bovine serum albumin (BSA) in PEM.
5. DNase: Dilute RQ1-RNA free DNase (Promega, Cat. No. 610A, concentration 1 U/μL) to 50 U/mL of PBS. Use fresh.
6. TUNEL Reaction Solution (In Situ Cell Death Detection Kit, Fluorescein: Boehringer Mannheim, Cat. No. 1684795): Remove 100 μL Label Solution from vial two and reserve for two negative controls. Add total volume of vial one (50 μL) to the remaining 450 μL Label Solution in vial two to obtain 500 μL TUNEL Reaction Solution. Mix well to equilibrate components. The TUNEL Reaction Solution should be prepared immediately before use and should not be stored.

2.3. Annexin V Assay

1. Citric acid: 0.5 M citric acid, adjust to pH 6.0 with NaOH. Autoclave before storing at 4°C.
2. Solution 1: 0.8 M ammonium sulfate (($NH_4)_2SO_4$), 100 mM citric acid. Autoclave before storing at 4°C.
3. Solution 2: 1% yeast extract, 2% sucrose, and 20 mM $MgSO_4$. Autoclave and store at 4°C.
4. Protoplasting Solution: In a 250 mL sterile flask, add 10 mL of Solution 1 and 20 mL of Solution 2. Dissolve 200 mg of Driselase, 300 mg of Glucanex, and 50 mg of Lysing enzymes in 5 mL of Solution 1 in a falcon tube and place on ice for 15 min. Because the enzymes dissolve poorly, they should be gently mixed every minute. In order to clarify, centrifuge at 2,000 rpm ($447 \times g$) for

5 min and then filter-sterilize through a 0.2 μm syringe filter directly into the same flask in which the other solutions were added (see Note 6). Dissolve 500 mg of BSA in 5 mL of Solution 1 and filter sterilize into the same flask (see Note 7).

5. Sucrose Solution: 1 M sucrose in autoclaved distilled water.
6. 1× Binding Buffer (Annexin V-FITC Apoptosis Detection Kit II, Calbiochem, Cat. No. CBA059): Dilute the 4× Binding Buffer from the kit to 1× in 1 M sucrose.
7. FITC-tagged Annexin V (Annexin V-FITC Apoptosis Detection Kit II, Calbiochem, Cat. No. CBA059).
8. Propidium iodide (PI) (Annexin V-FITC Apoptosis Detection Kit II, Calbiochem, Cat. No. CBA059) (see Note 8).
9. Items 4–8 from Subheading 2.1. Because protoplasts that have failed to adhere to slides can interfere with microscopy, we recommend the use of Gelatin Coated Adhesive or Colorfrost/Plus Microscope Slides (Fisher).

2.4. Evans Blue Staining

1. 1× PBS.
2. Evans Blue Solution: Dilute 1% Evans Blue (Sigma) in PBS.
3. Bright-field microscope.
4. Items 5–7 from Subheading 2.1.

2.5. ROS Detection

1. Solutions 1–6 from Subheading 2.3.

2.5.1. 2′,7′-Dichlorofluorescein Diacetate Staining

2. DCF Solution: Prepare 1 mM stock solution of 2′,7′-dichlorofluorescein diacetate (DCF, Molecular Probes) in DMSO. Divide into aliquots and store at −20°C. Prepare working solution fresh by diluting to 50 μM in water (see Note 9).
3. Items described in item 9 of Subheading 2.3.

2.5.2. Nitro Blue Tetrazolium Staining

1. MOPS: Prepare 5 mM solution of 3-(N-Morpholino) propanesulfonic acid (MOPS) in water. Adjust to pH 7.6 with NaOH.
2. NBT Solution: Make 2.5 mM solution of Nitro Blue Tetrazolium (NBT) in MOPS and store for up to 2 weeks at room temperature and protect from light.
3. Bright-field microscope.
4. Items 5–7 from Subheading 2.1.

3. Methods

Before performing the techniques described below, apoptosis should be induced. The conditions for induction of apoptosis, including dosage, exposure time, culture phase, temperature, media, etc., should be tested for each apoptotic inducer and fungal species. Ideally, the optimized conditions should induce apoptosis in the majority of the hyphae and in a dose-dependent manner. A control using the same volume of the solvent used to dissolve the apoptotic inducer should always be run in parallel with each experiment. Also, when performing the assays for the first time, it is advisable to use a known stimulator of apoptosis as a positive control (e.g., (3)).

3.1. Detection of Nuclear Condensation

1. Using sterile procedures, place up to four coverslips in the bottom of a 10 cm plastic Petri plate.
2. Inoculate 10^6 spores in 10 mL of appropriate liquid medium and gently pour into the plate containing the coverslips (see Note 10).
3. Incubate at temperature and time appropriate for each experiment (see Note 11).
4. Treat hyphae with apoptotic inducer diluted in fresh medium pre-incubated at the growth temperature. Include a parallel control sample that is not treated.
5. Transfer coverslips with forceps to a staining jar containing 10 mL of Fixing Solution. It is important to keep track of the side of the coverslip containing the growing hyphae. One suggestion to avoid confusion is to always keep the hyphal side facing you as you transfer the coverslips. Fix for 15 min at room temperature.
6. Transfer the coverslips to a new staining jar containing 10 mL of PBS and incubate for 5 min. Repeat this step two times (total of three washes) adding new buffer to the jar for each wash.
7. Transfer the coverslips to a new staining jar containing 10 mL of Nuclear Staining Solution, cover with aluminum foil to protect from light, and incubate for 5 min.
8. Rinse coverslips briefly in a beaker containing water. Dry the excess of water by carefully blotting the edge of the coverslip with a Kimwipe. Alternatively, the coverslips may be propped against a box resting on a Kimwipe to briefly drain. Make sure that the coverslips do not dry out (a few min should be enough). Always keep track of the side of the coverslip containing the hyphal growth.
9. Pipette 10 µL of Mounting Solution onto a microscope slide. Gently lay the coverslip on the Mounting Solution, with the hyphae facing down, avoiding the formation of bubbles.
10. Remove excess Mounting Solution that spread out from under the coverslip by carefully blotting with a Kimwipe (see Note 12).

11. Seal the edges of the coverslip with transparent nail polish.

12. Visualize samples using a fluorescence microscope, using a 60× objective and a UV filter set (see Note 13). Slides can be stored at 4°C in the dark for several months.

13. The proportion of hyphae with condensed nuclei should be determined and compared to the untreated control. Condensed nuclei are typically much smaller in size and lack the nucleolar shadow (i.e., small intranuclear region that does not stain) that is usually visible in normal nuclei. At least 200 hyphae should be counted in each sample.

3.2. TUNEL Assay

1. Follow steps 1–5 from Subheading 3.1 for growth and fixation of hyphae, but use the Fixing Solution described for the TUNEL Assay.

2. Wash coverslips three times with PEM for 5 min to remove Fixing Solution.

3. Cut a piece of parafilm and stretch it uniformly over an acrylic sheet or other smooth surface; the blunt side of a clean razor blade can be used to smooth out the parafilm and eliminate air bubbles. Pipette 200 µL of Digesting Solution such that it forms a drop on the parafilm. Gently lay the coverslip on the drop with hyphae facing down. Incubate at 28°C for 15–120 min (see Note 14). At each digestion time point (i.e., 15, 30 min, etc.), transfer one coverslip to a staining jar containing PEM and leave there until the final time point.

4. Wash coverslips three times with PEM for 5 min to remove the digestive enzymes.

5. Transfer the coverslips to a new staining jar containing 10 mL of Extraction Solution and incubate for 5 min at room temperature to permeabilize hyphal membranes.

6. Wash twice in PEM and once in PEM/BSA for 5 min.

7. Using parafilm as described in step 3, incubate the positive control with DNase at 37°C for 1 h. Meanwhile, keep the other samples in PEM/BSA.

8. Cover the bottom of a shallow plastic box with parafilm. Divide the TUNEL Reaction Solution into enough drops for each sample, including the positive control (use at least 50 µL for each one). Pipette 50 µL of Label Solution reserved for the negative controls (see Note 15). Lay coverslips on drops with hyphae facing down, keeping track of each different sample. Place 2 moistened cotton balls in the edges of the box and cover it with aluminum foil. Incubate at 37°C for 1 h. All following steps should be protected from light.

9. Wash coverslips three times with PEM/BSA for 5 min.

10. Follow steps 7–12 from Subheading 3.1.
11. The proportion of hyphae presenting TUNEL positive nuclei (green fluorescent nuclei) should be determined and compared to the untreated control (see Note 16). At least 200 hyphae should be counted in each sample. No TUNEL positive nuclei should be observed in negative controls and the majority of nuclei should be TUNEL positive in the positive controls (see Note 17).

3.3. Annexin V Assay

3.3.1. Generation of Protoplasts (See Note 18)

1. Inoculate 10^9 fresh spores in 50 mL of appropriate liquid medium in a 250 mL Erlenmeyer flask. Incubate with appropriate agitation, temperature, and time.
2. Treat hyphae with apoptotic inducer. Be sure to include a parallel control sample that is not treated.
3. Collect hyphae from each condition using centrifugation at 3,000 rpm ($1006 \times g$) for 3 min and discard supernatant.
4. Resuspend in 50 mL of Protoplast Solution and incubate at 30°C and 100 rpm. After about 3 h, monitor protoplast formation under the microscope. Protoplasts should appear slightly translucent when compared to undigested spores and hyphal fragments.
5. Filter the protoplast solution through sterile Miracloth into 50 mL falcon tubes. Centrifuge at 4,000 rpm ($1789 \times g$) for 10 min and discard supernatant.
6. Transfer flasks to ice (see Note 19) and wash protoplasts with 10 mL of cold Sucrose Solution. Gently re-suspend the pellet by pipetting up and down. Centrifuge at 4,000 rpm ($1789 \times g$) for 5 min and repeat the wash (see Note 20).
7. Resuspend protoplasts in 1 mL of cold Sucrose Solution and transfer to an Eppendorf tube (see Note 21).

3.3.2. Annexin V

1. Make different dilutions (1:5, 1:10 and 1:50) of resuspended protoplasts into 1× Binding Buffer and transfer 38 µL of each dilution to new Eppendorf tubes (see Note 22).
2. Add 2 µL of Annexin V and 2 µL of PI solutions to each tube.
3. Incubate at room temperature for 20 min in the dark.
4. Pipette 5 µL of Mounting Solution into a microscope slide. Add 10 µL of stained protoplasts to slides and place a coverslip on top.
5. Remove excessive Mounting Solution and seal the edges of the coverslip with transparent nail polish.
6. Analyze samples immediately using the fluorescein and rhodamine filters of a fluorescence microscope.

7. Count the number of Annexin V-positive (green fluorescent, especially at plasma membrane), PI-negative protoplasts (undetectable red fluorescence), which are scored as apoptotic.

3.4. Evans Blue Staining

1. Follow steps 1–4 from Subheading 3.1.
2. Transfer coverslips to a new staining jar containing Evans Blue Solution. Stain for 5 min at room temperature.
3. Wash coverslips three times with PBS for 5 min.
4. Mount slides as described in step 9 of Subheading 3.1 and analyze samples immediately using a bright field microscope.
5. Count the number of living hyphae (natural color) and dead hyphae (stained blue).

3.5. Detection of ROS

3.5.1. 2′,7′-Dichlorofluorescein Diacetate Staining

1. Generate protoplasts as described in Subheading 3.3.1 (see Notes 23 and 24).
2. Transfer 20 μL of resuspended protoplasts to a new Eppendorf tube and add 20 μL of DCF Solution.
3. Mount slides as described in step 9 of Subheading 3.1 and analyze samples immediately by fluorescence microscopy using the fluorescein filter set.
4. Count the number of green fluorescent protoplasts, which reflects an increased ROS production.

3.5.2. Nitro Blue Tetrazolium Staining

1. Follow steps 1–4 from Subheading 3.1.
2. Place coverslips on parafilm (prepared as in step 3 of Subheading 3.2) with the hyphae facing up.
3. Immediately and carefully add 500 μL of NBT solution to a corner of each coverslip without disturbing the hyphae.
4. Incubate in the dark at room temperature for 30 min.
5. Drain the extra solution using a Kim wipe and mount on slide using growth media instead of mount solution.
6. Observe immediately using light microscopy and count hyphal cells that contain blue-purple NBT formazan deposits.

4. Notes

1. Unless indicated otherwise, all solutions should be prepared in water that has a resistivity of 18.2 MΩ/cm and total organic content of less than five parts per billion.
2. Other chemical fixatives (i.e., glutaraldehyde, ethanol) could potentially be used as well.

3. The optimal concentration that results in good contrast between nuclear and cytoplasmic staining should be empirically determined for each stock solution.

4. Six-well microtiter plates can be used in place of staining jars. Transfer one coverslip to each well (hyphae facing up) and add 5 mL of solution.

5. Solution optimized for *A. nidulans*. Different fungi might require other enzymes that digest the components of their cell walls. Egg whites are obtained from fresh eggs that are cracked and filtered to remove yolks.

6. Even after the clarification the filter may clog. If that happens exchange the syringe filter.

7. Incubation at 37°C accelerates the process of dissolving the BSA.

8. PI is a potential carcinogen and must be handled with care.

9. The optimal concentration of DCF (25–50 μM) may vary between experiments and should be empirically determined.

10. Make sure that the coverslips are in contact with the plate to avoid hyphal growth on the underside of the coverslip. Sterilized forceps can be used to gently press the coverslips down.

11. The incubation period should be sufficiently long, enough for spores to break dormancy and form extending hyphae with at least eight nuclei. Note that this period should also include the incubation time with the apoptotic inducer.

12. Make sure all the excess of the Mounting Solution is removed, otherwise the nail polish will not stick well. In that case the coverslips may come off during microscopy or dry out from lack of sealing.

13. Although standard fluorescence microscopy is appropriate, the use of laser scanning confocal microscopy is preferable as it makes it possible to obtain images across multiple focal planes.

14. Because of variability between each batch of the Digestion Solution, it is advisable to set up several replicas for each condition and submit them to a range of digestion times (for example 15, 30, 60, and 90 min is recommended). The best digestion time will allow enough cell wall removal, so that TdT and FITC-dUTP penetrate the hyphae without destroying the fixed cells.

15. Reserve one untreated and one treated sample with the apoptotic inducer samples, digested for the longer incubation time, to be used as negative controls.

16. If all controls give the expected result and a low percentage of TUNEL positive nuclei are found in samples where apoptosis

17. was induced, it may be necessary to further optimize the conditions under which apoptosis is being induced (e.g., increase inducer concentration, increase exposure time).

17. If no TUNEL positive nuclei are found in the positive control, it is likely that the samples were not sufficiently digested or permeabilized.

18. The following Protoplast Generation protocol is optimized for *A. nidulans*.

19. From this step on, keep protoplasts on ice.

20. Be very careful manipulating the protoplasts since pipetting can disrupt the plasma membrane and increase the proportion of late apoptotic and necrotic cells.

21. Protoplasts can be kept on ice in a cold room overnight.

22. The purpose of using different dilutions is to have an adequate amount of protoplasts per field.

23. The same protoplasts can be used for the Annexin-V and ROS detection assays.

24. Cells treated with 1 mM of hydrogen peroxide, which induces oxidative stress, can be used as a positive control for the DCF staining.

References

1. Baehrecke EH (2002) How death shapes life during development. Nat Rev Mol Cell Biol 3:779–787
2. Gupta S (2001) Molecular steps of death receptor and mitochondrial pathways of apoptosis. Life Sci 69:2957–2964
3. Semighini CP, Hornby JM, Dumitru R, Nickerson KW, Harris SD (2006) Farnesol-induced apoptosis in *Aspergillus nidulans* reveals a possible mechanism for antagonistic interactions between fungi. Mol Microbiol 59:753–764
4. Semighini CP, Murray N, Harris SD (2008) Inhibition of *Fusarium graminearum* growth and development by farnesol. FEMS Microbiol Lett 279:259–264
5. Gorczyca W, Gong J, Darzynkiewicz Z (1993) Detection of DNA strand breaks in individual apoptotic cells by the in situ terminal deoxynucleotidyl transferase and nick translation assays. Cancer Res 53:1945–1951
6. Chapman RS, Chresta CM, Herberg AA, Beere HM, Heer S, Whetton AD, Hickman JA, Dive C (1995) Further characterisation of the in situ terminal deoxynucleotidyl transferase (TdT) assay for the flow cytometric analysis of apoptosis in drug resistant and drug sensitive leukaemic cells. Cytometry 20:245–256
7. Vermes I, Haanen C, Steffens-Nakken H, Reutelingsperger C (1995) A novel assay for apoptosis. Flow cytometric detection of phosphatidylserine expression on early apoptotic cells using fluorescein labelled Annexin V. J Immunol Methods 184:39–51
8. Cathcart R, Schwiers E, Ames BN (1983) Detection of picomole levels of hydroperoxides using a fluorescent dichlorofluorescein assay. Anal Biochem 134:111–116
9. Flohe L, Otting F (1984) Superoxide dismutase assays. Methods Enzymol 105:93–104
10. Beyer WF Jr, Fridovich I (1987) Assaying for superoxide dismutase activity: some large consequences of minor changes in conditions. Anal Biochem 161:559–566
11. Maly FE, Nakamura M, Gauchat JF, Urwyler A, Walker C, Dahinden CA, Cross AR, Jones OT, de Weck AL (1989) Superoxide-dependent nitroblue tetrazolium reduction and expression of cytochrome b-245 components by human tonsillar B lymphocytes and B cell lines. J Immunol 142:1260–1267

Chapter 21

Evaluation of Antifungal Susceptibility Using Flow Cytometry

Cidália Pina-Vaz and Acácio Gonçalves Rodrigues

Abstract

Flow cytometry has found wide applications in areas like haematology and immunology, but also presents great potential in microbiology. The susceptibility of clinical isolates of *Candida* and *Cryptococcus* to antifungal compounds can be assayed by flow cytometry using fluorescent probes like FUN-1, propidium iodide and JC-1, with several advantages. Following 1 or 2 h of incubation, depending on the antifungal compound, *versus* 48 h of the classical methods, it is possible to establish the different susceptibility profiles. Additionally, it provides information regarding the mechanisms of action and might infer about resistance mechanisms.

Key words: Antifungal susceptibility, Flow cytometry, FUN-1, Propidium iodide, JC-1, Azoles, Amphotericin B, 5-fluorocytosine, Caspofungin

1. Introduction

Although traditional methods in clinical microbiology usually provide an adequate sensitivity and specificity, the time required to obtain results is quite long, as they are invariably dependent on cell replication capacity, presence of biochemical subtracts for identification, or of drugs for the evaluation of the susceptibility profile. The availability of molecular biology techniques, particularly those based upon nucleic acid probes combined with amplification has provided considerable speediness and specificity to diagnostic microbiological laboratories. Although a revolutionary change in many of the traditional routine diagnostic procedures happened, reliable information about viability is null and only in very specific cases, information about susceptibility to drugs is made available. Flow cytometry had a considerable impact in areas of medical knowledge such as haematology or immunology, and it offers a broad growing range of potential applications in microbiology (1). Flow cytometric methods allow rapid measurement of light scattered and fluorescence emission by suitably illuminated cells. The cells, suspended in liquid, are evaluated individually but the results represent cumulative cytometric characteristics. The scattered light gives intrinsic cell information like size and complexity, but with the use of fluorochromes a wide range of cellular physiological or morphological parameters could be also evaluated like integrity of the membrane, pH, membrane potential, viability, etc. We have described several cytometric applications regarding the detection of *Mycobacteria* (2) and the evaluation of its drug susceptibility profile (3), the detection of *Cryptosporidium* (4) and *Giardia* (5), and the susceptibility of fungi (6–8). The increasing incidence of opportunistic fungal life-threatening infections has greatly enhanced the interest in novel methods for in vitro antifungal susceptibility testing. The standardized methodology recommended by the National Committee for Clinical Laboratory Standards (CLSI, formely NCCLS) M27-A2 (9), represented a significant step forward in the development of a reproducible standardized testing method. Nevertheless, this protocol is cumbersome and labor intensive. Additionally, at least 48 h are necessary to give results, and it does not provide any information about the fate of the fungal cells. Antifungal susceptibility testing, similar to antibacterial susceptibility testing, is becoming extremely useful in the selection of empirical treatment, particularly for testing of isolates from blood, from deep-seated infections, or from recurrent mucosal infections (10).

Flow cytometry allows a timely determination of antifungal susceptibility patterns and can provide additional information about cells physiology. We were also able to use it for clarification of the mechanism of action of drugs not classically considered antifungal like local anesthetics (11), ibuprofen (12) and some essential oils (13). As they induced primary lesion of the cell membrane, propidium iodide could enter after a short (few minutes) incubation. Those findings were confirmed using classic but rather hard working methods like electronic microscopy and assessment of potassium efflux. Flow cytometry also allowed us to infer about the presence of efflux pumps in *Candida* isolates justifying its antifungal resistance (7). Only after elucidating the underlying mechanism of antifungal resistance, it may be possible to develop meaningful strategies aiming its reversion. The blockade of efflux pumps made possible the reversion of resistant strains of *Candida* to a susceptible phenotype (14).

2. Materials

2.1. Fungal Culture and Preparation of Blastoconidia Suspensions

1. Sabouraud dextrose agar medium (Difco, USA) or Columbia agar medium (BioMérieux, Paris).
2. *Candida* and *Cryptococcus* strains in pure culture, 24–48 h old.
3. Newbauer chamber or McFarland scale.
4. Sterile distilled water.
5. Ethanol (70%).
6. Sodium azide.
7. Valinomycin (Sigma).
8. Phosphate-buffered saline (PBS) supplemented with 2% glucose.
9. Sodium deoxycolate.
10. HEPES solution, pH 7.2, supplemented with 2% glucose (GH solution, Sigma).

2.2. Storage and Preparation of Antifungal Compounds

1. Amphotericin B (Am B) (Sigma).
2. Fluconazole (Flu) (Pfizer, Groton, CT, USA).
3. Voriconazole (Pfizer).
4. 5-fluorocytosine (5-FC) (Sigma).
5. Itraconazole (Jansen, Beerse, Belgium).
6. Caspofungin (Merck, Rahway, NJ).
7. All antifungal compounds are prepared in dimethyl sulfoxide (DMSO) (at a final DMSO concentration below 2%) with the exception of fluconazole that is prepared in sterile water and store at −70°C, until use.

2.3. Fluorescent Probes

1. Propidium iodide (PI) (Sigma), prepared in HEPES solution, pH 7.2, supplemented with 2% glucose (GH solution), maintained in the dark, at 4°C.
2. FUN-1-(2-choro-4-(2,3-dihydro-3-methyl-(benzo-1,3-thiazol-2-yl)-methylidene)-1phenylquinoliniumiodide (Molecular Probes, Europe BV, Leiden, The Netherlands), prepared in GH solution, maintained in the dark, at 4°C.
3. JC-1 (5,5′, 6,6′-tetrachloro-1,1′,3,3′-tetraethylbenzimidazolcarbocyanine iodide) (Molecular Probes), prepared in PBS supplemented with 2% glucose, maintained in the dark, at 4°C.

2.4. Flow Cytometric Analysis

1. Flow Cytometer, FACSVantage SE Flow Cytometer (Becton Dickinson (BD), San Jose, California, USA), eqquiped with Clone-Cyte software (BD). The cytometer uses an argon laser regulated to 10 mW UV output, with a split 488 nm beam as the primary signal.

2. BDFacsFlow, BD Facs Rinse and BDFacs Clean solutions.

3. Rainbow Calibration Particles (six peaks), 3.0–3.4 µm, for calibration (BD).

2.5. Microscopic Analysis

1. Microscope glass slides and cover slips ($22 \times 40 \times 0.15$ mm).

2. Light microscope.

3. Epifluorescence microscope fitted with a mercury lamp, a BP 450–490 nm excitation filter and an LP 515 nm emission filter.

4. Vectashield Mounting Medium (Vector Laboratories, Burlingane, CA, USA), an antifading reagent.

3. Methods

Flow cytometric analysis using different fluorescent probes allows the appreciation of prelethal changes, either from a metabolic or a morphological point of view, which may predict accurately the susceptibility pattern to antifungal drugs. The performance of the following three fluorescent markers in terms of evaluation of susceptibility of *Candida* and *Cryptococcus* strains to standard antifungal agents was assessed: (1) propidium iodide (PI), a marker of cell death (only penetrates cells with severe membrane lesions), (2) FUN-1, a probe that is converted by metabolically active fungi from a diffuse cytosolic pool (green-yellow fluorescence) to red cytoplasmic intravacuolar structures (CIVS) (15). Metabolically disturbed cells show an increased intensity of green-yellow fluorescence, (3) JC-1, a fluorescent probe that penetrates the cytosol of eukaryotic cells and exhibits potential-dependent accumulation in mitochondria, which is indicated by a shift in fluorescence emission from green (529 nm) to red (590 nm). Its monomeric form emits green fluorescence but as the mitochondrial membrane becomes more polarised, JC-1 aggregates are formed (visualized as small intracellular red dots) and the color changes to red (16). Consequently, mitochondrial depolarization is indicated by a decrease in the red/green fluorescence intensity ratio.

3.1. Yeast Cells Suspensions and Antifungal Treatment

After picking one or two yeast colonies from an agar culture, prepare suspensions in PBS supplemented with 2% glucose, pH 7.0 (see Note 1). Adjust the blastoconidia concentration to 10^6 cells/ml using a Newbauer chamber, or using a 0.5 McFarland density scale and incubate with shaking (200 rpm) at 35°C with each antifungal agent. Incubation time is 1 h for all antifungal agents except 1 or 2 h in the case of amphotericin B (see below). Two concentrations of each drug (around the CLSI established or tentative MIC breakpoints) are used: amphotericin B, 1 and 8 μg/ml; fluconazole, 8 and 64 μg/ml; itraconazole 0.125 and 1 μg/ml; voriconazole, 1 and 4 μg/ml; 5-FC, 4 and 32 μg/ml; caspofungin 2 and 4 μg/ml. After incubation with the antifungal compound, yeast cells are washed in sterile water to prevent fluorochrome quenching and afterwards stained with the different fluorescent probes (see Note 2).

3.2. Staining of Yeasts Cells

1. To study the effect of all tested compounds except amphotericin B, cells are incubated for 30 min in the presence of 0.5 μM FUN-1/10^6 cells/ml, at 37°C in the dark (see Note 3). As control of the staining, yeast cells are incubated for 1 h with 1 mM sodium azide, an inhibitor of yeast metabolism. For each strain, distinct blastoconidia are prepared to be analyzed afterward by flow cytometry: (1) A nontreated (drug-free) and nonstained cell suspension (autofluorescence), (2) A nontreated but stained cell suspension (control of viable cells), (3) A suspension of cells treated with sodium azide and stained (control of metabolically disturbed cells), (4) A suspension of antifungal treated and stained cells.

2. Suspensions of cells treated with amphotericin B cannot be stained directly either with FUN-1 or IP (see Note 4). Two alternative protocols can be used: (a) after incubation for 1 h with the antifungal, stain yeast cells with 5.0 μM JC-1/10^6 cells/ml for 15 min, at 35°C, in the dark. As control of the staining, unstained cells (from the previous stage) should be incubated for 1 h with 100 μM of valinomycin (a K$^+$ ionophore known to reduce mitochondrial membrane potential). For each strain, distinct blastoconidia suspensions are prepared to be afterwards analyzed by flow cytometry: (1) A nontreated (drug-free) and nonstained cell suspension (autofluorescence), (2) A nontreated but stained cell suspension (control of viable cells), (3) A suspension of cells treated with valinomycin and stained (control of cells with reduced mitochondrial membrane potential), (4) A suspension of antifungal treated and stained cells. (b) After incubation for 2 h with the antifungal compounds, stain the cells with PI (1.0 μg/10^6 cells/ml) by incubating for 30 min, adding simultaneously sodium deoxycolate (a membrane detergent: 200 μL of 25 mM deoxycolate to 10 μL of PI), at room temperature, in the dark. Blastoconidia killed in 70% ethanol during 20 min are used as a control for the staining; wash twice with HEPES solution before staining with PI. For each strain, different suspensions are prepared to be analyzed by flow cytometry: (1) A nontreated (drug-free) and nonstained cell suspension (autofluorescence), (2) A nontreated and stained cell suspension (control of viable cells), (3) A suspension of ethanol killed cells and stained (control of unviable cells with severe lesion of the membrane), (4) A suspension of antifungal treated and stained cells.

3.3. Flow Cytometry Analysis

The distinct cell suspensions are analyzed on a FACSCalibur flow cytometer equipped with 3 PMTs, with standard filters (FL1: BP 530/30 nm; FL2: BP 585/42 nm; FL3: LP 650 nm), a 15 mW 488 nm Argon Laser, and a cell Quest Pro software (version 4.0.2, BD Biosciences, Sydney). Acquisition settings were defined using a nonstained sample (auto fluorescence) and adjusting the PMTs voltage to the first logarithmic (log) decade. At least 30,000 cells are analyzed. The cell scattergram (forward scatter (FS) and side scatter (SS)), the auto fluorescence (without fluorochrome) and the intensity of fluorescence at FL1 (green fluorescence), FL2 (yellow–green fluorescence) and FL3 (red fluorescence) are recorded using a logarithmic scale. The results are expressed as follows: for FUN-1, a staining index (SI) defined as the ratio between the mean fluorescence of treated cell suspensions and the corresponding value for the control (viable) cells at FL2; for PI, as a percentage of cells showing high fluorescence at FL3 or an increase on FL3 staining, resulting in clearly separated histograms of dead and viable cells; for JC-1, a ratio between the mean values of fluorescence at FL3 and FL1, calculated for each antifungal treated-suspension and compared with the corresponding value for the untreated cells. The equipment is controlled according to the manufacturer's recommendations.

3.4. Microscopic Analysis

3.4.1. Light Microscope Analysis

Blastoconidia suspensions are prepared in PBS and cell numbers are determined using a Newbauer chamber (at 400× magnification) and adjusted to 10^6 cells/ml. Alternatively, MacFarland scale can be used to adjust the suspension density to OD 0.5.

3.4.2. Epifluorescence Microscopy Analysis

For epifluorescence microscopy analysis, blastoconidia suspension (20 µL) is mixed with 20 µL of the antifading reagent (see Note 6) and visualized at 400× magnification. Regarding FUN-1 staining, cells showing red intravacuolar structures (CIVS) are viable cells, while metabolically unviable cells show a green diffuse staining without CIVS; for PI staining, diffuse red fluorescent cells represent dead blastoconidia with membrane lesions; with JC-1 we look for the presence of J-aggregates only present on untreated or resistant cells (see Note 7).

3.5. Susceptibility Analysis

1. Regarding azoles, 5-FC and caspofungin, the yeast strain will be considered susceptible if the incubation with the lower antifungal concentration tested result in an increase of the cell intensity of fluorescence, making the SI > 1 after FUN-1 staining. It will be considered resistant if SI is not >1 even at the higher concentration tested and S-DD (in the case of azoles) or I (in the case of 5-FC) if SI > 1 is observed only with the higher concentration tested and not with the lower concentration (see Note 5).

2. Regarding amphotericin B and JC-1 staining (protocol A) the strain will be considered susceptible only if the incubation with the lower concentration of the drug resulted in a decrease in the mean value of FL3/FL1 of cells stained with JC-1; with PI-sodium deoxycolate (protocol B) the strain will be considered susceptible if the lower concentration of the drug produced a 50% increase in intensity of fluorescence when compared to drug-free control of PI stained cells.

4. Notes

1. In previous flow cytometer analyses, we used yeast cells that were cultured with shaking (200 rpm) at 35°C in Sabouraud broth until late log phase (as determined by a growth curve constructed from absorbance readings at an optical density of 600 nm). However, antifungal susceptibility patterns are similar when suspensions of blastoconidia are prepared directly from the agar culture.

2. Quenching is the phenomenon of decreased fluorescence intensity due to chemical interactions between compounds. The mixing of drugs with fluorescent probes should be avoided or its mandatory exclude such possible interaction.

3. The use of phosphates with FUN-1 should be avoided as recommended by Molecular probes.

4. We have shown that the effect of amphotericin B could not be evaluated with PI. Although being a fungicidal and despite of the cells being dead, they do not stain with PI or with FUN-1. The reduction in staining did not result from interference of amphotericin B with the dyes, as deduced from the observation that the fluorescence spectra of both probes were not altered by the presence of amphotericin B (6). Amphotericin is a huge molecule that could thus turn the yeast cells impermeable to both probes. Some previous reports suggested that this drug can induce apoptotic cell death (17) which could explain the exclusion by PI. However, in that case FUN-1 staining should reveal considerable metabolic cell disturbance, resulting in an increase of the intensity of cell fluorescence and not in a reduction of the staining, as it did happen in fact. Chaturvedi et al. (18) used a detergent of the membrane to turn the cells permeable to PI, thus allowing its use to study of the effect of amphotericin B.

5. Yeast strains resistant to azoles often show a decrease of the intensity of fluorescence compared to nontreated cells (SI < 1), because FUN-1 is pumped out of the cells by efflux pumps,

similarly to antifungal compounds (7). The cells show CIVS, meaning that they are metabolically viable, but yield considerable less fluorescence.

6. An antifading is very useful when we want to examine in detail yeast cells under epifluorescence microscopy or take photographs. It prevents the fast decrease of fluorescence that occurs when we are performing the analysis.

7. Epifluorescence microscopy should be used as a control of flow cytometric analysis. While optimising a new cytometric protocol or assaying a new probe or whenever the result is not what you were expecting, always perform epifluorescence microscopic analysis; it provides insight to what is happening and will be a valuable aid to interpretation of FC results.

References

1. Álvarez-Barrientos A, Arroyo J, Cantón R, Nombela CA, Sánchez-Pérez M (2000) Applications of flow cytometry to clinical Microbiology. Clin Microbiol Rev 13:167–195
2. Pina-Vaz C, Costa-Oliveira S, Rodrigues AG, Salvador A (2004) Novel method using a laser scanning cytometer for detection of mycobacteria in clinical samples. J Clin Microbiol 42:906–908
3. Pina-Vaz C, Costa-de-Oliveira S, Rodrigues AG (2005) Safe susceptibility testing of Mycobacterium tuberculosis by flow cytometry with the fluorescent nucleic acid stain SYTO 16. J Med Microbiol 54:77–81
4. Barbosa JM, Costa-de-Oliveira S, Rodrigues AG, Hanscheid T, Shapiro H, Pina-Vaz C (2008) A flow cytometric protocol for detection of *Cryptosporidium* spp. Cytometry A 73:44–47
5. Barbosa J, Costa-de-Oliveira S, Rodrigues AG, Pina-Vaz C (2008) Optimization of a flow cytometry protocol for detection and viability assessment of *Giardia lamblia*. Travel Med Infect Dis 6:234–239
6. Pina-Vaz C, Sansonetty F, Rodrigues AG, Costa-Oliveira S, Tavares C, Martinez-de-Oliveira J (2001) Cytometric approach for a rapid evaluation of susceptibility of *Candida* strains to antifungals. Clin Microbiol Infect 7:609–618
7. Pina-Vaz C, Sansonetty F, Rodrigues AG, Costa-de-Oliveira S, Martinez-de-Oliveira J, Fonseca AF (2001) Susceptibility to fluconazole of *Candida* clinical isolates determined by FUN-1 staining with flow cytometry and epifluorescence microscopy. J Med Microbiol 50:375–382
8. Pina-Vaz C, Costa-de-Oliveira S, Rodrigues AG, Espinel-Ingroff A (2005) Comparison of two probes for testing susceptibilities of pathogenic yeasts to voriconazole, itraconazole, and caspofungin by flow cytometry. J Clin Microbiol 43:4674–4679
9. National Commitee for Clinical for Laboratory Standards (2002) Reference method for broth dilution antifungal susceptibility of yeasts. Approved Standard M27-A2, 2nd edn. CLSI, Wayne, PA, USA
10. Hospenthal DR, Murray CK, Rinaldi MG (2004) The role of antifungals susceptibility testing in the therapy of candidiasis. Diagn Microbiol Infect Dis 48:153–160
11. Pina-Vaz C, Rodrigues AG, Sansonetty F, Martinez-De-Oliveira J, Fonseca AF, Mårdh P-A (2000) Antifungal activity of local anesthetics against Candida species. Infect Dis Obstet Gynecol 8:124–137
12. Pina-Vaz C, Rodrigues AG, Costa-de-Oliveira S, Ricardo E, Mårdh P-A (2005) Potent synergic effect between ibuprofen and azoles on *Candida* resulting from blockade of efflux pumps as determined by FUN-1 staining and flow cytometry. J Antimicrob Chemother 56:678–685
13. Pina-Vaz C, Rodrigues GA, Pinto E, Costa-de-Oliveira S, Tavares C, Salgueiro L, Cavaleiro C, Gonçalves MJ, Martinez-de-Oliveira J (2004) Antifungal activity of *Thymus* oils and their major compounds. J Eur Acad Dermatol Venereol 18:73–78
14. Pina-Vaz C, Sansonetty F, Rodrigues AG, Martinez-De-Oliveira J, Fonseca AF, Mårdh P-A (2000) Antifungal activity of ibuprofen alone and in combination with fluconazole

against Candida species. J Med Microbiol 49:831–840

15. Millard PJ, Roth BL, Truong YH-P, Yue ST, Haugland RP (1997) Development of the FUN-1 Family of Fluorescent Probes for vacuole labeling and viability testing of yeasts. Appl Environ Microbiol 63:2897–2905

16. Cossarizza A, Baccarani-Contri M, Kalashnikova G, Franceschi C (1993) A new method for the cytofluorimetric analysis of mitochondrial membrane potential using the JC-aggregate forming lipophilic cation 5, 5′, 6, 6′-tretrachoro-1, 1′, 3, 3′-tetraethylbenzidazolcarbocyanine iodide (JC-1). Biochem Biophys Res Commun 197: 40–45

17. Phillips AJ, Sudbery I, Ramsdale M (2003) Apoptosis induced by environmental stresses and amphotericin B in *Candida albicans*. Proc Natl Acad Sci U S A 100:14327–14332

18. Chaturvedi V, Ramani R, Pfaller MA (2004) Collaborative study of the NCCLS and flow cytometry methods for antifungal susceptibility testing of *Candida albicans*. J Clin Microbiol 42:2249–2251

Chapter 22

Preparation of Fungi for Ultrastructural Investigations and Immunogoldlabelling

Gerd Hause and Simone Jahn

Abstract

Electron microscopic analysis of biological material requires optimal preparation of the samples. This is necessary to prevent degeneration processes and changes of the material during microscopic observation. Both would lead to artefacts. In this chapter, we present methods to prepare fungi or plant tissues infected with fungi for transmission electron microscopy. This includes chemical fixations as well as cryofixation and the subsequent embedding in suitable resins. For cryofixation, the high pressure freeze fixation is described in detail. Further on, protocols for freeze substitution of cryofixed samples and immunogoldlabelling of ultrathin sections are included.

Key words: Electron microscopy, Ultrastructure, High pressure freeze fixation, Freeze substitution, Immunogoldlabelling

1. Introduction

Structural investigations are more and more integrated in modern concepts of biological research. After a period of detailed descriptions of the morphology of cells and cellular organelles as well as of developmental processes, it is meanwhile very important to combine the possibilities of modern microscopic techniques with other methods and fields of biology.

Apart from some special techniques, for ultrastructural studies using electron microscopy, it is necessary to fix and prepare the material before observation. There are two main demands on the preparation methods. They should fix the material preferably very fast to immobilize all cellular structures and processes synchronously, and they should cause no or negligible artefacts. In general, there are two methods to obtain this: chemical fixation caused by cross-linking of proteins and physical fixation caused by freezing of the material. It was shown, however, that fast and nearly artefact-free fixation can be obtained by the application of cryotechniques only. The most severe problem of cryofixation is the formation of crystalline ice within the cells during freezing. Several techniques and apparatuses as cryo-plunging after infiltration with a cryoprotectant, fixation at a cold metal mirror, propane jet spray fixation and high pressure freeze fixation were developed to overcome ice crystal formation (see Note 1). Because of the large size of eukaryotic cells and tissues, they cannot be cryofixed by plunging or spray freezing, they have to be fixed using high pressure freeze fixation. Only this technique delivers frozen cells and tissues without freezing artefacts because a very high pressure (210 MPa) for a short time (~0.5 s) enhances significantly the viscosity of the cellular water during the freezing process and thus prevents ice crystal formation (for plant material see ref. (1–5)). But even for high pressure freezing, there are restrictions in the size of the samples. They should not be thicker than 200 μm and the diameter of the samples should be – depending on the manufacturer of the freezing apparatus – not larger than 1–2 mm due to the size of the specimen holders (planchets). One critical point during the procedure of high pressure freezing is the fast and careful preparation of the samples for freezing. There are specialities for all types of cells and tissues. Only a few samples have the right size to fit perfectly into the sample holders. For the majority of the objects of research, it is necessary to immobilize them (e.g., cell cultures in special tubes, (6)) or to cut them with appropriate tools.

In this chapter, we describe the methods as well as particularities of chemical and freeze fixation. Further, we add protocols for cryosubstitution of frozen material and immunogoldlabelling of proteins.

2. Materials

2.1. Chemical Fixation

1. SCB buffer: 0.1 M sodiumcacodylate (Plano, Wetzlar, Germany), pH 7.2.
2. Fixative 1:3% (v/v) glutaraldehyde (Sigma, Taufkirchen, Germany) in SCB.
3. Fixative 2:1% (w/v) osmiumtetroxide (Science Services, Munich, Germany) in SCB.
4. 4% (w/v) agar in SCB.
5. Ethanol.
6. 1% (w/v) uranyl acetate (Serva, Heidelberg, Germany) in 70% ethanol.
7. Embedding medium: epoxy resin according to Spurr (7).

2.2. High Pressure Freeze Fixation

1. Specimen holders: aluminium plachets – space for specimen 2 mm diameter and 0.2 mm deep, fitting into a HPM 010 high pressure freezer (BAL-TEC/Leica, Wetzlar, Germany).
2. Filler substances: hexadecene (Merck-Schuchardt, Hohenbrunn, Germany) or 20% (w/v) bovine serum albumine in PBS.

2.3. Freeze Substitution

1. Substitution vessels for the freeze substitution apparatus FSU 010 (BAL-TEC, Balzers, Liechtenstein) are 1.5 mL Eppendorf tubes.
2. Baskets for the frozen tissue fitting into the 1.5 mL Eppendorf tubes (see Note 2).
3. Acetone dried with KCl.
4. Substitution mixture A: 0.1% (w/v) uranyl acetate, 0.25% (v/v) glutaraldehyde in acetone.
5. Substitution mixture B: 0.1% uranyl acetate (w/v), 0.25% glutaraldehyde (v/v), 2% H_2O (v/v) in acetone.
6. Embedding medium: Lowicryl HM20 (Polysciences, Warrington PA).

2.4. Immunogold-labelling

1. Phosphate buffered saline (PBS, pH 7.2) containing 135 mM NaCl, 3 mM KCl, 1.5 mM KH_2PO_4, and 8 mM Na_2HPO_4.
2. Blocking reagent (BR): 1% (w/w) acetylated BSA (Aurion, Wageningen, NL), 0.1% (v/v) TWEEN® 20 in (PBS).
3. Staining solution 1: 1% (w/v) uranyl acetate in H_2O.
4. Staining solution 2: Ultrastain two containing lead citrate (Laurylab Saint-Fons Cedex; France).

3. Methods

A very important point for chemical as well as for freeze fixation is the fast transfer of the biological material into appropriate vessels to start fixation. To prevent degeneration and drying-out of the material during sampling, tissue should be prepared in buffer. Because the samples should be small for fixation, materials have to be dissected with tools (e.g., razor blades, biopsy punches, scalpels). These tools must be very sharp to avoid mechanical damage of the biological material. Further on, the dissection of the material must be carried out with a very soft pressure. In general, it is advantageous to train the sampling procedure intensively to be fast and careful.

3.1. Chemical Fixation of Tissues and Embedding in Epoxy Resin

1. Cut small (1–2 mm length of the edge) pieces of tissue with appropriate tools in buffer and transfer it into small containers filled with fixative 1. The samples should be small (1–2 mm) to allow fast diffusion of the fixative into the center of the sample.

Cell suspensions have to be centrifuged to sediment the cells. Thereafter, remove the culture medium, add fixative 1, and resuspend pellet in the fixative using a Vortex. During fixation and all subsequent steps, the containers with the samples should be placed on a rotator to improve the exchange of substances.

2. After 2 h (suspensions) to 4 h (tissue) of fixation at room temperature, wash with 3 mL SCB at least four times (15 min each). To avoid permanent centrifugation steps for the preparation of suspensions, they can be immobilized with agar. For this purpose, the suspension has to be centrifuged and the fixative has to be removed. Thereafter, the pellet of cells should be mixed immediately with one volume of prewarmed agar/SCB mixture. After cooling down, the solid mixture can be cut in small pieces and subsequently processed like pieces of tissue.

3. Postfixation with fixative 2. Incubate samples for 30 min (cells or thin tissues) up to 1 h (compact tissues) in fixative 2 (see Note 3).

4. Wash three times 10 min each with 3 mL H_2O. It is important to also wash the caps of the containers intensively to remove remnants of osmium.

5. Dehydrate the specimen in a graded series of ethanol. The procedure should be carried out stepwise by transferring the samples to 3 mL of 10, 30, 50, 70, 90%, and finally two times in absolute ethanol for 30 min each. To enhance the contrast of membranes, it is recommended to add an extra step between 50 and 70% ethanol by incubating for 1 h in 3 mL 1% uranylacetate in 70% ethanol. If the lab-time schedule demands a break of the procedure, it is recommended to store the samples in 70% ethanol at 4°C overnight or for some days to continue later with further dehydration.

6. Start with infiltration of the epoxy resin according to Spurr (7). The infiltration has to be performed stepwise with 2 mL each time; 4 h 25% resin/75% ethanol, 4 h 50% resin/50% ethanol, overnight 75% resin/25% ethanol, and subsequently at least two times pure resin for 8 h each time (see Notes 4 and 5).

7. Transfer of the samples in silicon embedding moulds filled with 0.3 mL of fresh epoxy resin.

8. Polymerization of the material at 70°C for at least 24 h (see Note 6).

3.2. High Pressure Freeze Fixation

3.2.1. Tissues

1. Trim tissues to fit into the chamber of the specimen holder (planchet) with an inner-diameter of 2 mm and a depth of 0.2 mm. Use biopsy needles, biopsy punches, or razor blades.

2. Gas within tissues (e.g., intercellular space in plants) will be compressed during the increase of the pressure and decompressed

after about 0.5 s during release of the high pressure. This very fast decompression of the gas can cause rupture of cells and tissues and has to be prevented. Therefore, the gas has to be replaced by degassing the leaf or root segments in 5 mL of 8% (v/v) aqueous methanol at 80 mbar for 5 min (8).

3. Fill the space in the cavity of the planchets that is not occupied by the biological material with either hexadecene or 20% BSA in PBS. This is necessary to avoid mechanical destruction of the cells during fixation because gas in the cavity has the same destructive effect during freezing as gas in intercellular space. Make sure to remove even small gas bubbles in the filling substance. Therefore, the loading of the specimen holder has to be carried out not only fast but also very carefully.

4. Finally, cover the cavity of the planchet containing the specimen with a second planchet.

5. Transfer of the specimen holder into a high pressure freezer and start fixation procedure. The freezing with liquid nitrogen has to start after reaching a pressure of about 210 MPa (~2,100 bar). This should occur within 0.2 s. The maintenance of high pressure is about 0.5 s. Thereafter, the frozen specimen holder has to be transferred rapidly into liquid nitrogen. The samples have to be stored in liquid nitrogen up to the start of freeze substitution.

3.2.2. Suspensions

1. To remove culture medium from cell suspensions, such as cultivated spores, the specimens can be centrifuged and the pellet can be transferred into the cavity of the planchet. Alternatively, cells can be concentrated at a sterile filter with appropriate pore size (9). The latter method is recommended, because changes of the cellular morphology caused by centrifugal forces can be excluded. Furthermore, for every freezing process, new cells can be concentrated quickly (see Notes 7 and 8).

2. Cover the specimen in the cavity of the planchet with a second planchet (flat side).

3. Transfer into a high pressure freezer and start the fixation procedure.

4. Rapidly transfer the frozen specimen holder in liquid nitrogen. Store samples in liquid nitrogen up to the start of freeze substitution.

3.3. Freeze Substitution

1. Transfer the frozen material in substitution tubes fitting into the substitution apparatus used (Fig. 1 and see Note 2). The tubes have to be filled with substitution solution and placed in holders cooled with liquid nitrogen. For tissues or mycelia grown in liquid culture, solution A is recommended, whereas

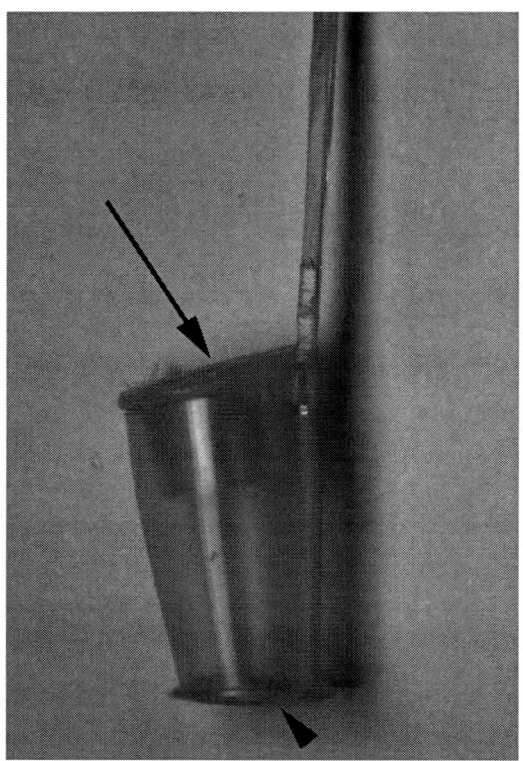

Fig. 1. Handmade basket for freeze substitution. The basket consisting of a trimmed 1 mL pipette-tip and nylon sieve (100 μm mesh) at the bottom (*arrowhead*) fits in a 1.5 mL test-tube. The cover (*arrow*) is also made from a piece of a pipette-tip and nylon-sieve.

for frozen cell cultures, solution B (contains 2% water) can deliver better results (contrast of membranes).

2. After a few seconds, the solution is frozen. Planchets containing frozen cell cultures are simply opened in liquid nitrogen and placed onto the frozen substitution solution (see Notes 9 and 10 for handling of frozen tissues). All handling of the samples have to be carried out in appropriate cooled vessels to keep the cells frozen at all times.

3. Transfer the frozen samples into the precooled substitution apparatus.

4. Perform cryosubstitution at −80°C for 2–4 days depending on the properties of the material. For thin tissues or frozen mycelia, 2 days of cryosubstitution should be sufficient whereas the cryosubstitution of pellets of cells or compact tissues should be extended up to 4 days.

5. Increase the temperature stepwise (6 h −60°C, 6 h −30°C) to −20°C. All subsequent steps will be carried out at −20°C.

6. Remove of the substitution solution and add 1 mL of precooled (−20°C) acetone. Fast addition of acetone is very important to prevent freeze-drying.

7. Frozen material still in the planchets during freeze substitution has to be taken out from the holders under −20°C conditions. Because of the restricted space in the Eppendorf tubes, the planchets should be transferred in a larger vessel filled with acetone and cooled to −20°C (e.g., within a bath of ethanol chilled with dry ice). The isolated frozen pellet is then transferred into a basket (Fig. 1), and the basket with the samples is placed in a tube with fresh cold acetone.
8. Wash twice with 1 mL cold acetone for 1 h each.
9. Perform a stepwise infiltration with Lowicryl: 4 h 25% Lowicryl in acetone, 4 h 50% Lowicryl in acetone, overnight 75% Lowicryl in acetone (see Notes 11 and 12).
10. Infiltrate with pure Lowicryl by incubating three times in 1 mL Lowicryl at −20°C for at least 8 h each time.
11. Polymerization. Add 1 mL of fresh Lowicryl to the samples and seal tubes carefully because oxygen would prevent polymerization. Place under a UV-lamp for 24 h at −20°C and then for 24 h at room temperature.
12. Following polymerization, wash the solid blocks containing the sample with ethanol to remove nonpolymerized Lowicryl.
13. Leave tubes open in a fume hood for at least 1 day.

3.4. Immunogold-labelling

For immunogoldlabelling material should not be fixed with osmiumtetroxide, because this compound causes changes of the protein structure. Therefore, the material prepared as described in 3.1 cannot be used for this purpose, whereas cryofixed and freeze-substituted material is suitable for this procedure (see Note 13).

1. Prepare ultrathin sections (50–90 nm) of the cryofixed/freeze substituted material and mount onto formvar coated nickel grids.
2. During all following steps (up to step 8), the grids have to be placed on a 20 µL droplet of the appropriate washing or incubation solution, which is favourably put on parafilm.
3. Block sections by incubating in 20 µL BR for 30 min at room temperature.
4. Incubate overnight in a humid chamber (e.g., a Petri-dish containing wet paper) with the primary antibody diluted in BR at 4°C (see Notes 14 and 15).
5. Wash four times with 20 µL BR at RT for 10 min each time.
6. Incubate with the secondary antibody diluted in BR at room temperature for 60–90 min. These antibodies are coupled with small gold-particles of defined size as reporter-structure.
7. Wash four times with 20 µL H$_2$O at RT for 5 min each time.

8. Contrasting of the sections in an EM-Stain (Leica, Wetzlar, Germany) for 1 h at 25°C with staining solution 1 and thereafter for 10 min at 25°C with staining solution 2. Alternatively, the sections can be contrasted in a Petri-dish at room temperature for 30 min with 1% aqueous uranyl acetate and thereafter for 5 min with 0.2% lead citrate.

4. Notes

1. Although there are obvious advantages of freeze fixation in comparison to chemical fixation concerning the preservation of subcellular structures (compare Fig. 2b, c), freezing techniques should only be applied if the appropriate technical equipment and methods are available. Otherwise freeze fixation can cause drastic artefacts because of ice crystal formation (Fig. 2a).

2. Freeze substitution baskets as sample containers fitting for the 1.5 mL Eppendorf tubes are made from 1 mL tips of pipettes and nylon sieves (100 μm meshes). An example is

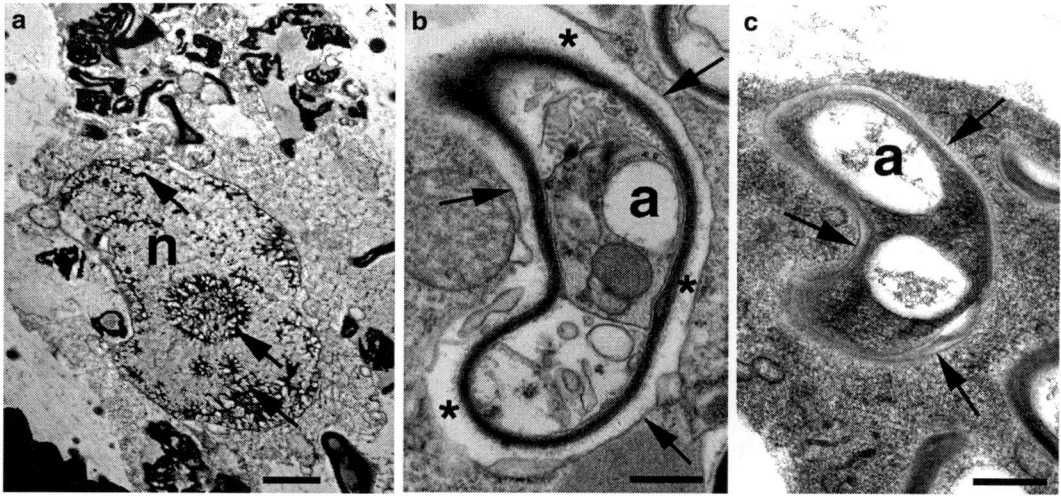

Fig. 2. Electron micrographs of root cells of *Zea mays* containing arbuscular structures of *Glomus intraradices*. (**a**): Root cortex cell with visible freeze artefacts caused by a failed freeze fixation. Especially in the nucleus (n) ice crystals of different size are visible (*arrows*). (**b, c**) arbuscular branches (a) in root cortex cells. Whereas the periarbuscular membrane (*arrows*) is tight-fitting to the arbuscule in high pressure frozen tissue (**c**), chemical fixation causes the detachment of this membrane from the arbuscule (**b**). This leads to the formation of a free space between membrane and arbuscule (stars in **b**), which is obviously a fixation artefact. Scale bars represent 2 μm (**a**) and 0.5 μm (**b, c**); (**b, c**) are reproduced from BIOspektrum, 2002 (2) by copyright permission of Spektrum, Akademischer Verlag GmbH, Heidelberg, Germany.

shown in Fig. 1. For freeze substitution in machines from other manufactures, it might not be necessary to use baskets.

3. Fixative 2 used for chemical fixation contains osmiumtetroxide. Because osmiumtetroxide is very toxic and has a high vapor pressure, it is strongly recommended to work in a fume hood. It is also important to be very careful during opening of the vial containing the solid osimiumtetroxide. The vials are ampoules, which are sometimes very difficult to open. The fixative should be freshly prepared and stored in a brown bottle.

4. Because the nonpolymerized resin according to Spurr is carcinogenic, these steps should be carried out in a fume hood. Use polyethylene-gloves (e.g., Manus PE-gloves, Dahlhausen & Co, Cologne, Germany) during handling of the material.

5. The epoxy-resin-mixture according to Spurr can be stored for longer time at −20°C.

6. It is important to preheat the oven to 70°C in order to avoid slow warming of the samples. If the oven is at room temperature and the epoxy resin with the samples warms up slowly, this can lead to brittle resin blocks inapplicable for sectioning.

7. The immobilization of cells in small cellulose capillary tubes is the gentlest method to collect cells for freezing (6), but it is not always advantageous because the density of cells can be quite low.

8. If a paste of concentrated cells is transferred into the planchet, it is not necessary to use filler substances.

9. Hexadecene will remain frozen at −80°C and thus represents a barrier for the substitution solution, which cannot diffuse into the biological material. Therefore, tissues frozen in planchets containing hexadecene as filler substance have to be taken out of the holder before substitution.

10. The most efficient method to take out the samples is to punch out the material in liquid nitrogen with a biopsy punch (2 mm, Stiefel, Offenbach, Germany) and transfer it in a basket fitting into the substitution tube (see Fig. 1). The basket has to be placed at the frozen substitution solution.

11. Because of the toxic and carcinogenic properties of Lowicryls, it is necessary to work during freeze substitution in a fume hood or to have very good local exhaust ventilation. Use polyethylene-gloves (e.g., Manus PE-gloves, Dahlhausen & Co, Cologne, Germany) during handling of the material. Because PE-gloves are not very tight on the fingers, the additional use of latex-gloves is recommended.

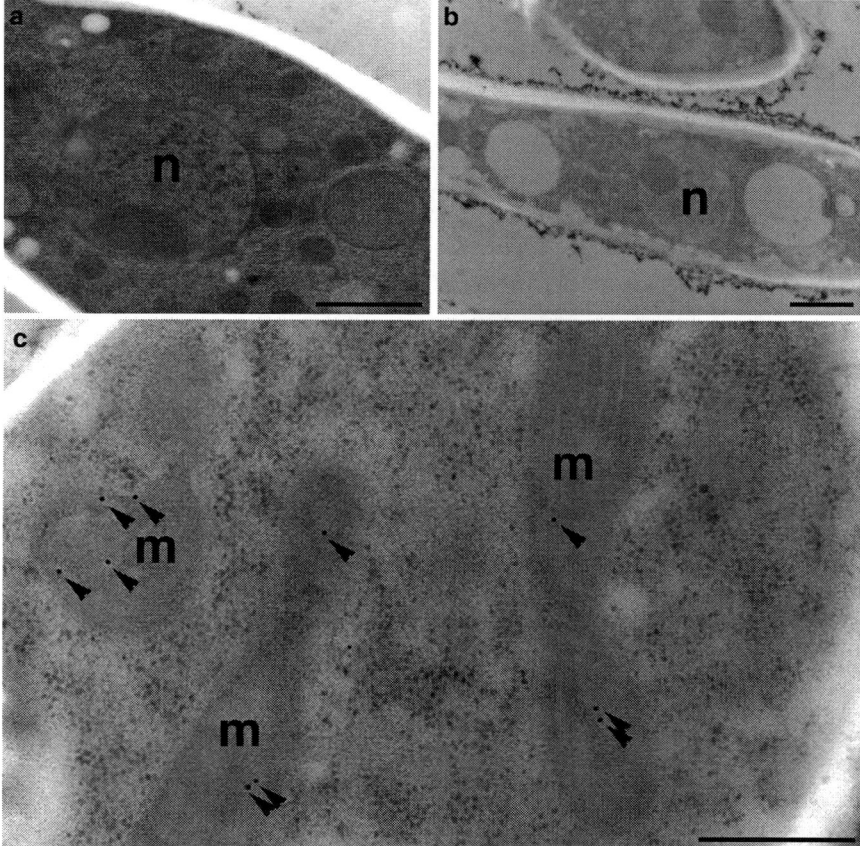

Fig. 3. Immunogoldlabelling of a protease in high pressure frozen hyphae of the plant pathogenic fungus *Rhynchosporium secalis*. (**a**) Part of a hyhpa containing a nucleus (n) and other organelles. This material was sectioned and contrasted without immunolabelling. (**b**) Overview of hyphae after immunogoldlabelling. Note the lost of contrast in comparison to **a**. Despite the labelling procedure causing some lost of contrast, this method is well suited for immunogoldlabelling. (**c**) shows the localization of the protease (arrowheads) in mitochondria (m). Scale bars represent 1 μm (**a** and **b**) and 0.5 μm (**c**).

12. The Lowicryl embedding mixture can be stored for a longer time if the empty space in the dark bottle is filled with nitrogen.

13. Immunogoldlabelling of high pressure frozen/freeze substituted material delivers suitable results, even though the contrast of the sections is somewhat diminished after the labelling procedure (Fig. 3).

14. The optimal dilution of the antibody for immunogoldlabelling has to be determined. In general, we start with the double antibody-concentration as used for Western-blotting.

15. The reaction with the first antibody can also be carried out at room temperature or at 37°C for 1 h. However, in our hands, incubation at 4°C delivers less background.

Acknowledgments

This work was supported by a grant of the Deutsche Forschungsgemeinschaft to G.H. (SFB 648 "Molekulare Mechanismen der Informationsverarbeitung in Pflanzen").

References

1. Craig S, Staehelin LA (1988) High pressure freezing of intact plant tissues. Evaluation and characterization of novel features of the endoplasmic reticulum and associated membrane systems. Eur J Cell Biol 46:80–93
2. Welter K, Müller M, Mendgen K (1988) The hyphae of *Uromyces appendiculatus* within the leaf tissue after high pressure freezing and freeze substitution. Protoplasma 147:91–99
3. Kiss JZ, Giddings TH Jr, Staehelin LA, Sack FD (1990) Comparison of the ultrastructure of conventionally fixed and high pressure frozen/ freeze substituted root tips of *Nicotiana* and *Arabidopsis*. Protoplasma 157:64–74
4. Kaneko Y, Walther P (1995) Comparison of ultrastructure of germinating pea leaves prepared by high-pressure freezing – freeze substitution and conventional chemical fixation. J Electron Microsc 44:104–109
5. Hess MW (1995) High-pressure freeze fixation reveals novel features during ontogenesis of the vegetative cell in *Ledebouria* pollen: an ultrastructural and cytochemical study. Biochem Cell Biol 73:1–10
6. Tiedemann J, Hohenberg H, Kollmann R (1997) High-pressure freezing of plant cells in cellulose microcapillaries. J Microsc 189:163–171
7. Spurr AR (1969) A low-viscosity epoxy resin embedding medium for electron microscopy. J Ultrastruct Res 26:31–43
8. Mendgen K, Welter K, Scheffold F, Knauf-Beiter G (1991) High pressure freezing of rust infected plant leaves. In: Mendgen K, Lesemann D-E (eds) Electron microscopy of plant pathogens. Springer-Verlag, Berlin, pp 31–42
9. Giddings TH (2003) Freeze-substitution protocols for improved visualization of membranes in high-pressure frozen samples. J Microsc 212:53–61

Chapter 23

Split-EGFP Screens for the Detection and Localisation of Protein–Protein Interactions in Living Yeast Cells

Emma Barnard and David J. Timson

Abstract

Proteomics aims to identify and classify the proteins present in a particular cell or tissue. However, we know that proteins rarely function alone and knowledge of which proteins interact with which other proteins is vital if we wish to understand how cells work. The budding yeast, *Saccharomyces cerevisiae*, is a well-established model for studying protein–protein interactions, and a number of methods have been developed to do this. A method for the in vivo detection and localisation of interacting pairs of proteins in living yeast cells is presented. The method relies on the ability of fragments of enhanced green fluorescent protein (EGFP) to reassemble if brought into close proximity. The reassembled EGFP regains the ability to fluoresce, and this fluorescence can be detected providing evidence of interaction and information about its location. *S. cerevisiae* is an ideal organism to apply this method to due to the relative ease with which its genome can be manipulated. The method described enables the modification of *S. cerevisiae* genes at the 3′-end with DNA encoding fragments of EGFP. Consequently, the expression levels of the proteins are unlikely to be affected and thus the method is unlikely to result in false positives. In addition to the protocol for labelling and detection of interacting pairs of yeast proteins, methods for simple tests for the effects of the labelling on the organism's function are presented.

Key words: Bimolecular fluorescence complementation assay, BiFC, Protein-fragment complementation assay, PCA, Enhanced green fluorescent protein, EGFP, Genomic modification

1. Introduction

The budding yeast, *Saccharomyces cerevisiae*, is now well-established as a model organism for studying protein–protein interactions. It has particular importance in the study of transcription, cellular signalling pathways and the cytoskeleton. The ease and safety of culture of the organism combined with a wealth of biochemical and genetic knowledge make the organism ideal for this role.

Consequently, there is a need to develop techniques to detect and analyse protein–protein interactions in *S. cerevisiae*.

The ideal method would enable reliable, rapid, in vivo detection of interactions, which did not affect the affinity of interaction. No single method meets this standard and it is generally necessary to probe interactions by a variety of complementary techniques in order to obtain sound information. The yeast two hybrid (Y2H) screen is rapid and takes place in living cells (1) but is notorious for generating false positives (i.e., detecting "interactions" that do not actually occur in the wild type organism) (2). These false positives are believed to occur, at least partly, because of the requirement to overexpress the proteins of interest thus resulting in inappropriate interactions being driven by the law of mass action. Tandem affinity purification (TAP) tagging seeks to overcome this problem (3). When employed in *S. cerevisiae*, it makes use of the relative ease with which that organism's genomic DNA can be modified and introduces sequence coding for two affinity tags at the 3′-end of genes of interest. By carrying out this modification away from the usual sites of regulation of gene expression, the genes are expressed at close to wild type levels. However, the lengthy, ex vivo affinity purification steps mean that short lived interactions are generally not detected (leading to false negatives).

Protein-fragment complementation assays (PCA) involve the fusion of inactive protein fragments to potentially interacting partners. If the partners interact, then the inactive fragments are brought together and function is restored (4, 5). A particularly useful type of PCA is the bimolecular fluorescence complementation (BiFC) assay in which the fragments of a fluorescent protein, e.g., enhanced green fluorescent protein, EGFP (6), are used. These assays provide a rapid and visual read-out and, if carried out on whole cells, also provide information about the localisation of the interaction. Recently, four separate groups have developed BiFC assays for use in living *S. cerevisiae* cells. Park et al. (7) described a plasmid-based assay which, although it lends itself to high throughput screening of potential interacting partners, potentially suffers the same overexpression drawbacks as the Y2H assay. A similar assay was also described by Blondel et al. (8). In an alternative version of the assay, Sung and Huh (9) introduced yellow fluorescent protein (YFP) encoding fragments into the *S. cerevisiae* genome. Successive modifications were achieved by mating two strains of opposite mating type. In a similar approach, Barnard et al. (10, 11) successively modified two genes in the same strain with DNA encoding N- and C-terminal fragments of EGFP. These hapto-EGFP fragments correspond to the first 157 amino acid residues (EGFP-N157) and residues 158 to the C-terminus (EGFP-C158). This protocol will result in labelling of every copy of the two proteins of interest whereas the method that relies on mating will only label 50%. Both these methods permit the simultaneous detection and localisation of interacting pairs of proteins, expressed at their wild type levels. Our method has been employed recently to detect interactions between cytoskeletal proteins at their wild type levels (12).

2. Materials

2.1. Oligonucleotides

1. Good quality oligonucleotide primers are required. We use high purity salt-free (HPSF) primers from Eurofins MWG Operon (Ebersberg, Germany) without any further purification. The freeze-dried oligonucleotide is dissolved in sterile, double deionised water (ddH$_2$O) to a final concentration of 100 µM. From this, aliquots are diluted to 10 µM in ddH$_2$O. Repeated freezing and thawing of these stocks is to be avoided.

2.2. Amplification of Insert DNA

2.2.1. Plasmids and Their Maintenance

1. Luria–Bertani (Miller) bacterial growth medium (LB): Pre-prepared solid LB medium should be dissolved to a final concentration of 25 g/L in ddH$_2$O. This should be autoclaved immediately and only dispensed using aseptic technique.
2. Ampicillin: This is conveniently stored as a 100 mg/mL solution in 50% (v/v) ethanol/water at –20°C.
3. LB-Ampicillin plates: LB medium should be made as above but supplemented with 1.5% (w/v) plant agar (Melford). Following autoclaving, the solution should be mixed gently (the introduction of bubbles should be avoided; rolling the bottle on the bench works well) and cooled under running water (the bottle should be rotated to avoid localised solidification of the agar). Ampicillin should be added to a final concentration of 100 µg/mL, the solution mixed gently and plates poured immediately using aseptic technique.
4. *Escherichia coli*: Any strain suitable for the long term maintenance of plasmids can be used (i.e., one deficient in recombinases). We use XL1-Blue (Stratagene, La Jolla, CA, USA).
5. Plasmids pEB1 and pEB2 (10): Small aliquots can be obtained from the Yeast Genetic Resource Center (Osaka University, Japan; http://yeast.lab.nig.ac.jp/nig/index-en.html/). Stocks should be prepared by DNA mini-prep from transformed *E. coli*. Any standard, commercial mini-prep kit can be used; alternatively, users can consult standard texts (13) for instructions on non-kit based methods (see Note 1). Inserts for labelling genes of interest with sequence encoding EGFP-N157 and a *TRP1* selectable marker can be produced using pEB2 as a template; pEB1 can be used to produce inserts encoding

EGFP-C158 and a *URA3* selectable marker. The marker genes are from *Kluyveromyces lactis* in order to minimise recombination into the *S. cerevisiae* genome.

2.2.2. PCR

1. Polymerase: KOD (Merck, Nottingham, UK).
2. Deoxynucleoside triphosphates (dNTPs): It is convenient to maintain a single solution of dATP, dGTP, dCTP, and dTTP all dissolved to a final concentration of 2.5 mM in ddH$_2$O. Repeated freezing and thawing of this solution should be avoided.
3. Thermal cycler: The protocols given here are appropriate for a Techne TC-312 machine. Minor modifications may be required for other machines.

2.2.3. Agarose Gel Electrophoresis and Extraction of DNA from Agarose

1. Agarose: Any reasonable quality agarose can be used.
2. TAE buffer: 40 mM Tris–acetate, pH 8.0; 1 mM EDTA. This is conveniently made by dilution of a 50× stock solution. For 1 L of this, the following should be mixed: 242 g of Tris-base, 57.1 mL glacial acetic acid, and 100 mL of 0.5 M EDTA (13).
3. Ethidium bromide: Dissolve to a final concentration of 1 mg/mL in TAE buffer. This compound is believed to be carcinogenic and mutagenic (14). Consequently, gloves should be worn at all times when handling solutions and gels that contain it. Care should be taken not to contaminate surfaces, door handles, etc (see Note 2).
4. DNA markers should be selected according to the predicted size of the products. 2 log DNA ladder (New England Biolabs) provides good size discrimination for the range 100–10,000 bp.
5. UV transilluminator and, preferably, some means of producing digital images of gels such as the ChemidocXRS from Bio-Rad (Hercules, CA, USA).
6. A kit for the purification of DNA from agarose slices (e.g., Sigma's Genelute DNA extraction kit).

2.3. Insertion of DNA into Yeast Genome

2.3.1. Yeast Strains and Growth Media

1. Yeast strains: Any *S. cerevisiae* strain, which is deficient in *TRP1* and *URA3*, can be used. We use JPY5 (MATα *ura3-52 his3Δ200 leu2Δ1 trp1Δ63 lys2Δ385*) (15).
2. YPD media: Pre-prepared solid YPD medium (Sigma, Poole, UK) should be dissolved to a final concentration of 50 g/L in ddH$_2$O. This should be autoclaved immediately and only dispensed using aseptic technique.
3. Minimal medium: Yeast nitrogen base without amino acids (Formedium, Hunstanton, UK), 6.7 g/L supplemented with

essential amino acids (tryptophan, 24 mg/L; histidine 24 mg/L; arginine 24 mg/L; methionine 24 mg/L; tyrosine 37 mg/L; lysine 37 mg/L; phenylalanine 61 mg/L; leucine 73 mg/L; aspartic acid 122 mg/L; threonine 244 mg/L), adenine (24 mg/L), and uracil (24 mg/L). The various components should be made up in 90% of the required total volume of ddH$_2$O, the pH adjusted to approximately 7.0 with NaOH and the solution autoclaved. For selection of recombinants uracil and/or tryptophan should be omitted as appropriate. A 20% (w/v) solution of glucose should be made up separately and autoclaved. After the solutions have cooled, the glucose solution should be added to the media aseptically to give a final concentration of 2% (w/v).

2.3.2. Transformation Solutions and Equipment

1. Boiled salmon sperm DNA (Sigma) acts as carrier DNA in the transformation: 2 mg/mL should be dissolved in sterile water.

2. Transformation mix: 33% (w/v) polyethylene glycol 3350; 100 mM lithium acetate; 0.07% (w/v) boiled salmon sperm DNA.

3. Centrifuge: A refrigerated centrifuge capable of spinning 50 mL falcon-style tubes up to at least $3,000 \times g$ such as the Biofuge Primo R from Heraeus.

2.4. Detection of Recombinants

1. PCR materials and thermal cycler as Subheading 2.2.2; Taq DNA polymerase (New England Biolabs).

2.5. Microscopy

1. Cell resuspension buffer: 50 mM Hepes–OH, pH 7.5; 150 mM sodium chloride, 10% (v/v) glycerol. The solution should be autoclaved and stored at 4°C.

2. Nuclear stain: Hoechst 33258 (Invitrogen, Paisley, UK) prepared as a 1,000× concentrated stock solution at 1 mg/mL in ddH$_2$O. The solution should be kept in a dark bottle or one covered in foil to minimise exposure to light. Since this dye binds strongly to DNA, it should be considered potentially mutagenic and gloves should be worn when handling solutions containing it.

3. Microscope: Leitz Laborlux D microscope with 450–490 nm filter for EFGP and 340–380 nm filter for Hoechst. Microscopes of similar specification should also be appropriate. Ideally, the microscope should be fitted with image capture and analysis software.

2.6. Further Tests

2.6.1. Cell Size and Morphology

1. Microscope as above and either a graticule system for estimating cell dimensions or software to do this.

2. Software to calculate means and standard deviations and to carry out statistical tests (e.g., Microsoft Excel).

2.6.2. Growth Kinetics

1. Spectrophotometer and plate reader; sterile 96-well plates.
2. Multichannel pipette.
3. Software for non-linear curve fitting, e.g., GraphPad Prism (GraphPad Software, San Diego, CA, USA).

3. Methods

In order to modify two genes sequentially, *S. cerevisiae* cells should be labelled first using one hapto-EGFP encoding construct (Subheadings 3.1, 3.2, and 3.3). PCR methods should be used to verify that the cells have been correctly labelled (Subheading 3.4) and then these cells should be labelled with the second hapto-EGFP encoding construct using essentially the same protocol (Subheadings 3.1, 3.2, and 3.3). Following verification of the second successful labelling, the cells can be imaged to determine if fluorescence can be detected from reassembled EGFP (Subheading 3.5) and subjected to further analyses, as required (Subheading 3.6). A suggested workflow is outlined in Fig. 1.

3.1. Primer Design

Primers should be designed such that the final PCR product will be flanked by sequences from the 3′-end of gene of interest (not including the stop codon) and the sequence immediately 3′ to the stop codon.

1. Both pEB1 and pEB2 contain common priming sequences and so the same primers can be used to generate insert DNA from both plasmids.
2. The forward primer should be designed such that it incorporates at least 45 bases from the 3′of the gene of interest, including the codon immediately prior to the stop codon. This sequence should be followed immediately by the forward common priming sequence (ggaggatctggaggg).
3. The reverse primer should be designed so as to include the reverse complement of at least 45 bases immediately 3′ to the gene of interest followed immediately by the reverse common priming sequence (tacgactcactataggg).

3.2. PCR Amplification of Insert DNA

3.2.1. Maintenance of Plasmid DNA

Stocks of pEB1 and pEB2 should be maintained by transformation into a suitable *E. coli* strain and isolation by DNA mini-preparation. Both plasmids confer ampicillin resistance on *E. coli* cells, and so this antibiotic should be included in plates for selection of recombinants.

1. Transform *E. coli* with the plasmids. Any method for the production and transformation of competent cells is appropriate.

Split-EGFP Screens for the Detection and Localisation 309

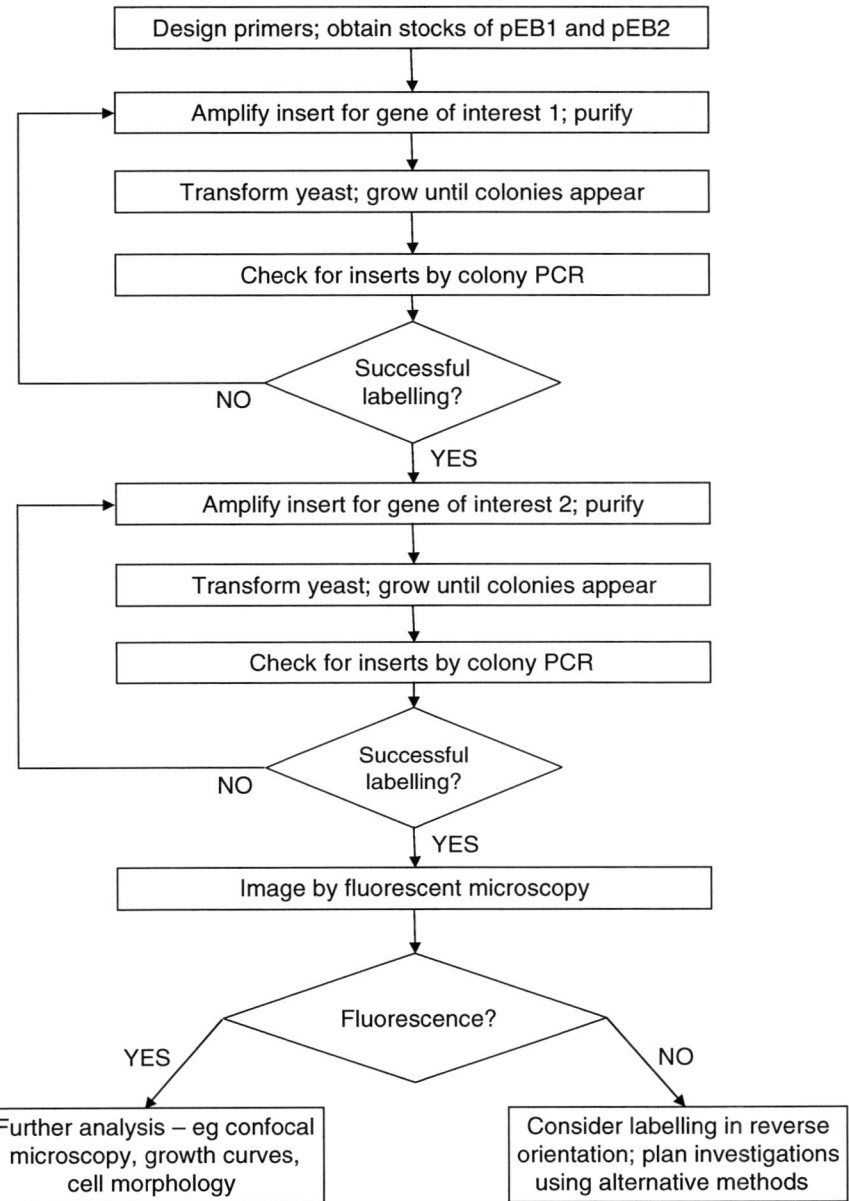

Fig. 1. A workflow outlining the strategy for labelling a pair of proteins suspected of interaction. If no problems are encountered, the process can be completed in 2–3 weeks.

2. Once transformed colonies have been produced, prepare plasmid DNA by mini-preparation using a suitable commercial kit.

3. Dissolve the plasmid DNA in 5 mM Tris–HCl, pH 8.0 (not TE buffer as this contains EDTA, which can interfere with subsequent steps) and store at −20°C.

3.2.2. Amplification of Insert DNA

1. Prepare the following mix: Sterile 34.5 μL ddH$_2$O, 5.0 μL 10× KOD buffer, 4.0 μL dNTP mix (2.5 mM), 2.5 μL Forward primer (10 μM), 2.5 μL Reverse primer (10 μM), 1.0 μL pEB1 or pEB2 (as prepared above in Subheading 3.2.1), 0.5 μL KOD polymerase. It is convenient to begin by adding the ddH$_2$O and then add the remaining components. The polymerase should be added last.

2. Perform PCR reaction according to the following program: 35 cycles of {98°C 15 s, 50°C 30 s, 72°C 20 s}, 72°C 5 min, then 4°C until required.

3.2.3. Detection and Purification of Insert DNA

1. The entire 50 μL volume should be resolved by agarose gel electrophoresis. For most product sizes (100–3,000 bp), 1% agarose gel electrophoresis in TAE buffer (13) is adequate. If only combs suitable for smaller volumes are available, then several wells can be carefully taped together to achieve this. Sellotape™ (or a similar product) is preferred to autoclave tape for this purpose as the latter often contains fluorescent powder, which can interfere with the analysis of the gel in UV light. Markers should be loaded to enable an estimate of the size of products to be estimated.

2. Following electrophoresis, visualise the gel by UV transillumination. If no band of an appropriate size is observed, repeat stages 1–4 altered thermal cycling parameters (e.g., extension time and annealing temperature).

3. Carefully excise bands corresponding to the desired product using a clean scalpel or a razor blade. Care should be taken to avoid exposure to UV light: wear a full face mask along with gloves and a lab coat. The excised band can be stored at −20°C at this stage if required.

4. Extract the DNA from the gel slice using a suitable commercial kit.

5. Analyse a small aliquot (5 μL) of the purified DNA on an agarose gel to check purity and yield.

3.3. Transformation of Yeast and Selection of Recombinants

1. Grow an overnight culture (5 mL) of the yeast strain to be transformed in suitable media (YPD for unmodified yeast; selective media for strains, which have been modified).

2. Transfer the culture into 100 mL of media and grow until $A_{600\ nm} = 0.8–1.0$. This typically takes 17–24 h.

3. Pellet the culture by centrifugation (2,000 × g for 5 min) and resuspend in 25 mL ddH$_2$O.

4. Pellet the cells by centrifugation as above and resuspend in 1 mL of sterile ddH$_2$O. Repeat this step one more time. It is more convenient to use a bench-top centrifuge for the second wash.

5. For each transformation, pellet a 100 μL aliquot of the suspension produced in step 3 by centrifugation and resuspend in 360 μL of transformation mix.
6. Add insert DNA. Typically, 2–5 μL of the linear DNA as produced in Subheading 3.2.3 above is required, but some experimentation may be required to determine the optimal level for the particular insert (see Note 3).
7. Incubate the mix at 40°C for 40 min (see Note 4).
8. Pellet the mix by centrifugation and resuspend in 500 μL of ddH$_2$O.
9. Plate the cells onto selective media. Take care to ensure that the cells are evenly spread and do not form aggregates at specific locations (see Note 5).
10. Incubate the plates inverted at 30°C, until colonies appear (typically 3–4 days).

3.4. Detection of Recombinants

Detection of correct insertions is carried out by colony PCR (see Note 6). This allows the simultaneous analysis of several colonies. The basic principle for the detection method is that two primers (one corresponding to sequence within the gene of interest and one to sequence in the 3′ non-coding region) will amplify a short fragment in non-labelled cells and a longer one in cells in which the hapto-EGFP and marker encoding sequences have been successfully inserted (see Fig. 2).

3.4.1. Design of "Check" Primers for Colony PCR

1. These should be designed such that the forward (5′) primer is located within the coding sequence of the gene of interest and the reverse is in the 3′ non-coding region of the gene.

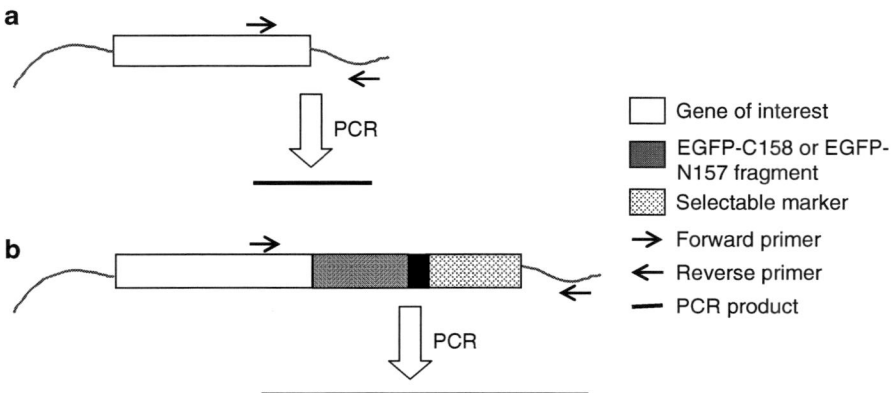

Fig. 2. The strategy for checking if genes of interest have been successfully modified. Two primers, one corresponding to sequence within the gene of interest and immediately 3′ to the gene, are designed and used to amplify genomic DNA. In unmodified yeast (**a**), a relatively short fragment is amplified. However, if labelling has been successful (**b**) a larger fragment is produced. Modification with sequence encoding EGFP-N157 (and its corresponding selectable marker) results in a size increase of 1,650 bp. Modification with EGFP-C158 results in a 1,980-bp increase.

Ideally, they should be located approximately 300–500 bp apart. Online resources such as the *Saccharomyces* Genome Database (SGD, http://www.yeastgenome.org/) can be consulted to assist in the design of primers, and bioinformatics resources such as BLAST (16) can be used to check for similarity to other regions of the yeast genome and potential non-specific binding of the primer.

3.4.2. Colony PCR to Detect Recombinants

1. Label each colony with a unique identifier.
2. For each colony to be screened, prepare the following mix: 17.8 μL sterile ddH$_2$O, 2.5 μL 10× Taq buffer, 2.0 μL dNTP mix (2.5 mM), 1.3 μL Forward primer (10 μM), 1.3 μL Reverse primer (10 μM), 0.3 μL Taq polymerase.
3. Pick individual colonies and mix directly into the PCR mix immediately prior to PCR amplification. Leave some of the colony behind on the plate to enable further analysis should it prove to result from a positive transformant.
4. Perform PCR reaction with the mix according to the following program: 98°C 50 s, 35 cycles of {98°C 10 s, X°C* 30 s, 72°C 80 s}, 72°C 10 min, 4°C until required. *Where X = the lower melting temperature of the two primers plus 3°C. Generally, this is in the range 45–50°C.
5. Analyse the products by agarose gel electrophoresis and estimate the sizes of the products by reference to DNA markers of known size. If insertion has been successful, proceed to the next stage. If not, screen additional colonies or repeat the transformation (see Note 7).

3.5. Visualisation of EGFP-Labelled Proteins

Use cells from either liquid culture or from colonies. One advantage of the technique is that it permits visualisation of interactions in living cells. However, on occasions (for example, following nuclear staining), it may be necessary to fix the cells prior to viewing. Typical positive results from in vivo imaging of *S. cerevisiae* cells expressing reassembled EGFP are shown in Fig. 3.

3.5.1. Visualisation of EGFP Fluorescence in Live Cells

1. Pick colonies from plates and resuspend in the 100 μL of cell resuspension buffer (or PBS, if preferred).
2. Place approximately 10–20 μL of this suspension (or a liquid culture) onto a microscope slide and cover with a coverslip. If necessary, the suspension can be spread out by pressing gently down on the coverslip with the flat end of a pencil (or similar implement).
3. Observe under the microscope using a ×100 fluotar oil immersion lens (see Note 8).

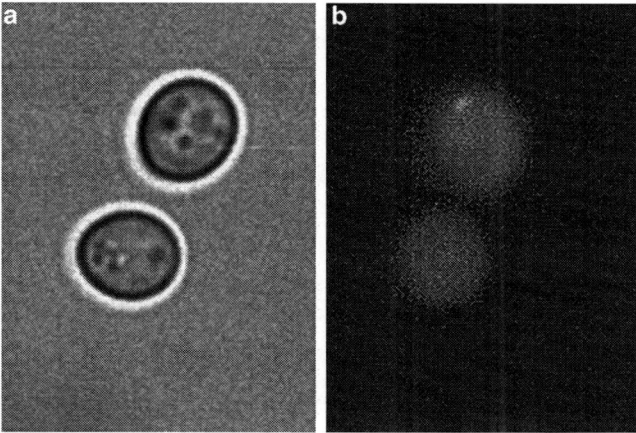

Fig. 3. An example of a successfully detected interaction. The yeast phosphofructokinase is a highly expressed heterooctomeric glycolytic enzyme composed of four subunits of Pfk1p and Pfk2p (19, 20). Here, Pfk1p is labelled with EGFP-N157 and Pfk2p with EGFP-C158. The cells were imaged under white light (**a**) and fluorescent conditions (**b**). Fluorescence was clearly detected throughout the cytoplasm of the living yeast cells.

3.5.2. Nuclear Staining and Visualisation

1. Stain with Hoechst 33258 at a final concentration of 1 μg/mL.
2. Wash the cells 2–3 times by pelleting followed by resuspension in cell resuspension buffer (or PBS). This is necessary as Hoechst tends to adhere to the external membranes of the cells resulting in the appearance of uniform blue staining if the cells are not washed.
3. Visualise using the same microscope and procedures, except that a filter block of 340–380 nm should be introduced to observe blue light.

3.5.3. Data Handling and Subsequent Manipulation

Capturing images digitally enables them to be stored for subsequent analysis and publication. However, care must be taken to avoid excessive or inappropriate manipulation of the images. Recently, recommendations have been published (17) suggesting that the only acceptable manipulations are: cropping, resizing and adjustments to the brightness and contrast. If such manipulations are undertaken, they should be applied to the whole image and should not be used to suppress data. Original files should always be archived in order to deal with queries. Good practice would suggest that they should be made available to *bone fide* researchers to carry out their own analyses.

3.6. Further Tests and Troubleshooting

In many investigations, it may be sufficient to demonstrate the existence of the interaction and localise it within the cell. Further detail on the localisation of the interaction may be obtained through more sophisticated imaging techniques such as confocal microscopy. However, in other cases, it may be necessary to test the consequences of the modification on the function of the proteins of interest. In such cases, specialised assays may already exist, which are particular to these proteins. However, the following tests are suggested for all interactions.

3.6.1. Cell Size and Morphology

1. Under direct (white light) illumination, compare the morphology of cells to unmodified ones. Any differences should be noted.
2. For both modified and unmodified cells, pick a sample ($n=100$) of cells at random and measure their diameters using either a graticule or visualisation software. If the cells are not spherical, the longest axis should be chosen. For the same cells record the number of buds per 100 cells.
3. Repeat step 2 at least twice.
4. Record the mean diameters and budding frequencies and compare using a t-test (or other suitable statistical test). Significant ($p<0.05$) differences should be noted.

3.6.2. Growth Kinetics

1. Streak a labelled strain onto an appropriate selective minimal medium plate. At the same time, streak unmodified cells onto a minimal medium plate.
2. Incubate the plates inverted at 30°C until distinct colonies appear.
3. Pick a colony from each plate and grow, shaking at 30°C overnight in appropriate medium.
4. Estimate the extent of the growth by measuring the absorbance at 600 nm using fresh medium as a blank (see Note 9). It may be necessary to dilute the culture into fresh medium to obtain a meaningful value. Note that the linear response of most spectrophotometers is between approximately 0.1 and 1.5 in absorbance mode.
5. Dilute the cultures into 5 mL (final volume) of medium such that the final $A_{600\,nm}$ is 0.05. Mix well the fresh culture by vortexing.
6. Divide the cultures into 150 µL aliquots in a 96-well plate. At least three aliquots of each strain should be used. Set up an identical number of media only controls on the same plate.
7. Read the absorbance at 600 nm immediately using a plate reader (e.g., Labsystems 352) to obtain the zero time values (see Note 10).
8. Incubate the plate, shaking at 30°C for at least 24 h (see Note 11).
9. Read the absorbance at convenient intervals (at least every hour). If the cells have settled in the wells between readings, it is important to resuspend them by gentle pipetting before measuring.

10. Calculate the mean of the absorbencies for each strain at each time point and plot against time. The standard deviations of these means should be used to provide error bars. The graph can be conveniently fit to the Gompertz growth equation, $A = A_0 + C \cdot \exp(-\exp((2.718 \cdot G_{max}/C) \cdot (L-t) + 1))$ where, A_0 is the initial absorbance, C is the difference between the initial and maximum optical densities, L is the lag time, G_{max} is the maximum specific growth rate (18) using non-linear curve fitting programs, such as GraphPad Prism. Curves can be contrasted by eye or more formally by comparing parameters derived from this equation. Modifications that cause minimal perturbation to cellular physiology would be expected to result in little or no change to the growth kinetics. However, it should be noted that the *TRP1* marker gene appears to confer the ability to grow to slightly higher cell densities in a 24-h period (10).

4. Notes

1. Some users may prefer to make glycerol stocks of transformed *E. coli*.
2. A number of alternatives to ethidium bromide are available (e.g., Sybr Green). It is claimed that they are not as dangerous as ethidium bromide.
3. The amount of DNA appears to be critical in some cases. Too little DNA and there will be insufficient for transformation; too much and there is a risk of insertion at non-specific sites. We have observed both effects. Using a number of different volumes of purified DNA may be advisable. If this is done, colonies resulting from the transformation with the lowest volume of DNA should be screened first.
4. The optimum length of time required for a particular strain may vary. If a strain other than JPY5 is used, then optimisation of this time may be required. If information is available in the literature about similar transformation procedures (e.g., for TAP-tagging or epitope tagging), then this should be used as a starting point.
5. It can be tricky to plate 500 μL onto standard, 90-mm diameter plates. This problem can be overcome either by spreading the mix across more than one plate or by using larger (e.g., 150 mm diameter) plates.
6. In theory, detection could also be achieved by western blotting using appropriate antibodies. In order to do this, however, it would be necessary to obtain antibodies whose epitopes were solely located in the N- and C-terminal fragments or to

the specific proteins under investigation (if available). It would also be necessary to ensure that the proteins of interest were expressed under the growth conditions being used.

7. Although, in theory, this screening procedure should result in either a positive or a negative result, we have seen several instances of mixed populations. At present, we do not know if this results from contamination or from rejection of the insert by a subset of the cells. If a mixed population is detected, repeated streaking of the colony onto selective plates can sometimes yield a pure population.

8. A positive signal is strong evidence in favour of an interaction, but the absence of a signal is not sufficient to prove lack of interaction. Not all interactions occur under all growth conditions. (For example, many "global" screens miss well-characterised interactions which only occur under specific circumstances such as with different carbon sources or under particular stresses.) So the growth conditions may need to be altered. It is also worth sampling the cells at different stages in their growth as the biochemistry of *S. cerevisiae* varies between lag, exponential and stationery phases. Furthermore, stronger evidence for lack of interaction is provided if both possible labelling reactions are carried out – i.e., if the interacting proteins are X and Y, testing *both* X-EGFP-N157/Y-EGFP-C158 and X-EGFP-C158/Y-EGFP-N157. Sometimes only one labelling configuration will result in the restoration of fluorescence (10).

9. The use of absorbance measurements to estimate cell growth relies upon the light scattering effects of cells. Thus, the precise wavelength used is not critical but should be kept constant throughout the experiment. It may be necessary to adjust the wavelength at this stage if a 600-nm filter is not available on the plate reader to be used in later stages. Often, plate readers are fitted with a 590-nm filter (for measurement of protein concentration by the Bradford assay). This filter works perfectly well for the growth assay.

10. If a plate reader is not available, the assay can be carried out in a spectrophotometer (or colorimeter) instead. If this is done, it is recommended that a larger culture volume is set up initially (e.g., 100 mL) and samples (1 mL) are taken, measured for each reading and disposed of.

11. If it is not convenient to read every hour over a 24-h period, then the initial overnight culture should be retained and diluted again 8–10 h after the first plate is set up. The second dilution should be used to make up a second plate, which can then be incubated overnight and read for a further 8–10 h the following day alongside the first plate to give a total timeframe of 24–30 h.

Acknowledgments

We thank Drs John Nelson, Neil McFerran and Alan Trudgett (Queen's University, Belfast) for their assistance in the development of the method. EB acknowledges a PhD studentship from the European Social Fund.

References

1. Fields S, Song O (1989) A novel genetic system to detect protein–protein interactions. Nature 340:245–246
2. Serebriiskii I, Estojak J, Berman M, Golemis EA (2000) Approaches to detecting false positives in yeast two-hybrid systems. Biotechniques 28:328–336
3. Rigaut G, Shevchenko A, Rutz B, Wilm M, Mann M, Seraphin B (1999) A generic protein purification method for protein complex characterization and proteome exploration. Nat Biotechnol 17:1030–1032
4. Remy I, Galarneau A, Michnick SW (2002) Detection and visualization of protein interactions with protein fragment complementation assays. Methods Mol Biol 185:447–459
5. Barnard E, McFerran NV, Nelson J, Timson DJ (2007) Detection of protein–protein interactions using protein-fragment complementation assays (PCA). Curr Proteomics 4:17–27
6. Wilson CG, Magliery TJ, Regan L (2004) Detecting protein–protein interactions with GFP-fragment reassembly. Nat Methods 1:255–262
7. Park K, Yi SY, Lee CS, Kim KE, Pai HS, Seol DW et al (2007) A split enhanced green fluorescent protein-based reporter in yeast two-hybrid system. Protein J 26:107–116
8. Blondel M, Bach S, Bamps S, Dobbelaere J, Wiget P, Longaretti C et al (2005) Degradation of Hof1 by SCF (Grr1) is important for actomyosin contraction during cytokinesis in yeast. EMBO J 24:1440–1452
9. Sung MK, Huh WK (2007) Bimolecular fluorescence complementation analysis system for in vivo detection of protein–protein interaction in *Saccharomyces cerevisiae*. Yeast 24:767–775
10. Barnard E, McFerran N, Trudgett A, Nelson J, Timson DJ (2008) Detection and localisation of protein–protein interactions in *Saccharomyces cerevisiae* using a split-GFP method. Fungal Genet Biol 45:597–604
11. Barnard E, McFerran N, Trudgett A, Nelson J, Timson DJ (2008) Development and implementation of split-GFP based bimolecular fluorescence complementation (BiFC) assays in yeast. Biochem Soc Trans 36:479–482
12. Pathmanathan S, Barnard E, Timson DJ (2008) Interactions between the budding yeast IQGAP homologue Iqg1p and its targets revealed by a split-EGFP bimolecular fluorescence complementation assay. Cell Biol Int 32:1318–1322
13. Sambrook J, Fritsch EF, Maniatis T (1989) Molecular cloning, a laboratory manual. Cold Spring Harbor Laboratory Press, New York
14. Perlman PS, Mahler HR (1971) Molecular consequences of ethidium bromide mutagenesis. Nat New Biol 231:12–16
15. Wu Y, Reece RJ, Ptashne M (1996) Quantitation of putative activator-target affinities predicts transcriptional activating potentials. EMBO J 15:3951–3963
16. Altschul SF, Madden TL, Schaffer AA, Zhang J, Zhang Z, Miller W, Lipman DJ (1997) Gapped BLAST and PSI-BLAST: a new generation of protein database search programs. Nucleic Acids Res 25:3389–3402
17. Rossner M, Yamada KM (2004) What's in a picture? The temptation of image manipulation. J Cell Biol 166:11–15
18. Gompertz B (1825) On the nature of the function expressive of the law of human mortality, and on a new mode of determining the value of life contingencies. Philos Trans R Soc Lond 115:513–585
19. Kopperschlager G, Bar J, Nissler K, Hofmann E (1977) Physicochemical parameters and subunit composition of yeast phosphofructokinase. Eur J Biochem 81:317–325
20. Clifton D, Fraenkel DG (1982) Mutant studies of yeast phosphofructokinase. Biochemistry 21:1935–1942

INDEX

A

Actin
    F-actin 226
    microfilaments 226
    patches 227
    staining 228, 229, 231, 232
Actin cables 226, 227
Analysis of fungal transformants 61–64, 70–71
Annexin V 270–273, 276–277, 279
Antibodies 192, 193, 199, 203, 204, 212,
    214, 218, 220–223, 225–231, 241, 255, 297, 315
Antifading reagent 284, 286
Antifungal drugs
    amphotericin B 283–285, 287
    caspofungin 283
    fluconazole 283
    5-fluorocytosine (5-FC) 283, 285, 286
    itraconazole 283, 285
    voriconazole 283, 285
Apoptosis 269, 270, 273, 274, 278, 279

B

Bimolecular fluorescence complementation (BiFC) 304
Biotechnology 6

C

Calcofluor white 229
cDNA
    library 116
    synthesis 105, 108, 111, 142, 144, 145, 159, 162
Cell death 269–279, 284, 285
Cell lysis 26, 39, 168–169
Cell wall
    degrading enzymes 22, 227, 228, 230
    permeabilization 140–142, 144, 228, 230
    staining 229
Chemiluminescence (ECL) 194, 199, 204, 206
Chromatin immunoprecipitation (ChIP) 190–192,
    196–199, 211–223

Chromosome painting 236–238, 242–244, 250–252
Clonase 56, 57, 59, 64–66, 74
Co-Immunoprecipitation (CoIp),
Coupled enzyme assay 207–208
Cross linking 161, 196, 197, 226, 227, 291
Cryosubstitution 292, 296
Cytoskeleton 225–232, 303

D

DelsGate 55–76
Digital image analysis 125, 126
Direct fluorescent in situ RT-PCR
    (DIFIS-RT-PCR) 137–150
Disease quantification 126
DNA
    fragmentation 270
    multiple displacement amplification
        (MDA) 175–185, 238, 242, 243, 250, 251
    purification 37, 39, 158–160

E

Evans blue 271, 273, 277

F

Fiber FISH 236–238, 244–245, 252–254
Fixation 139–140, 143–144, 229, 231, 232, 292–295
Flow cytometer 283
Flow cytometry 281–288
Fluorescence in situ hybridization (FISH) 235–256
Fluorescent dyes
    FITC 232, 236–237, 241,
        243, 249, 270, 271, 273, 278
    FUN-1 284–287
    H_2DCFDA 271
    JC-1 283–287
    propidium iodide (PI) 236–237,
        241, 243, 271, 273, 276, 278, 282–287
    rhodamine (TRITC) 228, 232,
        236–237, 241, 243, 249, 276
    rhodamine phalloidin 232

Formaldehyde 81, 89, 97, 105, 109, 196, 197, 212, 214, 216, 227–229, 232, 271
Freeze substitution 293, 295–299
Fungal infection 125–135
Fungal transformation
　agrobacterium 22–26, 34, 76
　electroporation 24–29, 31, 34, 40, 71, 74, 79, 84, 85, 92, 96
　protoplasts 3–18, 21, 22, 34, 46–47, 52, 53, 59–61, 64, 65, 68–69, 75, 232, 238, 244, 252, 255, 271–273, 276, 277, 279
Fungi
　Ashbya gossypii 226
　Aspergillus nidulans 3, 12, 36, 37, 42–43, 45, 49, 78, 96, 226, 270, 278, 279
　Botrytis cinerea 104, 107, 108, 110–112, 116, 235–236
　Candida 79, 282–284
　Cochliobolus heterostrophus 3, 4, 7–8, 12–15, 116, 118, 121, 236, 238
　Colletotrichum gloeosporioides 22, 28
　Cryptococcus 78, 283, 284
　Erysiphe cichoracearum 154
　Fusarium graminearum 43, 52, 202, 204–207, 270
　Fusarium oxysporum 41
　Fusarium verticillioides 76, 202, 207–209
　Glomus intraradices 139–143, 145–149, 298
　Magnaporthe oryzae 78, 116
　Neurospora crassa 3, 21, 33–40, 77–79, 82, 84, 86, 87, 91–96, 110, 189–199, 259–268
　Saccharomyces cerevisiae 3, 33, 34, 37, 79, 166, 211–223, 226, 303, 304, 306, 308, 312, 316
　Trichoderma reesei 166, 167
　Ustilago maydis 3, 56, 59–60, 64, 68, 70, 75, 79, 226, 227, 231, 232, 236
　Verticillium 56, 76

G

Gateway 56, 57, 59–60, 64
Gene
　cloning 56, 57, 64, 71, 83, 96, 122
　deletion 14, 16, 17, 33–40, 55–76
　expression 77–98, 103–113, 115, 116, 138, 139, 144–149, 154–156, 165–172, 189–199, 304
　knockout 33
　silencing 77–98, 189
　tagging 41–53
　transcription 139, 197, 220
GFP. *See* Green-fluorescent protein
Glucanex 5, 6, 17, 44, 46, 272
Glucokinase 202, 204, 207–209
Green-fluorescent protein (GFP) 29, 30, 42–43, 162, 226, 227, 259, 261–266, 268

H

High pressure freeze fixation 292–295
His-Myc tag 194

I

Immunofluorescence 138, 225–232
Immunogoldlabelling 291–300
Immunoprecipitation 190, 191, 195–199, 211–223
In situ hybridization (ISH) 138, 235–256
ISH. *See* In situ hybridization

L

Laser microdissection 153–162
Live Cell imaging 259–268

M

MAPK. *See* Mitogen activated protein kinase
MDA. *See* Multiple displacement amplification
Media
　carboxymethylcellulose (CMC) 203, 205
　Columbia agar 283
　Emerson's YpSs (EMS) 22, 24, 25, 27
　Gamborg B5 .... 104, 107, 108
　minimal medium agar 24, 43, 45
　potato dextrose broth (PDB) 57, 68, 73, 208, 246
　Sabouraud dextrose agar (SDA) 283
　SC 37, 38
　Srb's micronutrients 6, 7, 116, 117
　trace element solution 167, 260, 261
　Vogel's 80, 191, 260–262
　YENB 23, 25, 26
　YEPS 57, 61, 68, 69, 231
　YPD 36, 38, 215, 306, 310
Microscopy
　confocal microscope 138, 140, 142, 145, 231, 260, 278, 313
　electron microscope 282, 291
　fluorescent microscope 240, 242
　light microscope 31, 94, 240, 277, 284, 286
　total internal reflection fluorescence (TIRF) 260, 262–264
Microtubule
　astral microtubules 226, 227
　depolymerization 260, 262, 264
　dynamics 225, 226, 259–268
　polymerization 260, 262, 263
　α tubulin 225
　β tubulin 95, 107, 225, 259, 261–266, 268
　γ tubulin 226
Mitochondria
　membrane potential 269, 285
　staining 270

Mitogen activated protein kinase
(MAPK) 115, 202–207, 209
Multiple displacement amplification
(MDA) 175–185, 238, 242, 243, 250–251
Multiplex gene expression 165–172
Mycorrhiza 116, 137–145, 236

N

NBT. *See* Nitroblue tetrazolium
Nitroblue tetrazolium (NBT) 271, 273, 277
Northern blot 79–82, 87, 88, 90, 91, 94, 95, 97, 98, 118, 119, 121
Novozyme 17, 61, 75, 228
Nuclear condensation 270–272, 274–275
Nuclei
 DAPI (4′,6-diamidino-2phenylindole) 229, 231, 237, 239–241, 243, 245, 247, 249, 255, 270
 Hoechst 229, 231, 270, 271, 307, 313
 propidium iodide (PI) 237, 241, 243, 271, 273, 276–278, 282–287
 staining 229, 270, 271, 274, 307, 312, 313
 Sytox green 229, 231

P

PCR
 DOP-PCR (degenerate oligonucleotide-primed PCR) 176, 238, 242–243, 250, 251
 real time PCR 97, 103–113, 199, 223
 RT-PCR 90, 119, 121, 122, 137–150, 223
Phosphatidylserine 270, 271
Phosphorylation assay 203–204
Plasma membrane
 permeabilization 227, 270
 staining 271
Protein-protein interaction..................................... 303–316
Proteins
 blotting 206
 cross-linking 196, 197, 226, 227, 291
 extraction 189–194, 196, 203, 205–206, 208, 209
 immunoprecipitation (IP) 190, 191, 195–196, 199, 285
 purification 189–190, 196, 212
 quantification..205–206, 212
 separation 192, 206
Protoplasts...................... 3–18, 21, 22, 34, 46–47, 52, 53, 59–61, 64, 68–70, 75, 232, 238, 244, 252, 255, 271–273, 276, 277, 279

R

Reactive oxygen species (ROS)....... 270, 271, 273, 277, 279
REMI. *See* Restriction enzyme-mediated integration
Restriction enzyme-mediated integration
(REMI) 4–5, 12–13, 18
RNA
 dsRNA..78, 79, 82, 83, 92
 isolation 86, 117–119, 122, 154, 155, 157–158, 162
 linear amplification... 158
 mRNA..78, 86, 91, 94, 95, 97, 119–122, 166, 170–172, 225
 quantification..118, 170–171
 reverse transcription.........................97, 104, 145, 162
RNAi.. 77–98
ROS. *See* Reactive oxygen species

S

SDS-PAGE....................191–193, 195, 196, 203, 206, 207
Software
 imageJ..80, 94, 206
 MetaMorph 6.0/6.1.. 262, 263
 Scion Image.. 125–135
Southern blot...27, 29, 49, 50, 71, 97, 182
Split-EGFP.. 303–316
Split-marker 10–12
SSH. *See* Suppression subtraction hybridization
Suppression subtraction hybridization
(SSH) 115–122
Syber Green..104, 110, 113

T

Texas Red 140, 142, 144
TRAC. *See* Transcript analysis with aid of affinity capture
Transcript analysis with aid of affinity capture
(TRAC) 165–172
Transposons.. 4, 41–53
TUNEL (TdT-mediated dUTP nick
end labeling) 270, 272, 275–276, 278, 279

V

Valinomycin.. 283, 285

W

Western blot 189, 190, 192–196, 199, 204, 206, 300, 315
WGA. *See* Whole genome amplification
Whole genome amplification (WGA) 176, 182